Hard Facts about Soft Machines

Hard Facts about Soft Machines:
The Ergonomics of Seating

Edited by
Rani Lueder
Humanics ErgoSystems, Inc.
PO Box 17388
Encino, CA 91416-7388, USA

and

Kageyu Noro
Waseda University
School of Human Sciences
Tokorozawa 359, Japan

UK Taylor & Francis Ltd, 4 John St., London WC1N 2ET

USA Taylor & Francis Inc., 1900 Frost Road, Suite 101, Bristol PA 19007

British Library Cataloguing in Publication Data

A catalogue record for this book is available from the British Library

ISBN 0 85066802 6

Cover design by Amanda Barragry

Typeset in Great Britain by Santype International Limited.
Printed in Great Britain by Burgess Science Press, Basingstoke
on paper which has a specified pH value on final paper
manufacture of not less than 7.5 and is therefore 'acid free'.

Contents

Preface

A well-known British radio programme for small children (*Listen with Mother*) some years ago, would start with the storyteller saying, 'Are you sitting comfortably, then we'll begin'. It became a catchphrase, still in use, but begged a lot of questions! Not least amongst those questions is what 'sitting comfortably' really means.

Much of the work on seating has dealt with sitting at work, very little has looked at what in Britain are called 'easy chairs', the large heavily upholstered furniture which is observable in most homes. Domestic furniture generally 'design' emphasizes appearance over comfort which is equated with softness of upholstery. There is little evidence that the many years of seating research has had much effect on the design of personal furniture, or on the understanding of the public about sitting requirements.

Not least to be criticized in this is the concept that designers, a word which is imprecise in its coverage, are not necessarily the appropriate people to design furniture. Amongst those in this category have been architects for example, who have produced many striking designs of chairs, but at the same time chairs which are inadequate, unstable and downright dangerous. Until some ergonomics is a required component of an architect's training they should recognize the necessity of taking some good ergonomics advice.

Although work furniture has had so much more attention a visit to any office or factory will show that there is a long way to go before this work is translated into the wider world. Perhaps sitting is too easy an action, just bending at the hips and knees, and there you are. But the reactions of the body, over time, to this simple action are yet matters for investigation. The impact of Branton's view of sitting as an unstable posture, (Branton 1969), with its consequent requirement to study seated subjects over long periods and observe their behaviour in detail, still does not illuminate many studies. In particular the approach of attempting to interpret from behaviour what relevant model is operating, rather than working from a theoretical model is seen in the minority of investigations. There is nothing wrong with theoretical models, of course, providing they are representative of reality. Unfortunately our knowledge of the realities of sitting are still uncertain, and a dependence on theory without the support of empirical evidence will lead us into false, indeed dangerous, conclusions.

The revival of an international symposium on seating is long overdue, and these papers, developed and expanded from the Tokyo symposium are a welcome indication that the interest may now be maintained. Increased international concern in the health and safety of people at work is one of the driving forces behind this revival. The contributions in this book are a significant step in the science of seating and should lead to a better understanding of the mechanics, the dynamics and the effects of seating on the sitter. It will also, no doubt, lead to the recognition that technology can permit the increased personalisation of a seat, so that it is truly the seat that adapts to the sitter rather than the opposite.

There is still a long way to go in the understanding of seats and sitters, and a lot of work to do in the design of better seats. This exploration will require seat models which are truly ergonomic, taking account not only of the biomechanics and physiology of sitting but the task requirements and environment of the sitter as well. Then we will begin to see a closer identity between theory and practice in the applications of ergonomics to seating.

E N Corlett

Reference

Branton, P. (1969), Behaviour, Body Mechanics and Discomfort. *Ergonomics* **12**, 316–27. Reprinted (1993) in *Person-Centred Ergonomics*, (Eds D. J. Oborne, R. Branton, F. Leal, P. Shipley and T. Stewart) Taylor & Francis, London.

*Notes on Contributors**

Rani Lueder, CPE, is President of Humanics ErgoSystems, Inc., an ergonomics consulting firm in Encino, California. She has consulted in ergonomics and workplace safety since 1982, and is a member of the ANSI committee revising American National Standards for Computer Workstations. She also edited and co-authored *The Ergonomics Payoff: Designing the Electronic Office* (*HRW*). She is a member of the Human Factors and Ergonomics Society (US), the Ergonomics Society (Europe), and is Certified with the Board of Certification in Professional Ergonomics.

Kageyu Noro is the Associate Dean of School of Human Sciences, Waseda University. With the support of many staff members and visiting professors, his laboratory carries out varied studies, such as participatory ergonomics, kansei, seating, and virtual reality as high-technology ergonomics. He is known as the person who originated the concept of participatory ergonomics in 1984. The handbook *Illustrated Ergonomics*, developed with his concept of product-oriented ergonomics, edited by him, and contributed by 118 authors, was awarded one of the top prizes for Japanese books on science and technology in 1990. The 15-instalment TV educational programme 'Ergonomics for Everyday Life' he edited as invited professor of the University of the Air, a Japanese broadcast university, is broadcasting on a Japan-wide channel. He is Chair of the Seating Research Committee (Zaken). He is the first chair of the International Working With Display Units (WWDU) Group.

Marvin J. Dainoff, PhD, is Professor of Psychology and Director, Center for Ergonomic Research, Miami University, Oxford, Ohio. He is also President of Marvin Dainoff Associates Inc., Consultants in Ergonomics. He received his PhD in Experimental Psychology from the University of Rochester in 1969.

Paul Cornell works at Behavioral and Environmental Research, Steelcase Inc., Grand Rapids, MI, USA.

Keiichi Ohno graduated from Nihon University in 1985. He joined Uchida Yoko Co., Ltd and took charge of the R&D for work stage furniture. He is now taking charge of marketing the chair with the R&D Department.

John Roebuck began his interest in human factors while working on his Bachelor of Science degree in mechanical engineering at the University of California at Berkeley. From UCLA he obtained a Master of Science degree in engineering with a biotechnology option. He has been employed at Douglas Aircraft Co., Collins Radio Co., Rockwell International Corporation and McDonnell Douglas. His experience has included engineering design, human factors research and consulting, technical editing, and management positions as proposal manager, project engineer and supervisor on contract studies and

* First-named authors only

principal investigator for internal research and methods development. His 25 years at Rockwell included work on Apollo, Space Shuttle and Space Station as well as many IR&D projects. Since February 1987 he has been self-employed as a consultant, doing business as Roebuck Research and Consulting. Since early 1988 he has worked part time at the Douglas Aircraft component of McDonnell Douglas developing anthropometric requirements for a proprietary computer–human model. He is the author or co-author of over 40 technical documents, including the text book titled *Engineering Anthropometry Methods* (John Wiley & Sons in 1975), and is currently writing a book on anthropometric estimation methods.

Max Vercruyssen, PhD, Assistant Professor of Gerontology and Psychology (University of Hawaii) and Assistant Professor of Human Factors (University of Southern California), is a gerontologist and ergonomist whose basic research has focused on the quantification and explanation of central nervous system changes, particularly in integrity measures like speed of behaviour, caused by environmental stressors, neurotoxins, and endogenous phenomena like organism ageing and practice (i.e., repeated exposures). His applied research consists mainly of laboratory investigations, the findings of which have been instrumental in developing federal and international standards for protective equipment (e.g., respirators and clothing ensembles). As a private consultant, Dr Vercruyssen has advised clients on seating arrangements for optimal workstation design, but posture became an experimental manipulation in his laboratory research when he discovered that it could alter the rate of information processing in humans and that such changes showed potential for explaining age-related slowing of behaviour.

Yutaka Haruki graduated from Graduate School of Letters, Waseda University in 1961, where he became a lecturer of School of Letters in 1966. He became an associate professor in 1969, and in 1974 became a professor. Since 1987, he has been professor of School of Human Sciences. His research targets the animal and the child, and has researched conditioning and human learning. Recently, he has been interested in the relationship between nonverbal behaviour and emotion, especially the effect that nonverbal behaviour has on emotion.

Tom Bendix, MD, PhD, works at Laboratory for Back Research, 7601 State University Hospital, DK-2200 Copenhagen, Denmark and Rehabilitation Engineering Center, University of Vermont, USA.

Steven Reinecke, Biomedical Engineer MS, has been affiliated with the Department of Orthopedics at the University of Vermont since 1980. In 1983 he joined the Vermont Rehabilitation Engineering Center for low back pain to study seating. In 1987 he became Principal Investigator for low back pain seating studies. He has consulted for a number of chair manufacturers in product development, product evaluation, and seating research. His research studies have ranged from bicycle seats, office chairs, and motor vehicle seats to wheelchair seating. His work has concentrated on the biomechanics of the body while in the sedentary position.

A. C. Mandal, MD, is Chief Surgeon, Taarbek Strandvej 49, 2930 Klampenborg, Denmark.

Karl H. E. ('Eb') Kroemer's interests are particularly in research and application aspects of anthropometry, biomechanics, and work physiology. Since 1981, he has been a professor at Virginia Tech (VPI&SU) where he directs the Industrial Ergonomics Laboratory. His academic degrees are in mechanical engineering from the Technical University Hannover in Germany. He worked as a research engineer at the Max Planck Institute for Work Physiology and as director of the Ergonomics Division of the BAU (the German NIOSH). In the USA, he was a research engineer with the USAF Human Engineering Division, and professor of ergonomics and industrial engineering at Wayne State University. He was a UN ergonomics expert in Romania and India. He is the author of more than 130 publications, among them several books.

Joseph A. Sember III, gained a Bachelor of Science degree in Civil Engineering from the University of Arizona. From 1973 to 1976, he was Project Manager on large military construction projects. In 1976, he became Manager of applications engineering at P. L. Porter Co., which is a manufacturer of seat recline systems. Since 1986, he has been Executive Vice President of Jasco Products, a manufacturer of pneumatic seat adjustment systems. In 1990 a company called Numotech Inc was formed by five medical and technical professionals to develop special seating systems for the wheelchair bound population. Joe Sember III is President of Numotech Inc. Together, Numotech, as the Research and marketing arm, and Jasco Products as the engineering and manufacturing arm have, over the last two years conducted an extensive research and development programme resulting in the systems and studies referenced in the chapter.

Clifford M. Gross, PhD, is President and Chief Executive Officer of Biomechanics Corporation of America. He has provided consulting, training and research services worldwide, across a wide number of industries and has published more than 50 articles and abstracts on workplace ergonomics, biomechanics and injury prevention. As guest speaker, he has addressed professional audiences around the world. Dr Gross has established corporate ergonomics programmes at many major US corporations including EI DuPont de Nemours, Dow Corning Corporation, Ford Motor Company, General Motors, Procter & Gamble, Michelin Tire Corporation, Consolidated Edison, Cooper Industries, Mobay Corporation, Union Carbide Corporation, Steelcase, Lear Seating and Thomson Consumer Electronics. The company which he founded and directs is the first public ergonomics firm in the United States.

Am Cho, is professor of the Department of Industrial Engineering, Dongguk University, Korea. Since 1986, he has been a member of executive committee of the Ergonomics Society of Korea. His research interests are virtual reality, working environment ergonomics, and seating.

Tomoko Hibaru completed the school nurse training programme of the Faculty of Education, Kumamoto University in 1974. After the work of the junior high school of three years, she is working at Shirono elementary school of Kitakyushu municipal.

Tadao Koga, Dr Eng., joined Citizen Watch Co. in 1957 after graduating from the Department of Craft, Faculty of Art, Tokyo University of Art. He became the lecturer of the Department of Industrial Design, Faculty of Design, Kyushu Institute of Design in 1969. He is researching the design methodology concerning physically handicapped persons and the aged's environment.

Narumi Hirao, MD, BS, graduated from the Department of Physics, Faculty of Science, Kobe University in 1976. Afterwards, she entered University of Occupational and Environmental Health, School of Medicine and graduated in 1986. She worked for Fujidenki Hospital as a doctor of internal medicine and the health doctor after working with Kantou Rousai Hospital as a resident. She has been working for Hyogo Hospital since 1992. She is interested in women's work problems.

E. Nigel Corlett is Professor Emeritus of the University of Nottingham and Scientific Adviser to the University's Institute for Occupational Ergonomics. Prior to joining the University at Nottingham he was Professor of Industrial Ergonomics at the University of Birmingham. Professor Corlett studied mechanical and production engineering and worked in industry as designer, factory manager, and head of design and development in a major domestic equipment company. His research has embraced biomechanics including seating, workspace and equipment design and organization design. He has edited or authored several books. He is Honorary Fellow of the Ergonomics Society, a Fellow of the Fellowship of Engineering and holds the DSc of the University of London.

Youichi Suzuki graduated from the Department of Industrial Design, Kanazawa Art and Technology College in 1964. He joined Kokuyo Co., Ltd in the same year. At present, he is Director of the Products Development Section of the Furniture Department, with responsibility for planning and development of office furniture.

Mitsuaki Shiraishi, graduated from Chiba University in 1983, having completed a masters course in Architecture. He was then employed by the Joint Okamura Corporation (Office Furniture maker) Field Ergonomics for Interior space and Industrial Design.

Kozi Morooka, Doctor of Engineering, is Head Professor of Tokai University Graduate School of Management Engineering. He was born in 1930. After many years of study he graduated from a Doctors' course at Keio University Japan, and went on to become Assistant Professor of Administration Engineering at Keio University, Honorable researcher of Wisconsin University of Canada, Professor of the Engineering Department, Tokai University, Lecturer of the Medical Department, Tokai University, and Visiting Professor of Windsor University, Canada, he is also a former member of the Japan Medical Association.

Christin Grant received her PhD in Urban, Technological, and Environmental Planning from the University of Michigan, where she specialized in environmental psychology. In the past 18 years, she has done numerous human factors research projects affecting the design of furniture, machines, and environments. A Research Fellow at the Center for Ergonomics at the University of Michigan, and a member of the Human Factors Society and the Environmental Design Research Association, she lives in Ann Arbor, Michigan.

Ida Festervoll is a consultant and a Physical Therapist at Håg a.s. She has consulted for 10 years on the development of ergonomic work chairs and seats for transportation, industry, banks and offices, and is Norwegian delegate to the European Standardization Committee CEN, TC 207 Office Furniture and TC 122 Ergonomics. She has 10 years of developing working chairs, and serving as consultant to one of Norway's biggest banks, where 90% of the employees have their own computer.

Hector Serber is President of American Ergonomics Corporation, a research and development firm dedicated to advanced ergonomic seating system design. Serber has been granted three US Patents #4,650,249, Ergonomic Seating Assembly (1987), #4,832,407, Variable Posture Chair and Method (1989) and for vehicular safety, Seat Assembly and Method (1993), #5244,252 as well as numerous international patents and patents pending in Europe and Asia. He holds a Tecnico Mecanico degree from Collegio Industrial de la Nacion, BA, Argentina, and studied Mechanical Engineering at California State University at Sacramento.

Mark C. Volesky is a Senior Product Engineer at Steelcase Inc. During the writing of this chapter, he was a senior research engineer at Steelcase. His responsibilities included monitoring ergonomics and VDT related legislation at all levels of government. He has served as consultant to the Canadian Standards Association (CSA) during the drafting of the Office Ergonomics Standards and has taught classes on ergonomics. He is a senior member of the Institute of Industrial Engineers and a past member of the Human Factors Society. Volesky earned his Bachelor of Science degree from North Dakota State University.

L. B. Kruk works at Marketing and Research, Proformix, Whitehouse Station, New Jersey, USA.

PART I

Introduction

Introduction

Kageyu Noro

Rani Lueder

The genesis of chairs and sitting is not known. A primitive version of the chair was discovered in the Toro ruins in Japan, circa 100 AD. During the Middle Ages, chairs were also used by high-level samurai who commanded field operations, and by priests during religious ceremonies.

Given this, it is surprising that chairs did not predominate until recently. In Japan, citizens began to use chairs during the Meiji era (1868–1912). In the West, seats were primarily used by royalty, and only gradually infiltrating the upper classes. Stewart (1986) describes a 1458 painting, in which robed justices of the French high court sprawled across the floor in session. Chairs became available to the emerging leisure classes during the reign of Louis XIV. 'Commoners' reconciled themselves to sitting on baskets, benches, and such.

Over time, technological innovations, first directed towards the achievement of specific ends, were incorporated into general purpose seating. For example, swivel-tilt seat capabilities were probably developed to facilitate navigation at sea. Today, the evolution of chair design is frequently driven by specialized requirements, such as to accommodate car driving or the disabilities.

Recent advances in the variety, complexity, and rapid pace of change of the materials and related sciences continue to accelerate innovations for all the various stages, associated with product design, manufacture, assessment and (ultimately) recycling. For example, intelligent computer systems, using computer-aided design and large-scale databases, are extending our ability to take advantage of the almost infinite variety of materials and composites now available to allow chairs to perform in new ways.

These advances continue to promise breakthroughs that allow seats to become more comfortable, yet easier to use and adjust. Some models can already flex in new ways, improving both postural support and freedom of movement. Cushioning composites may impart both firm support and 'cushy' comfort; features that have historically been at odds. Seat tensioning and contours may respond to the specific user, and absorb shocks. Chairs are increasingly becoming lighter, safer, stronger, and more stable, and at less cost and with less waste. Some chairs have become 'intelligent', and impart 'memories' to accommodate different users, tasks, and positions.¶

These possibilities arrive not a moment too soon. Never before has there been such a recognition of the health consequences that may accompany seated work. A review of research over the last decade underscores the prevalence of physical discomforts experienced by computer users, as well as general office workers (Kroemer, 1988; Lueder, 1992; National Research Council, 1983; Sauter and Schleifer, 1991).*

An increasing recognition of the health hazards incurred from lack of movement (Kilbom, 1987; Lueder, 1992) has accentuated concerns regarding the future impact of a technology-driven work processs that systematically reduces the potential for movement – to a greater extent every year.

In Japan, attention to seating has also been heightened by a shortfall of workers since 1990. Employers have become increasingly focused on how to attract good workers with ergonomic seating. Concurrently, greater employee expectations in 'quality of work life' extended to their expectations of seating comfort.

Within Japan, this demand became so acute that there arose a severe shortage of ergonomic seats during 1990 and 1991. It also provided the impetus for 'The New Office Campaign' project by Japan's Ministry of International Trade and Industry, which accelerated corporate interest in ergonomic seating.

Zaken (a Japanese abbreviation of 'Posture and Chair Study Committee') was established in April of 1991 through financial support from the Japan Institute of Posture Research. The objectives of this committee are: to help organizations evaluate chairs; to provide users with information to improve their seated comfort; and, to promote an exchange of information between users and manufacturers.

This book is an outgrowth of this initiative. Originally based on the Proceedings of the 1989 Second International Symposium on the Science of Seating (Tokyo), it included fifteen papers from participants in seven countries†. This was the first such symposium since the 1968 International Symposium on Sitting Posture, organized by Dr Etienne Grandjean, and held at the Swiss Federal Institute of Technology.

---

* It should be noted that this attention is presently centered in the West. The frequency and types of physical problem of computer users in countries such as Japan are not well understood, but appear less pronounced at this time.

† This symposium was jointly sponsored by Waseda University and the Japan Society for the Promotion of Science. It was conceived and organized by Professor Noro, with the collaboration of Rani Lueder, MS (US).

¶ My thanks to Mr Bill Marks, Dupont, for his insightful comments. (RL)

The Symposium's objective was to address a few of the myriad of questions that remain regarding how to design ergonomic seats. Although a vast body of research on seating in recent years has greatly advanced our understanding, we are left with many gaps in how to address users' discomforts and support their activities. The answers that we have gleaned have only underscored how much remains to be discovered.

- Some issues are ignored, because little research attention has been directed to them. Such gaps may result, for example, from the difficulty of performing this research or an absence of sources of research funding. A sampling of such issues may include:

  What is the relationship between comfort and physiologic well-being?
  Do users have an intuitive sense of how they should sit?
  How does posture affect our emotions and awareness?
  How does one reconcile the various trade-offs implicit in designing a seat (e.g., the interaction of seat elements; the benefits of increasing support at the cost of promoting movement)?
  How can tomorrow's technologies, such as artificial intelligence, be applied to address seating requirements?

- Some issues have 'fallen through the cracks', between the questions commonly investigated by ergonomists, and the answers needed by manufacturers and users. Such gaps may result, for example, from dogma (which besets all disciplines), resistance to paradigm shifts, or a limited understanding of the holistic interplay among seat considerations. A sampling of such subjects may include:

  How does one begin to prioritize the costs and benefits associated with different seat features?
  How does one compare highly adjustable seats with 'dynamic' versions, which ostensibly reduces the need for user intervention?
  What kinds of chair do special populations need (e.g., pregnant women, school children)?
  What user, task, and seat factors affect the kind and level of seat tention?
  What are the implications of the wide variability in the population in the symmetry and shape of the spine (particularly the degree of lumbar lordosis), and other individual factors on seating design?

- Frequently, research has been fragmented, and reflected national orientations. Correspondingly, our attention is often limited to the perspectives imparted by our society.

The contributions we offer here, all fully revised and updated, cannot provide conclusive answers to many of the broad spectrum of questions that need to be answered, but we believe that they can stimulate new areas of research, new design applications, and further the international exchange of information.

Kageyu Noro
and Rani Lueder
Editors
Tokyo/Los Angeles
August 1994

References

Kilbom, A., (1987), Short- and long-term effects of extreme physical inactivity: a brief review, in Knave, B., Wideback, P. G., (eds.) *Work with Display Units*, Amsterdam, Elsevier Science, 219–28.

Kroemer, K. H. E., (1988), VDT work station design, in Helander, M. (ed.) *Handbook of Human–Computer Interaction*, Amsterdam: Elsevier Science, 521–39.

Lueder, R., (1992), Seating, posture, and ergonomics, in Sweere, J. (ed.) *Chiropractic Family Practice: A clinical manual*, Gaithersburg, Maryland: Aspen Publishing. Section 21.

National Research Council, (1983), *Video displays, work and vision*, Washington, DC: National Academy Press.

Sauter, S. L. and Schleifer, L. M., (1991), Work posture, workstation design, and musculoskeletal discomfort in a VDT data entry task, *Human Factors*, **33**(2), 151–67.

Stewart, D., (1986), Modern designers still can't make the perfect chair, *Smithsonian*. March, 97–105.

PART II

Adjustability

1

Sashaku: a user-oriented approach for seating

K. Noro

Creating an information loop for seating

Introduction

This chapter describes a model for chair adjustability. First, *sashaku,* as used traditionally in Japan, is introduced and sashaku for workstations using VDT (called VDT sashaku) is described. Second a parametric model which includes VDT sashaku is introduced and 'Foot overplus posture' and 'Foot float posture' are defined. Third, the mathematical interpretation of those postures is described. Finally, as only adjustment by anthropometric dimensions is known to date, a new adjustment method is proposed.

Information required by users to adjust their chairs

Chair guidelines are available from many sources; These include the Labour Standards Bureau, Ministry of Japan (1985), Human Factors Society of the United States (1988), Personnel Department of Waseda University (1987), and other organizations. These guidelines often refer to the need for adjustability. The Occupational Health Guidelines for VDT Work (Labour Standards Bureau, Ministry of Labour, 1985) recommend that seat pan heights should be adjusted. Such guidelines are applied to workplaces, but the adjustment is made by individual users. Frequently, the user understands why adjustability is important but not how to adjust their seat correctly. To accomplish this, they must understand both what is the correct height, and how to adjust the chair. Information required for seat pan height adjustment should flow between both the user and chair manufacturer. To collect such data, the user must have correct information and a sufficient seat adjustment range.

One problem with the conventional method of chair adjustment is that there is no algorithm for customizing workstations for individual users; even if the dimensions of specific workstation components accommodate individual users, dimensional conflicts between components may preclude an optimal arrangement of the workstation for the user (Povlotsky and Dubrovsky, 1988). Povlotsky and Dubrovsky (1988) and Cornell and Kokot (1988) reported a poor correlation between seat pan height and anthropometric dimensions of the user. These findings indicate that it is difficult for users to adjust workstations to their preferences without outside assistance.

Proposal of new concept

Noro (1987) proposed that a technological information loop would bridge the gap between user behaviours and ergonomic data on workstation adjustment. A revised version of this information loop is shown in Figure 1.1.

An integral component of such a loop would consist of software, which incorporates health and work management data, and hardware for automating the adjustment. The core of this system is the Sensor Workstation, an experimental adjustment apparatus. Noro, Togami and Onomura (1985), and Noro (1986, 1987) designed this Sensor Workstation to preclude dimensional conflicts of individual components and accommodate users. Togami and Noro (1987) used this apparatus to assess chair and desk dimensions for the aforementioned Occupational Health Guidelines for VDT Work (Labour Standards Bureau, 1985). It was also used by Hirasawa, Noro and Togami (1990) to establish VDT Work Guidelines (Personnel Department, Waseda University, 1987). The Sensor Workstation is limited to laboratory research, however, rendering it too cumbersome for large-scale data collection (exceeding 30 subjects). Prior to beginning the experiment, users were trained in adjustment and use of their chair (see section on Implementation of participatory ergonomics and analysis of obtained data) to increase participation and facilitate communication with ergonomists. The first phase of this study involved the investigation of the relationship between seat pan height settings

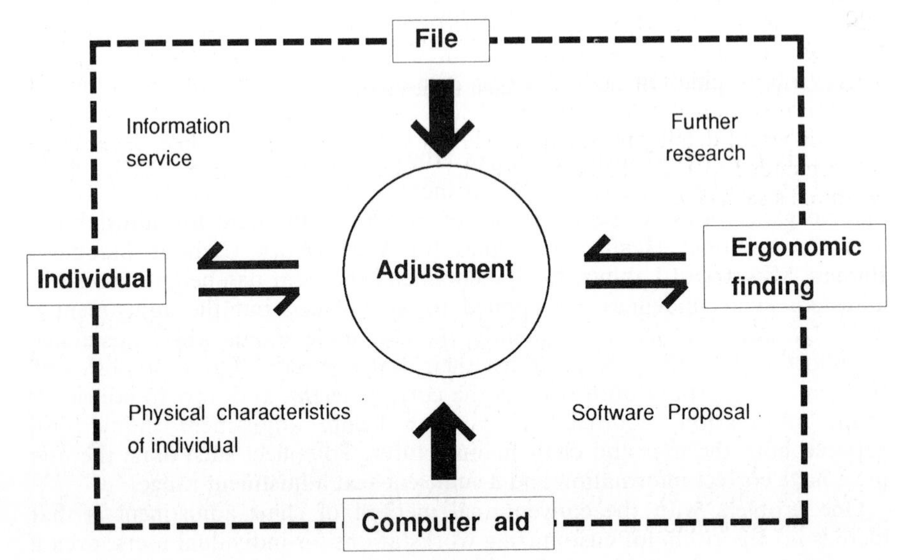

Figure 1.1 A revised information loop between users and ergonomics research

and anthropometric dimensions and user preferences. These were conducted under experimental conditions independent of conventional chair adjustable ranges.

Analytical tools

The parametric model described in the following section depicts the software for communicating with users (see the information loop depicted in Figure 1.1). Contrarily, user preferences are communicated to the adjustment software to refine seat adjustability. This chapter describes data and information from two investigations; the investigations comprise two experiments. The first is a series of experiments using office workers as subjects. The experiments were conducted in the chair usage class of 1988 and 1989, and the chair usage class of the office workers at Waseda University. The purpose of the second experiment is to observe and measure workstations and workers' postures. A method for office workers to easily measure the height of the seat pan of a chair does not exist. A possible solution for office workers is a method of measuring seat pan height without seated. The theoretical value which is able to refer when seat pan height was measured by the above method is proposed in this chapter. The seat pan height without sitting is called seat cushion height or uncompressed seat pan height. The seat pan height with seated users is referred to as compressed seat pan height.

Theoretical framework

Sashaku

Sashaku, developed by Mishima in 1893, is a Japanese term that refers to the distance between the desk (work surface) and seat pan (Ohnishi, 1977). Mishima recommended the use of sashaku for allocating chairs to elementary school pupils. The Ministry of Education subsequently used this criterion throughout elementary schools in Japan. Of course, sashaku does not equally apply to both children and adults and to all tasks. Togami and Noro (1987), and Hirasawa *et al.* (1990) proposed that Mishima's guidelines are inappropriate for visual display terminal (VDT) work surface heights. These researchers recommended VDT sashaku (see following section) which differs from Mishima's sashaku.

Parametric model

A model for VDT work was developed that represents the interrelationship between office workers and their environment (Hirasawa *et al.*, 1989). The parametric model is designed to determine workstation dimensions to accommodate individual operators, as shown in Figure 1.2. Two submodels, shown in Figure 1.3, have been developed to date. The viewing axis submodel describes the relationship shown in Figure 1.4. When the display is located below the sitting eye height of the operator, the following relation holds among the parameters a, b, and q:

$$a + b + q = 180^\circ \quad (1)$$

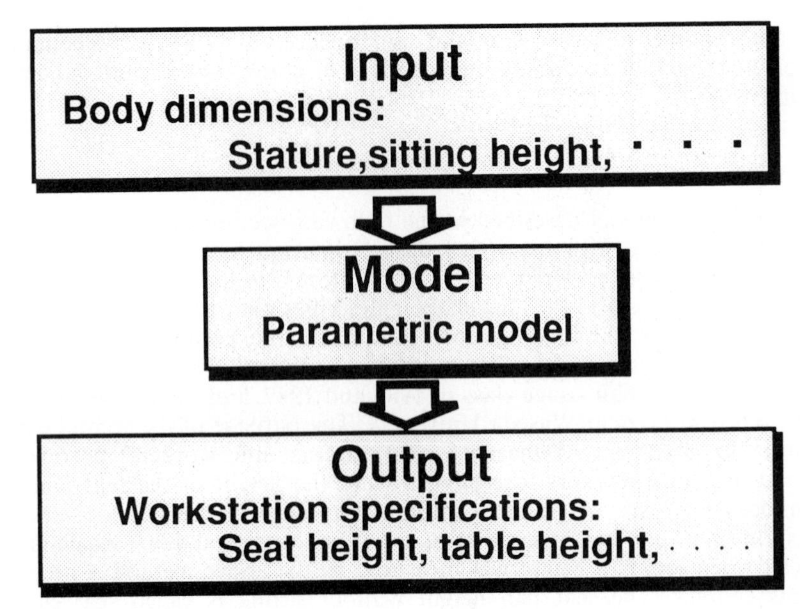

Figure 1.2 Proposed parametric model. Stature and sitting height of a person are used to develop suggested workstation settings of user

where the ranges of *a*, *b*, and *q* are as follows:

$$\begin{aligned} 0 \le a \le 180^\circ \\ 0 \le b \le 90^\circ \\ 0 \le q \le 90^\circ \end{aligned} \tag{2}$$

The submodel for determining the seat pan height, work surface height, and display height follows. The seat pan height may be set by two methods:

1. Sitting on a chair put both feet on the floor (the height of the seat is calculated from popliteal height).

The seat pan height affects the distribution of seated pressures. Equation (3) represents a formula for calculating the optimum distribution; the value obtained from this equation represents the theoretical seat pan height setting (described later). To customize the calculated seat pan height for a particular user, it is finely adjusted to suit his or her preference, and is discussed subsequently.

$$\underset{\text{(seat cushion height)}}{\text{Seat pan height}} = \text{popliteal height} - 2.5\text{ cm} + \underset{\text{height}}{\text{cushion}} + \underset{\text{shoe heel}}{\text{height of}} \tag{3}$$

where 2.5 cm is introduced, based on the assumption that the seat pan height should be slightly lower than popliteal height to prevent impacting the thighs (Noro and Hirasawa, 1988).

2. To make keyboarding easy (seat pan height is calculated from work surface height).

When the work surface height is fixed, the seat pan height can be determined by subtracting sashaku from the work surface height, as denoted by equation

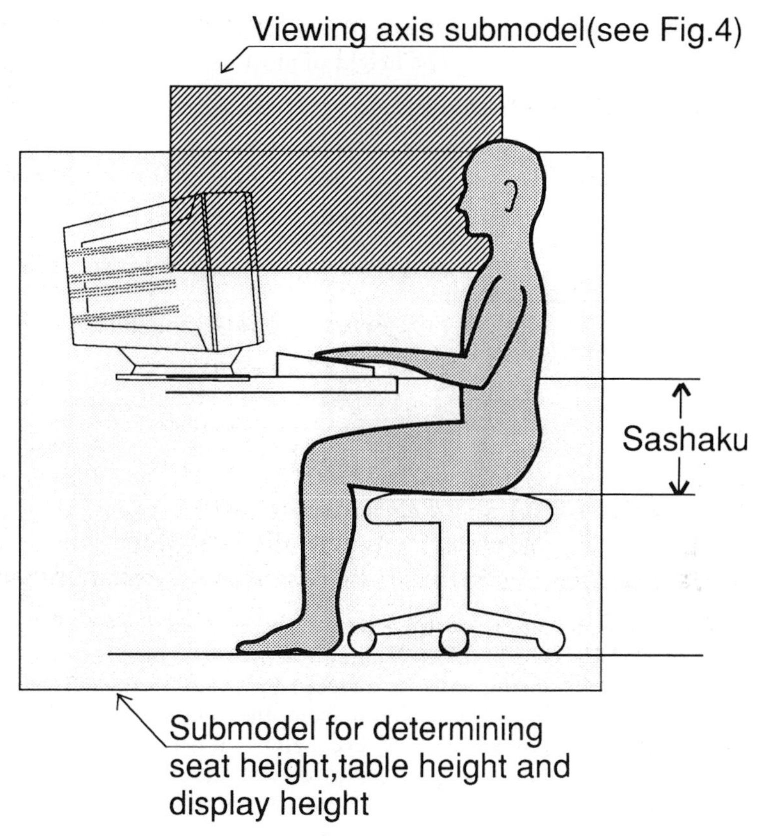

Figure 1.3 Division of parametric model into two submodels for upper parts of VDT workstation. Two submodels may be connected as required: (1) distance from floor to top of thigh; (2) leg room below desk; (3) support thickness, A: Sashaku, B: distance from floor to seat pan

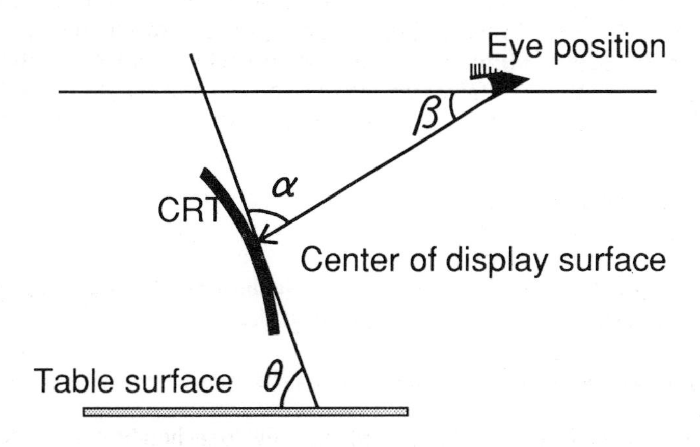

Figure 1.4 Viewing angle submodel

(4). JIS S 1010 (Japanese Standards Association, 1983) specifies two office work surface heights of 67 and 70 cm. The height of most desks in Japanese offices is 70 cm.

$$\text{Seat pan height (seat cushion height)} = \text{work surface height} - \text{sashaku} + \text{cushion height} + \text{height of shoe heel} \quad (4)$$

where sashaku for VDT work depends on the relationship between the screen centre height and sitting eye height.

When the elbow angle is 90°, the muscular load is minimized. When the operator sits upright, sashaku is denoted as

$$\text{Sashaku} = \frac{\text{sitting height}}{3 - 5\ \text{cm}} \quad (5)$$

This sashaku is called VDT sashaku.

The term '–5 cm' of the above equation is added for keyboarding easily. For VDT work, the seat pan height is determined by subtracting sashaku from the work surface height. When a short person sits at the VDT at a standard height, equations (4) and (5) show that a very high seat and foot support is needed. Results of calculation by equations (4) and (5) correspond to Figure 22 on page 54 of ANSI/HFS 100–1988 (Human Factors Society, 1988).

A submodel of the viewing angle is denoted by equation (6). The two submodels are connected by VDT sashaku from the relationship between the VDT screen centre height and sitting eye height. When the operator faces the display screen, the centre of the display screen should correspond to the natural viewing angle of the operator. VDT sashaku in this case represents:

$$\text{VDT sashaku} = \text{sitting eye height} - \text{height to display screen centre} - l \cdot \sin \beta \quad (6)$$

where l is the viewing distance and β is the viewing angle.

Equation (6) is important in that it defines the geometrical relationship between the lower and upper body parts of the seated person performing a visual task. For example, these relations can be used to define desk and chair dimensions to accommodate recommended viewing distance and angles for individual users (Noro and Hirasawa, 1988).

Implementation of participatory ergonomics and analysis of obtained data

This section describes the case of sitting on a chair with the feet touching on the floor, equation (3) alone is used in the calculations that follow. The results of calculation by Equation (4) are discussed in the conclusions section.

Participatory ergonomics

A participatory approach was employed to encourage users to adjust their chairs. It is based on the assumption that nonexperts should be encouraged to become actively involved in ergonomics. This involvement is called 'participatory ergonomics' or 'participatory approach' (Noro, 1991).

Chair usage class

A preliminary survey of the chair usage class was conducted, involving the measurement of body dimensions of 30 female and 69 male, university office workers (Hirasawa *et al.*, 1989). This measurement process was conducted by expert office workers using the sensor workstation described earlier. The chair usage class was held once in 1988 and in 1989. For the 1988 class, only one chair was used, and its seat pan height was adjusted to suit each participant. For the 1989 class, 10 chairs were used, and each was preset to a seat pan height ranging from 35 to 50 cm. Since the minimum seat pan height of chairs on the market in Japan is 39 cm, participants were able to use heights not commercially available.

The experimental scene in the chair usage class is shown in Figure 1.5.

Data analysis

Data were analysed as follows:
1. Statistical analysis of collected anthropometric data.

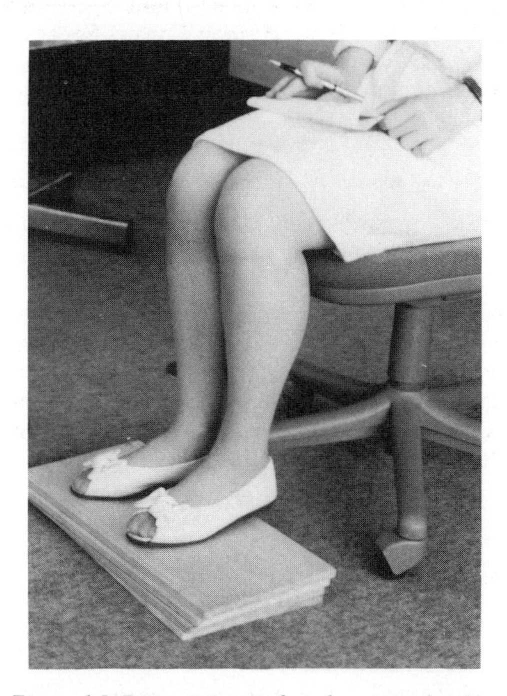

Figure 1.5 Determining preferred seat pan heights

Participants are each seated with seat pan height set according to theory, and asked what they think about their chair. If they are not satisfied with the theoretical seat pan height setting of the chair, they can try another chair, or different seat pan height setting. The foot rest is used to supplement this setting, to the effective seat pan height

2. Comparison of calculated and preferred values of seat pan height for individual participants by statistical analysis as well as of statistics from participants in the chair usage classes and from office workers.
3. Analysis of preferred values.

Body dimensions of participants were measured with Martin instruments and a sitting height metre during the practical education phase. The 1988 class was attended by 101 women, aged between 18 and 20 years, who applied for the class. The objective of the investigation was to determine whether seat pan height ranges of chairs on the market accommodated the spectrum of female users. These results were reported by Hirasawa *et al.* (1989).

A scatter diagram of stature versus calculated seat pan height (popliteal height −2.5 cm) is shown in Figure 2.6. Stature is generally correlated with seat pan height, variabilities of calculated seat pan heights amounted to 7–8 cm among participants of the same stature. Povlotsky and Dubrovsky (1988) and Cornell and Kokot (1988) observed a poor correlation between selected seat pan heights and anthropometric data in the United States of America

An analysis was made of the differences between the theoretical seat pan height settings and those preferred by class participants and office workers.

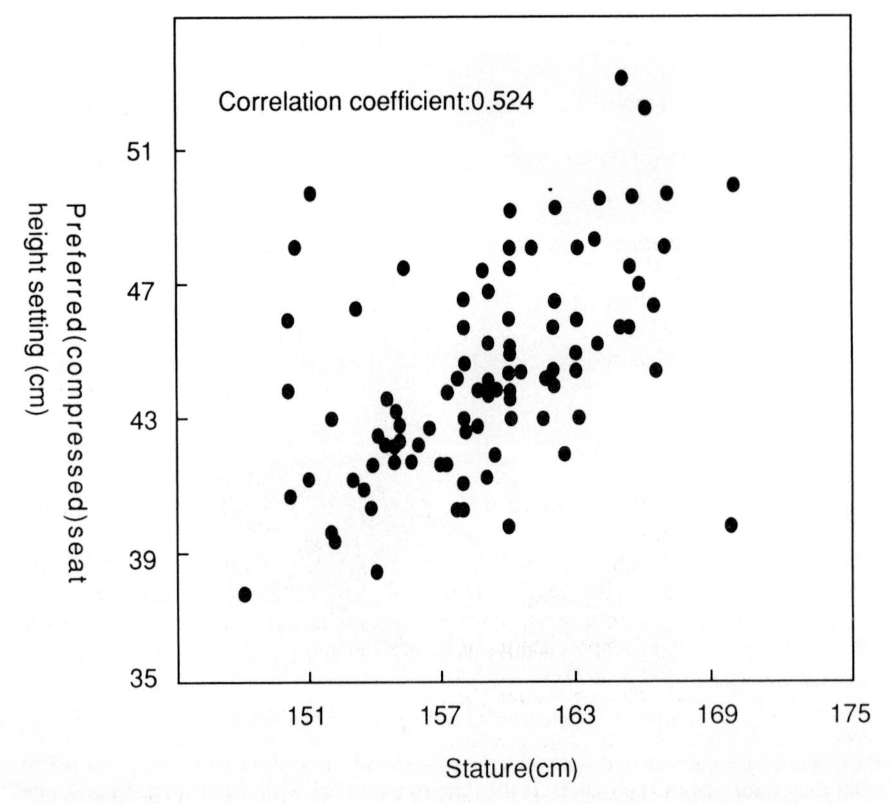

Figure 1.6 Scatter diagram of calculated seat pan height (popliteal height minus 2.5 cm) versus stature

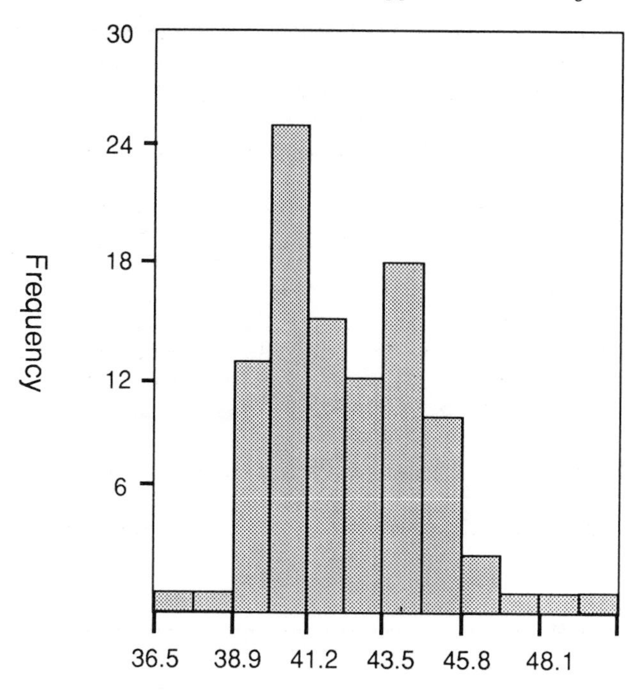

Figure 1.7 Frequency distribution of preferred seat pan height settings of 101 women

Participants were asked by the experimenter to adjust the seat pan height to their liking. Figure 1.7 shows the frequency distribution of selected seat pan heights of the 1988 class. It can be seen that the distribution exhibits slight peaking at 41 and 45 cm. Also, the distribution drops steeply near the lower limit. This 'drop off' is related to insufficient adjustability ranges of available chairs; if seats had been able to adjust to lower positions the distribution would have been more normal.

A scatter diagram of preferred seat pan height versus stature is shown in Figure 1.8. The two variables are generally correlated to each other, but variabilities in preferred seat pan height up to 8 cm were evident with users approximately 160 cm in stature.

In the 1989 chair usage class of 106 people, 10 chairs were preset at different seat pan heights to avoid some deviations resulting as users adjusted their seat pan height, as was the case with the 1988 class. The coefficient of correlation between the theoretical and preferred seat pan height was 0.90 for class participants, and 0.96 for office workers.

Figure 1.9 shows the frequency distribution of difference between preferred and theoretical seat pan height of male and female office workers investigated in 1987.

This distribution is normal (PL. 01). The 90 per cent confidence interval of difference between theoretical and preferred seat pan height was ± 2.5 cm. When the confidence interval is added to the theoretical seat pan height (45

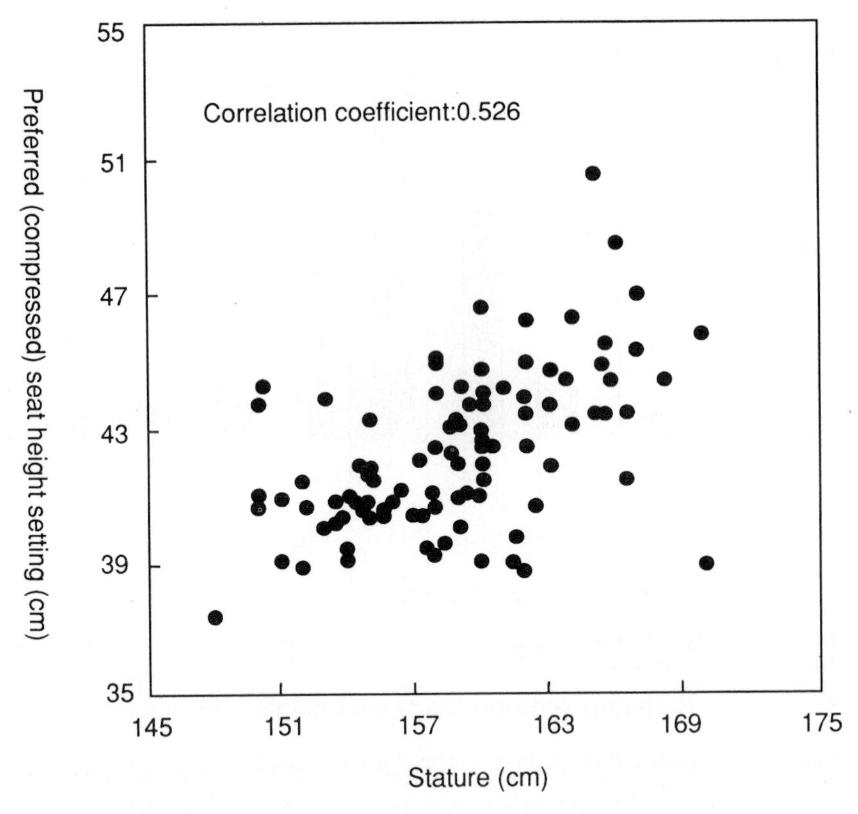

Figure 1.8 Scatter diagram of preferred seat pan height versus stature

cm) of the 5th percentile Japanese woman (stature of 147 cm) estimated from anthropometric data, the minimum adjustable seat pan height is calculated as 37 cm. The equivalent compressed seat pan height is about 35 cm. In Japan, no chairs are commercially available that can be adjusted to the seat pan height of 37 cm. The seated participants in the 1989 chair usage class were asked whether they preferred seat pans higher or lower than theoretical setting. The results of the questionnaire survey are provided in Table 1.1.

A very few participants (5.7 per cent) agreed to the theoretical seat pan height setting. In the 1989 class, the participants first sat on chairs set at the theoretical seat pan height, and the seat pan height was then adjusted to suit

Table 1.1 Results of questionnaire survey in which participants in 1989 chair usage class were asked whether they preferred higher or lower seat pans than predicted

Prefer seat pan height lower than predicted	39/106 (36.8%)
Prefer seat pan height higher than predicted	61/106 (57.5%)
Agree with predicted value	6/106 (5.7%)

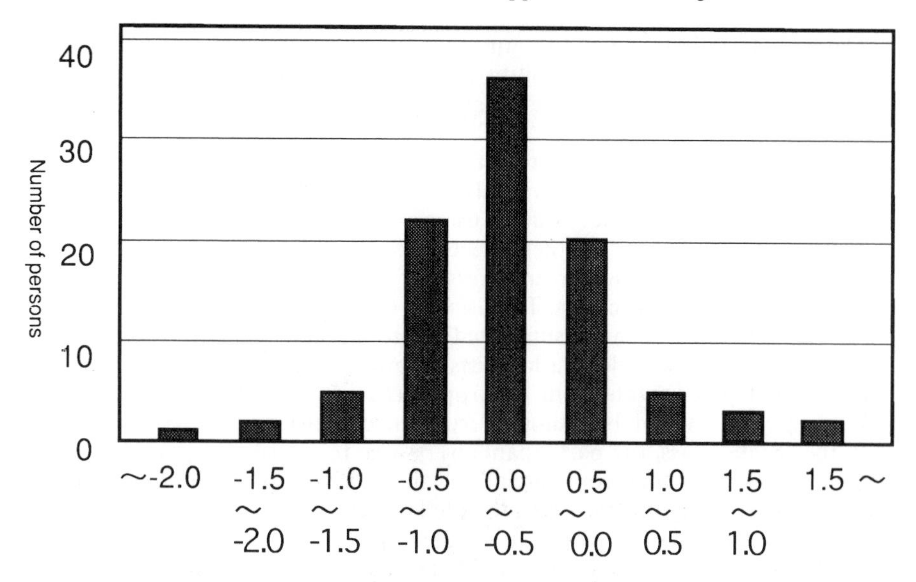

Figure 1.9 Distribution of difference between preferred and theoretical seat pan height settings of office workers

them. Consequently, findings may have been influenced by an order effect. The average change from the theoretical seat pan height was 2.2 cm when the seat pan height was decreased, and 1.2 cm when the seat pan height was increased.

Participants cited 87 reasons why they preferred seat pan height lower, and 110 reasons why they preferred higher seat pan heights larger than the theoretical setting. Some of the frequently cited reasons are listed in Table 1.2.

The findings obtained in the preliminary survey and the two chair usage classes may be summarized as follows:

Table 1.2 Relatively frequent reasons cited for low chair and high chair, respectively (participants were allowed to make multiple replies)

I prefer low chair because:	
'can place feet on floor'	20
'no thigh compression'	14
'feel comfortable'	15
I prefer high chair because:	
I am tall	4
I usually like high chairs	
I am tired in legs with low chair when I wear high-heeled shoes	
'have long knee length'	
'must bend legs when I sit on low chair'	
I feel uncomfortable when I sit on low chair	13

1. The variability between stature and theoretically calculated seat pan height is substantial, as is the variability between stature and preferred seat pan height. Thus, the method of determining seat pan height from stature alone is insufficient.
2. Although seat pan height predictions were statistically correlated with selections, in many cases individuals of a given stature selected very different seat pan heights. One group of users selected seat pan heights that were higher than predicted; another selected seat pan heights that were lower. The users in the former group normally prefer high seat pans, and feel uncomfortable on low chairs. The users in the latter group prefer low chairs because their feet remain supported or to minimize thigh compression. There users may be called 'rationalists'. However, the extrapolation of theoretical seat pan heights from anthropometric data may serve as a useful starting point and can be estimated from the dimensions of individual users.
3. In the second class, the participants were able to adjust seat pan heights of their chairs lower than commercially available chairs. The minimum adjustable seat pan height, based on user preferences, was 37 cm. The equivalent compressed seat pan height is about 35 cm.
4. The chair usage class allows the participants to exchange information among themselves, allows for the valuable collection of research data, and enables the participants to determine their preferred seat pan height. Increase in user knowledge about chairs is also likely to improve seating design.

Office investigation

In the chair usage class described above, the desk and the VDT screen were not used. Individual preference of seat pan height was examined under this condition. In this section, seat pan height at the time of working with a desk and VDT screen was measured. All the office workers were experienced VDT workers; they were male and female subjects aged 18–59-years-old. The number of males was 45, the number of females 20. Measured seatpan height is considered as the preferred seat pan height. Those measurement values were compared with the theoretical values derived by equations (3) and (4). Their working posture was recorded by VCR at the same time. A minimum number of items of the workstation were measured. Namely, the heights of the work desks and chairs were measured. And all office workers' popliteal heights and the heights of the heel of their shoes were measured. The measured values of the seat pan height are shown in Figure 1.10.

The distribution of the measured value of the seat pan height of the chair ranged from 38 to 51 cm as shown in Figure 1.10. The centre of the distribution shifts to higher position than the distribution in Figure 1.7; and the distribution in Figure 1.10 has two peaks. The distribution ranged from 46 to 51 cm and corresponds to the setting based on the desk. These facts differ from the result of the experiments discussed above (Figure 1.7). However, subjects did not use the desk and the VDT screen at that experiment. The height of the seat was measured under the condition. On the other hand, subjects used the desk and the VDT screen during the office investigation in this section. Subjects who use the desk and the VDT screen tend to raise the seat. What is the

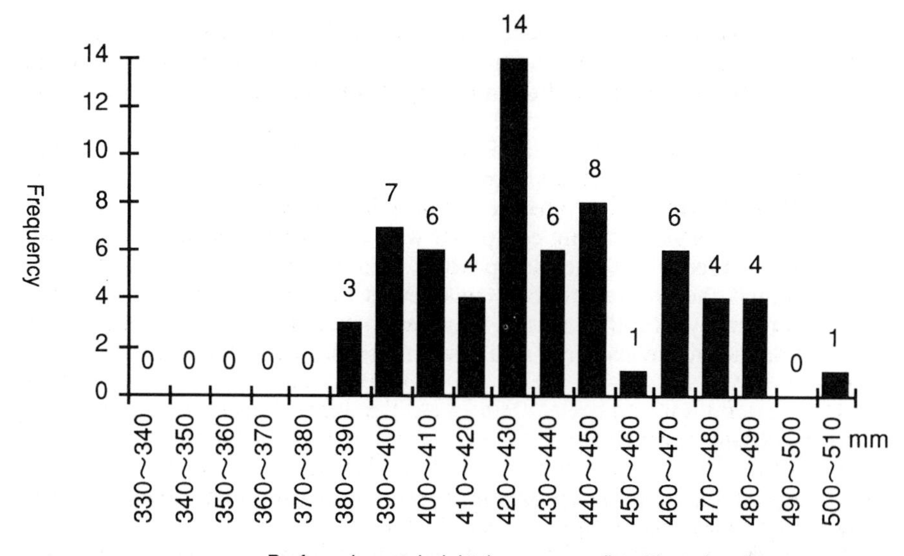

Figure 1.10 Frequency distribution of preferred seat pan height settings measured in the office investigation

reason for this fact? It may be that the operation of the keyboard becomes easier, therefore, this fact shows the basic necessity of the theoretical value which depends on the height of the desk.

Thirty-four observed heights (53 per cent) of the 64 subjects differed considerably from the theoretical values calculated by equation (3). Office

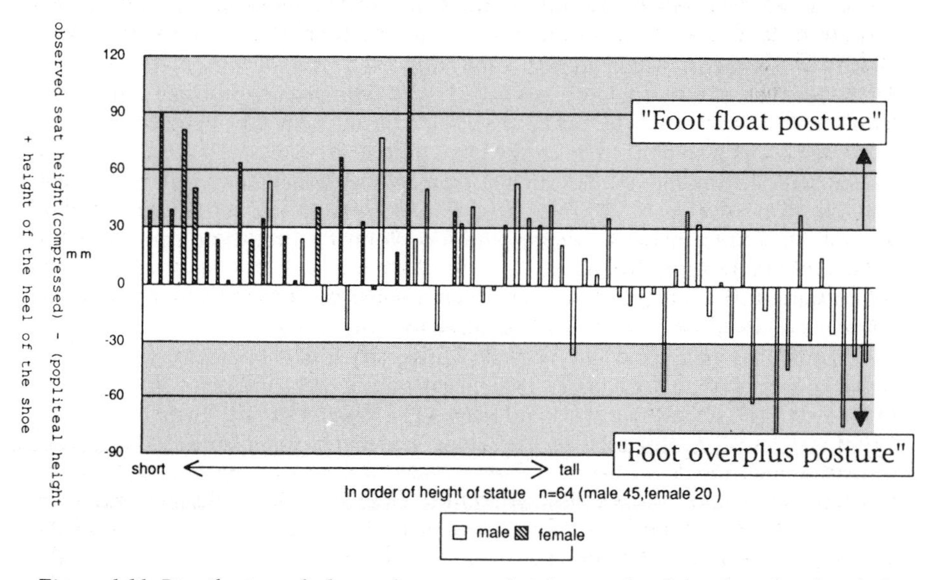

Figure 1.11 Distribution of observed seat pan height—popliteal height + height of the heel of the shoe. The office investigation

workers' way of sitting was classified by referring to the difference between the observed seat pan height and the (popliteal height + height of the heel of the shoe). The differences were plotted in Figure 1.11.

The allowable range was assumed to ± 3 cm of 0 point of the vertical axis of Figure 1.11 as a work-hypothesis. People whose difference value is over $+3$ cm are people whose foot floats 3 cm or more above the floor. This state will be called 'Foot float posture'. People whose difference value is over -3 cm are people whose legs are too long compared with the height under the desk. This state will be called 'Foot overplus posture'. 'Foot float posture' might be the posture which the worker chooses to operate the keyboard and to see the screen more easily. Another reason may be worker's preference or habit. The above explanation is considered as a mathematical interpretation of those postures.

Conclusions

Multi-function office chairs are becoming more prevalent. Office workers want to know how such chairs should be used. This chapter attempts to address this need with a participative approach which includes means of predicting generalized workstation settings to accommodate individual users.

This study suggests that user dimensions are insufficient for predicting selected seat pan heights. User seat pan height preferences depend on a number of factors, such as seating comfort, task, and health. Users need appropriate information and chairs with sufficient adjustment ranges to allow them to decide for themselves how to sit. Such information should comprise a system that is a 'soft technology' for seating. Users or nonexperts in ergonomics need a mechanism to communicate their perceptions and needs about chairs.

VDT sashaku and a parametric model were developed as theory necessary to implement the proposed information loop. The validity of the theory cannot be fully described within the space constraints of this chapter. The validity of the theory is considered to be supported by equation (3), high correlation between theoretical and preferred seat pan height, and centre of distribution in Figure 1.9. The necessity of calculating seat pan height based on desk height was also confirmed in the last section.

Seat pan heights may be determined from popliteal height or by subtracting sashaku from the work surface height. Observation of actual workplaces shows the existence of foot overplus and foot float posture; these postures may be the preferred posture patterns.

Consequently, seat research must discriminate between optimum seat pan height in relation to floor and in relation to work surface and screen. In the VDT guidelines established by Waseda University for its operators (Personnel Department, Waseda University 1987), seat pan height guidelines are presented for fixed and adjustable work surfaces VDT users, ranging from 151 to 179 cm in stature, in tables relating the chair and desk dimensions to the body dimensions of users.

Product-oriented ergonomics, including chairs, is often difficult because users must depend solely on the existing products. 'Ergonomic' chairs should not be forced upon users albeit designed according to anthropometric data. Users should be provided with a selection of chairs to choose from that allow them to be comfortable, and accomplish their work.

References

ANSI/HFS 100-1988, (1988) *Guidelines for Human Factors Engineering of Visual Display Terminal Workstations*, Human Factors Society, Santa Monica.

Cornell, P. and Kokot, D., (1988), Naturalistic observation of adjustable VDT stand usage, in *Proceedings of the Human Factors Society's 32nd Annual Meeting*, 496–500.

Hirasawa, N., Noro, K. and Cho, A., (1989), Accommodation of office chair to female workers, *Human Sciences*, **2**, 23–31.

Hirasawa, N., Noro, K. and Togami, H., (1990), Study on adaptation of VDT components to individual operators, *Japanese Journal of Ergonomics*, **26**, 87–93. (In Japanese.)

JIS S 1010, (1983), *Standard Size of Writing Desks for Office*, Japanese Standards Association, Tokyo. (In Japanese.)

Kohara, K., Ouchi, K. and Terakado, (1967), A study on height differences between desk and feet, *Japanese Journal of Ergonomics*, **13**, 159–65. (In Japanese.)

Labour Standards Bureau, Ministry of Labour, (1985), *Occupational Health Guidelines for VDT Work*, Japan Industrial Safety and Health Association, Tokyo. (In Japanese.)

Noro, K., Togami, H. and Onomura, T., (1985), Concepts of computer technology utilization for health management of VDT work, in *Proceedings of 1st Human Interface Symposium*, 255–58. (In Japanese.)

Noro, K., (1986), New concepts of computer technology utilization for health and work control of VDT operator, in Knave, B. (Ed.) *Proceedings of International Scientific Conference: Work with Display Units*, Stockholm, 473–76.

Noro, K., (1987), Prediction of high-technology development and problems of man-machine systems, in Noro, K. (Ed.) *Occupational Health and Safety in Automation and Robotics*, London: Taylor & Francis, 70–72.

Noro, K. and Hirasawa, N., (1988), Construction of multi-parametric model for workstation and operator (part 1), description of model, and example of engineering workstation (part 2), in *Proceedings of Annual Meeting of Kanto Chapter, Japan Ergonomics Research Society*, 23–31. (In Japanese.)

Noro, K., (1990), Participatory ergonomics: Concepts, advantages and Japanese cases, in Noro, K. and Brown, Jr. O. (Eds) *Human Factors in Organizational Design and Management-III* (Amsterdam: North-Holland, 83–86.

Noro, K. and Imada, A., (1991), *Participatory Ergonomics*, London: Taylor & Francis, 3–29.

Noro, K., (1992), Construction of parametric model of operator and workstation, *Ergonomics*, **35**(5/6), 661–676.

Ohnishi, T., (1977), *Practice of Healthy Life* Kyoto: Higashiyama Shobo. (In Japanese.)

Personnel Department, Waseda University, (1987), *VDT Work Guidelines*, Waseda University, Tokyo. (In Japanese.)

Povlotsky, B. and Dubrovsky, V., (1988), 'Recommeded' versus 'preferred' in design and use of computer workstations, in *Proceedings of the Human Factors Society's 32nd Annual Meeting*, 501–505.

Togami, H. and Noro, K., (1987), Optimum height of VDT work table, *Japanese Journal of Ergonomics*, **23**, 155–162. (In Japanese.)

2

Adjustability in context

Rani Lueder

Many computer users in today's offices experience high levels of stress. The reasons are not new, and have been evident in work situations throughout history. What is recent, however, is the widespread attention to seat and work station adjustability to address such problems.

Although much has been learned in the process, many related considerations have often been ignored. These include:

- Should seats adjust?
- Under what conditions?
- Are adjustments used?
- Do users set them correctly?
- What is their impact on health and productivity?
- What about so-called 'Dynamic seating', which supposedly obviates some adjustments?

Discrepancies are often apparent between assumptions held by ergonomists, claims within industry, and reports of end users (sf. Lueder 1983, 1986*a*; Miller and Suther, 1983; Sauter and Arndt, 1984; Cornell and Kokot, 1988, 1993).

The Ergon Chair, designed by Bill Stumpf in the 1970s for Herman Miller Furniture, may be the earliest recognized example of adjustable and ergonomic seating. However, a wide-spread attention to such seats did not take place until concerns emerged in the early 1980s with the introduction of computers.

Problems were erupting that we did not understand. Some of these concerns have since been largely discounted. Other considerations are still controversial, such as relating to psychosocial, organizational, and job characteristics.

Kroemer (1988) notes that musculo-skeletal and visual discomforts and pain constitute between 50 per cent and 80 per cent of objective and subjective symptoms in North America and Europe. In particular, many physical problems, although not unique, are compounded by intensive computer operation. Posture is inherently constrained during this work; typically, hands are on keyboard, mouse or numeric keypad, eyes on the document or screen, neck (perhaps) cradling the phone, and with feet on the floor.

We now recognize that constrained postures are associated with discomforts, a host of long-term (chronic) physical disorders, and psychological problems (sf. reviews by Kroemer, 1988; Lueder, 1986*b*; 1992). Such problems are compounded by poor working postures, inappropriate lighting, stress, and other factors often found in automated offices.

Does adjustability improve performance?

McLeod *et al.* (1980) found that students made fewer grammatical errors when seats were set in the mid-position. Other research has studied seat and work station adjustability in conjunction. For example, Dainoff (1990) and Springer (1982) each found that the combination of adjustable seats and computer work stations increased keystroke rates.

Springer's research found that an adjustable ergonomic seat alone increased keystroke rate between 4 and 8 per cent during word processing and data entry (respectively).

Hozeski (1986) compared several types of adjustable seat, finding that although subjects preferred the synchro-tilt with automated adjustability, their performance was better at a standard secretarial chair. He suggested this was because the seat adjustments were too complex. Subsequently, Hozeski and Rohles (1987) found the synchro-tilt was associated with increases of 7.5 per cent in the composite score of typing productivity and work-space attitudes; they did not describe how the score was attained.

Does adjustability improve health and comfort?

Seating affects the postures people sit in, their restlessness, and patterns of movements (for review see Lueder, 1983). People often evidence strong territoriality towards their seat and exhibit strong individual preferences. Turf wars may erupt, and devious tactics applied, centering on seating.

Adjustability also contributes to comfort and well-being. Researchers Shute and Starr (1984) of American Telephone and Telegraph found that when people sat at adjustable seating, they reported less shoulder and upper back pain and less intense pain of the lower back.

Do people use adjustments?

The term ergonomic seating is often equated with adjustability. Yet the two are not equivalent. Various studies have found that people often do not use their adjustable furniture (Webb *et al.*, 1984; Shute and Starr, 1984). Out of one hundred fully adjustable work stations at Ontario Hydro, less than 5 per cent had ever been adjusted; none on a daily basis. Only 12 per cent of the users were aware that they could be adjusted (Schneider, 1983).

A survey of over 2000 air traffic controllers indicates that these workers seldom adjusted their seats (Kleeman and Prunier, 1980). Only about 10 per cent of the controllers would adjust their seats during the day; more than half were not even aware of some of the adjustments. Their response was reminiscent of the game musical chairs. At the moment of truth, each person grabbed the best seat available at the start of their shift; the last worker in gets no choice in what the seat is set at. (The author has also witnessed such behaviour within organizations.)

An article in the Wall Street Journal (Freedman, 1986) describes a female with short legs struggling at a high chair seat. Rather than using the controls, 'when my legs get kind of numb, I kind of scoot down so they touch the floor. Then, my lower back gets numb'. And so on.

Are users aware of adjustments?

A survey of 23 managers and facility managers suggests a predominant perspective that users do not know that their seating adjusts and when they do, it is seldom used (Lueder, 1983).

Adjustability alone does not suffice. People need to know that seats adjust. They must recognize the basis for performing adjustments. Controls must be easy to use. The adjustments and ranges must be acceptable. Hands on instruction helps (Dainoff, 1984).

McLeod *et al.* (1980) found that users who were provided with full manufacturer's literature adjusted their seat less often; too much information may actually serve as a disincentive to using controls.

Which adjustments are used most frequently?

Adjustments are not equally desired or used. A survey at State Farm Insurance (Springer, 1982) indicated that ability to adjust furniture and ease of adjustment was rated third and fifth among work station factors; more important than the amount of light, privacy, and noise control. Further, seat and backrest height were valued more highly than other controls. Arm height adjustment and seat slope* were not considered important (see Figure 2.1).

The survey by Kleeman and Prunier (1980) found most adjustments were (1) seat height, (2) backrest height and (3) backrest tilt. Among seat characteristics, adjustability was rated third in importance after comfort and safety and far above durability and appearance (see Figure 2.2).

Figure 2.1 User ranking of importance of chair adjustment features (Springer, 1982)

Chair Characteristics	Average Valence	Rank
Seat height adjustment	8.9	1
Back rest height adjustment	8.2	2
Ability to turn while seated	7.8	3
Back tilt adjustment	6.6	4
Arms	5.6	5
Seat tilt adjustment	5.2	6
Ability to lean back	5.2	6
Carpet casters	4.8	8
Foot rest	2.8	9

* Such findings should be tempered with situational factors. That is, subjects lacked familiarity with this new feature, training, and understanding of the basis for using controls.

Figure 2.2 Rank order of preferred adjustments (Kleeman and Prunier, 1980)

Rank	Adjustment	Percentage agreeing with this ranking or higher
1	Seat height	73.3
2	Back height	59
3	Back tilt	64.4
4	Seat depth	63.8
4	Seat pitch	57.7
4	Arm height	57.9
4	Seat width	50
5	Back width	56.7
5	Back lateral	51
6	Arm	63.6

Similarly, De Groot and Vellinga (1984) found that seat height was adjusted most frequently, with backrest height in second place. Seat pan angle adjustments were used the least often (see Figure 2.3).

Shute and Starr (1984) compared a standard seat with an improved ergonomic and adjustable version. Staff trained in using the standard secretarial and ergonomic chairs adjusted both seat pan height and backrest height of both chair types frequently. This high rate of adjustment may have resulted from the attention received during the experiment as well as training. Again, seat pan height was adjusted more frequently than backrest height. The biggest discriminators between seats were in ratings of ease of adjustment and ability to adjust while seated.

Launis and Lehtela (1985) studied 154 computer users of shift work. Figure 2.4 shows that seat height controls were used more often than any of the work station elements, with backrest height in second place.

It is interesting that adjustments were considered more important by staff than they were used. For example, although 17 per cent of them usually adjusted their seat height, 42 per cent consider this control very useful correspondingly, 51 per cent never or seldom adjusted their seat height, but only 7 per cent considered them useless. Thirty per cent usually or sometimes adjusted their backrest height, but 90 per cent considered them useful or very useful.

Figure 2.3 Frequency of readjusting controls (De Groot and Vellinga, 1984)

	No. of readjustments per hour during the shift
Seat height	0.18
Table (front part)	0.28
Table (back part)	0.26
Seat slope	0.05
Seat depth	0.07
Backrest height	0.16

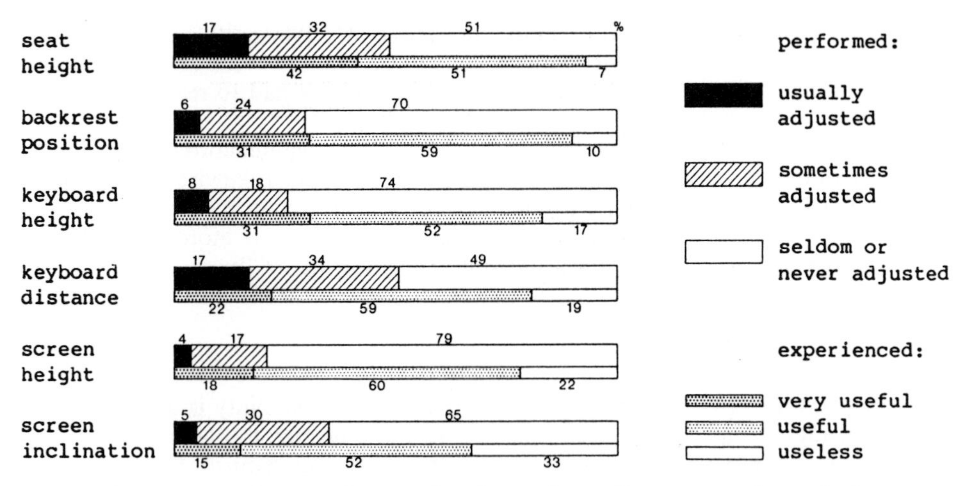

Figure 2.4 Per cent of users indicating that they perform adjustments, and perceived usefulness (*n = 143*) (*Launis and Lehtela, 1984*)

People consider adjustability important and will use controls frequently if circumstances are good. Seat height is rated the most important element in the work station and is used the most often.

Backrest height adjustment is rated second in importance, and is considered more important than backrest angle controls. Interestingly, this appears to contradict physiological research, which indicates that backrest inclinations (particularly up to 25° or 30° from the vertical) are very effective at reducing disc pressure on the spine and muscle activity. On the other hand, backrest height possibly need not be exact (Andersson, 1980; Boudrifa and Davies, 1984).

Adjustable seat depth was not considered very important, although this feature might have received higher ratings from particular individuals, such as small females. Arm rest height controls were not considered important. Seat pan angle controls* were not rated highly, although this may have been an artifact of experimental conditions.

How are the adjustments used?

A number of factors affect the use of adjustments. In addition, the presence of adjustments affects how seats are used. Burandt and Grandjean (1963) noted that adjustable backrests were used 58 per cent of the time, while fixed backrests only 35 per cent of the time. This represents an increase in usage by two-thirds.

The initial setting affects usage as well. Cushman (1984) of Eastman Kodak noted that people who adjust the seat from the lowest position, raised it less than people who adjusted it downwards from the highest position.

---

* This primarily refers to forward slope adjustments, where the front of the seat is lower than the rear.

At the University of Connecticut (McLeod *et al.*, 1980), the original height of the setting also affected how much users liked the seat. If the original setting was at mid-point, they liked the chair less than if they had to reset it from an extreme position.

Correspondingly, these researchers suggest that manufacturers ship their seats out with an initial extreme adjustment position. Likewise, salespeople might ask potential clients to try out the seat when set at an extreme.

Finally, people with back problems want adjustable seats more than others, although they seem to like the same kinds of seat (Hall, 1972).

Ease of adjustability

The previous discussion underscores that ease of adjustability is as important as the adjustments themselves. Controls must also be easy to understand.

Users do not always recognize that their seat can adjust. For example, although most of the staff of a television newsroom had expensive ergonomic seats, not a single staff member recognized that the backrest height adjusted. This was because, in the name of aesthetic purity, the designer made the control indistinguishable from the seat backrest.

Even furniture salesmen are sometimes unable to adjust their furniture. On one occasion furniture representatives could not decipher a manual; on another, a tool that was required to set the seat height was lacking.

Some general characteristics of easy adjustments are given below (*Lueder, 1985*):

1. Controls can be used from the standard seated work position.
2. Labels and instructions on the use of the seat are readily understandable. Literature should not be too extensive, or users will not bother to read it.
3. Controls are evident and understandable. Seat height controls that are activated by seat swivel typically go unnoticed. Backrest height adjustments are often not recognized. Tension controls confuse.
4. No tools!
5. Immediate feedback is provided. Delay (such as caused by swivelling the seat to set height; feedback is delayed because the user must intermittently get up, rotate, sit and so forth).
6. The controls adjust in a logical and consistent manner (e.g., a lever raises a seat when pulled up).
7. A minimum of motions is needed, and little effort.
8. One-handed adjustment only.
9. The act of adjustment is intrinsically reinforcing (maybe even fun!).

When do people adjust their seat?

Apparently, in most situations people do not commonly use adjustments. One likely reason for this is inadequate seat design.

However, under certain circumstances, people will adjust their seats continuously, and their discomforts are correspondingly reduced. For example, Dainoff (1984) found that it is not enough to train users in how to adjust the seat; they must also be taught why they should do so, and receive hands-on instruction. This has been the author's experience as well.

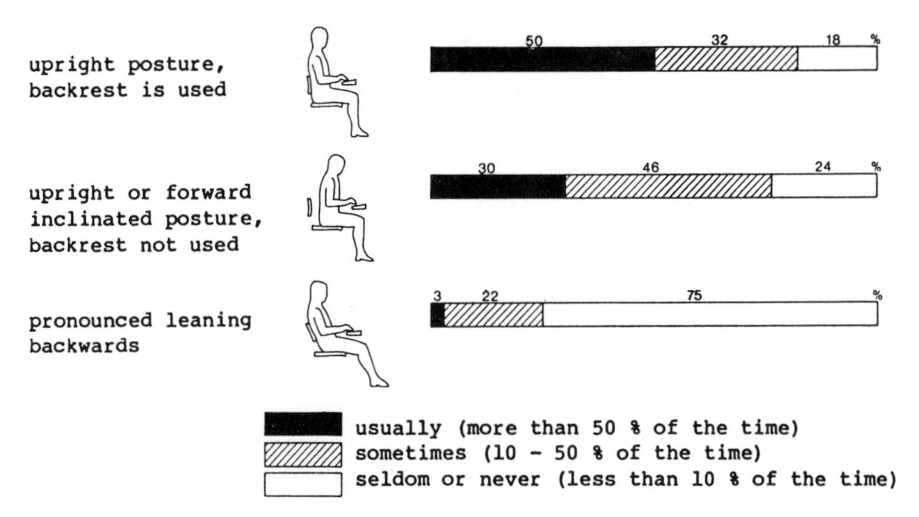

Figure 2.5 The reported percentual time distribution of each of three sitting postures at VDT (n = 144) (Launis and Lehtela, 1985)

Such training presupposes knowledge of what postures to assume. A general consensus has emerged that no single posture is ideal because posture would then be constrained; however, disagreement remains about the physiological acceptability of different positions.

People also sit differently than we thought they should (see Figure 2.5). Several studies have found virtually no correlation between users' anthropometric dimensions and adjustment settings (Chisvin, 1983; Cornell and Kokot, 1993; De Groot and Vellinga, 1984; Grandjean, 1980; Verbeek, 1991; Ong *et al.*, 1988; Grandjean *et al.*, 1983; Lueder, 1986*a* and 1986*b*). After training, the correlations between body sizes and settings did not increase, but adjustments became more intercorrelated (De Groot and Vellinga, 1984). Verbeek (1991) found that users sat remarkably similar with and without training.

The author believes that the biomechanics of posture is, to some extent, specific to the individual as well as the task. There is considerable variability within the population in spinal curvature, tendon length and flexibility, muscle strength, etc. How can we presume to dictate what is correct?

Potkin (1990) has developed a technique called Posturcise to help people establish their natural working postures. This involves teaching them to locate the extremes of their ranges of motion during both static positioning and movement, and find their natural positions within this range. Another approach is the use of surface EMG biofeedback in order to help individuals to increase awareness over postures and movements. This is achieved by re-learning the neuromuscular components of voluntary muscle control (Ettare, 1993).

Further, Winkel and Oxenburgh (1990) suggest that many of the adverse consequences of sitting might be attributed to a lack of physical variation, rather than by bad postures *per se*. If so, furniture adjustments should not only support good postures, but encourage continuous changes of settings over the course of the day.

Task length and adjustabiltiy

The longer one works, the more important adjustability becomes. The Canadian Labour Congress (1982) found the largest difference in complaints occurred between those employees working at a computer for four hours or less and those working for five hours or more. Muscle problems were three to four times more common in workers using the VDT for between seven and eight hours compared with operators with four or fewer hours of daily exposure.

Likewise, among air traffic controllers (Kleeman and Prunier, 1980), most were comfortable up to five hours, but the rate of discomforts grew quickly after that.

Apparently, many employees working at a computer for more than four hours a day are dramatically more disposed to discomforts. If their hours of operation are not reduced, adjustable seating and work stations will not eliminate the potential for, but may mitigate the extent of such discomforts.

Dynamic seating

In recent years, there has been a heightened recognition of the health consequences associated with lack of movement (for a review of the literature, see Lueder, 1992). This has caused some researchers to suggest that comfort and movement may sometime conflict, and that movement is more important than comfort (Winkel and Oxenburgh, 1990).

Further, a growing recognition that many users do not bother to adjust their seats has contributed to the impetus of seats that obviate the need for some of the adjustments. This may be enacted in at least two ways:

1. Seat motions are synchronized automatically. For example, the seat pan angle is automatically positioned in direct ratio to the angle of the backrest.
2. The seat floats by following and supporting the users' movements. For example, as the user leans forward, the backrest angle supposedly follows suit*.
3. The seat flexes and pivots with changes in weight and position.

The theoretical basis for this approach is intriguing and deserves investigation. The author uses a version of dynamic seat at her computer. However, these seats are not innately ergonomic; nor do they bypass a need to conduct and apply ergonomic design-based research. The primary contention is not standard adjustable *vs.* dynamic. Rather, it is how each seat is designed and the context in which it is used.

Many dynamic seats only adjust in seat height, tension, and perhaps include seat/back angle locks. In the author's experience, if the seat height is adjusted by rotating the seat, they will be used infrequently at best. Further, in order to comply with ANSI's† recommendations for minimum seat height, many

---

* Often not to the extent claimed; for example, see Dainoff *et al.* (1986).

† The American National Standards Institute provides seating and work station guidelines for the USA. For a copy of ANSI/HFS 100-1988 send $25 to Human Factors Society, P.O. Box 1369, Santa Monica, CA 90406, USA. Its revision is anticipated in April 1994.

manufacturers have been forced to reduce the upper end of the adjustability ranges in their pneumatic seats‡.

As a result, seats often do not adjust high enough. Research consistently indicates that people adjust their seat height to afford a comfortable working arm posture (Burandt and Grandjean, 1963; Floyd and Roberts, 1958; Floyd and Ward, 1964; Langdon, 1965; Lueder, 1986*a*). As a result, this problem is particularly acute for small females working at fixed height desks.

If not told, most people are not aware of seat tension adjustments and do not understand why or how to use them. Even at its lowest setting, the extent of seat tension is often excessive for the elderly or small females.

Dainoff *et al.* (1986) found users with floating seat/back supports would often select the extreme forward and extreme back position. Apparently the 'float' did not provide sufficient support, so users limited their postural ranges instead.

Often, the backrest heights of dynamic seats do not adjust. It may be that the specific location of the lumbar support of the backrest does not need to exactly match the curve of the spine, as long as it is in the general lumbar area (Andersson, 1980; Andersson, 1986; Boudrifa and Davies, 1984). Very tall and short people may not be fully accommodated at a fixed height backrest, but the extent of this discrepancy is less at a single (fixed) backrest angle.

However, due to pelvic rotation, the height of our lumbar curve changes as the backrest inclines. As a result, the backrest displaces relative to the person; because the pivot point of the seat is different from the pivot point of the user, a discrepancy occurs between the backrest and the user's back during inclination.

At least two manufacturers presently make seats in which the backrest lowers (vertically) as it inclines. This benefits because it helps the backrest follow and support the contours of the spine as it leans rearward. However, little research exists to define the boundaries of these necessary seat displacements. Further, the relationship between the seat and back displacements would depend on such seat factors as its relative pivot location.

References

Andersson, G. B. J., (1980), The load on the lumbar spine in sitting postures, in Oborne, D. J. and Levis, J. A. (eds) *Human Factors in Transport Research*, New York: Academic Press, 231–9.

Andersson, G. B. J., (1986), Loads on the spine during sitting, in Corlett, N., Wilson, J. and Manenica, I. (eds) *The Ergonomics of Working Postures*, London: Taylor & Francis., 309–18.

Boudrifa, H. and Davies, B. T., (1984), The effect of backrest inclination, lumbar support and thoracic support on the intra-abdominal pressure while lifting, *Ergonomics*, **27**(4), 379–86.

Burandt, U. and Grandjean, E., (1963), Sitting habits of office employees, *Ergonomics*, **6**, 217–28.

‡ Due to cost and structural limitations, pneumatic cylinders typically afford fixed ranges of adjustability.

Canadian Labour Congress, (1982), *Towards a More Humanized Technology: Exploring the impact of video display terminals on the health and working conditions of Canadian office workers*, Ottawa, Canada: Education and Studies Centre.

Chisvin, S. G., (1983), *Video Terminals: An ergonomics evaluation of work station and operating postures*, Unpublished thesis, University of Windsor, Windsor, Canada.

Cornell, P. and Kokot, D., (1988), Natural observation of adjustable VDT stand usage, *Proceedings of the Human Factors Society 32nd Annual Meeting*, Santa Monica, 496–505.

Cornell, P. and Kokot, D., (1994), Use of adjustable VDT stands in the office, in Lueder, R. and Noro, K. (eds) *Hard Facts about Soft Machines: The ergonomics of seating*. London: Taylor & Francis.

Cushman, W. H., (1984), Data entry performance and operator preferences for various keyboard heights, in Grandjean, E. (ed.) *Ergonomics and Health in Modern Offices*, London: Taylor & Francis, 495–504.

Dainoff, M. J., (1984), Ergonomics of office automation–A conceptual overview, *Proceedings of the International Conference on Occupational Ergonomics*, Toronto, 72–80. Volume 2 (Reviews).

Dainoff, M. J., Mark, L. S., Daley, R. and Moritz, R., (1986), Seated posture in the dynamic mode: Preliminary results, *Proceedings of the Human Factors Society 30th Annual Meeting*, Dayton, Ohio., 197–201.

Dainoff, M. J., (1990), Health and performance effects of ergonomical improvements in VDT workstation design, in Sauter, S. L., Dainoff, M. J. and Smith, M. J. (eds) *Promoting Health and Productivity in the Computerized Office: Models of successful intervention*, London: Taylor & Francis, 49–67.

De Groot, J. P. and Vellinga, R., (1984), Practical usage of adjustable features in terminal furniture, *Proceedings of the International Conference on Occupational Ergonomics*, Toronto, 308–11.

Ettare, D., (1993), Personal Communications, June 15, 1993, Biofeedback Associates of California, San Jose, CA.

Floyd, W. F. and Roberts, D. F., (1958), Anatomical and physiological principles in chair and table design, *Ergonomics*, **2**(1), 1–16.

Floyd, W. F. and Ward, J. S., (1964), Posture of school children and office workers, in *Proceedings of the 2nd International Conference on Ergonomics, Dortmund*, 351–60.

Freedman, A. M., (1986), Today's office chair promises happiness, has lots of knobs, *Wall Street Journal*, Wednesday, June 18.

Grandjean, E., (1980), Ergonomics of VDTs: Review of present knowledge, in Grandjean, E. and Vigliani, E. (eds), *Ergonomic Aspects of Visual Display Terminals, Proceedings of the International Workshop in Milan*, London: Taylor & Francis, 3–12.

Grandjean, E., Hunting, W. and Piderman, M., (1983), A field study of preferred settings of an adjustable VDT work station and their effects on body postures and subjective feelings, *Human Factors*, **25**, 161–75.

Hall, M. A. W., (1972), Back pain and car seat comfort, *Applied Ergonomics*, **15**, 82–91.

Hozeski, K. W., (1986), Subjective preference and use of work station adjustability features, *Proceedings of the Human Factors Society 30th Annual Meeting*, Dayton, Ohio, 890–3.

Hozeski, K. W. and Rohles, R. H., (1987), Subjective evaluation of chair comfort and influence on productivity, in Knave, B. and Wideback, P. G. (eds) *Work with Display Units*, Amsterdam: Elsevier Science Publishers, B. V. (North-Holland).

Kleeman, W. and Prunier, T., (1980), Evaluation of chairs used by air traffic controllers of the US Federal Aviation Administration, in Easterby, R., Kroemer, K. H. and Chaffin, D. (eds) *Nato Symposium on Anthropometry and Biomechanics: Theory and Application*, London: Plenum Press.

Kroemer, K. H. E., (1988), VDT workstation design, in Helander, M. (ed) *Handbook of Human–Computer Interaction*, Amsterdam: Elsevier Science Publishers B. V. (North-Holland), 521–39.

Langdon, F. J., (1965), The design of card punches and the seating of operators, *Ergonomics*, **8**, 61–8.

Launis, M. and Lehtela, J., (1985), Adjustable VDT work stations as experienced by infrequent users. *Ergonomics International 85: Proceedings of the Ninth Congress of the International Ergonomics Association*, Bournemouth, England. E2/2, 394–6.

Lueder, R., (1985), The economics of adjustability, *Proceedings of the Office Automation Conference*, Atlanta, Georgia, February 4–6, 1985, 105–12.

Lueder, R., (1983), The ergonomics of furniture adjustability, *The Ergonomics Newsletter*, **2**(1), 2–10.

Lueder, R., (1992), Seating for a new worker, in Sweere, J. (ed) *Chiropractic Family Practice*, Rockville, Maryland: Aspen Publishers, Inc.

Lueder, R. K., (1986*a*), Seat height revisited, *Bulletin of the Human Factors Society*, Santa Monica: The Human Factors Society, November, 3–5.

Lueder, R., (1986*b*), Workstation design, in Lueder, R. (ed) *The Ergonomics Payoff: Designing the electronic office*, New York: Nichols Publishing.

McLeod, W. P., Mandel, D. R. and Malven, F., (1980), The effects of seating on human tasks and perceptions, in Poydar, H. R. (ed) *Proceedings of the Symposium of Human Factors and Industrial Design in Consumer Products*, Medford, Massachusetts: Department of Engineering Design, Tufts University, 117–26.

Miller, W. and Suther, T. W., III, (1983), Display station anthropometrics, *Human Factors*, **25**(4), 401–8.

Ong, C. N., Koh, D., Phoon, W. O. and Low, A., (1988), Anthropometrics and display station preferences of VDU operators, *Ergonomics*, **31**(3), 337–347.

Potkin, M., (1993), Personal Communications, June 1, 1993, M. Potkin, Dr. of Chiropractic in Tarzana, CA.

Sauter, S. L. and Arndt, R., (1984), Ergonomics in the automated office; gaps in knowledge and practice, in Salvendy, G. (ed) *Human Computer Interaction*, Amsterdam: Elsevier Science Publishers, 411–4.

Schneider, M. F., (1983), *Assessment of Video Display Work Stations*, Toronto: Ontario Hydro (internal report of their research).

Shute, S. J. and Starr, S. J., (1984), Effects of adjustable furniture on VDT users, *Human Factors*, **26**(2), 157–70.

Springer, T. J., (1983), Personal communication.

Springer, T. J., (1982), Visual Display Terminals: A comparative evaluation of alternatives. State Farm Mutual Automobile Insurance Company, Bloomington, Illinois, March.

Verbeek, J., (1991), The use of adjustable furniture: Evaluation of an instruction programme for office workers, *Applied Ergonomics*, **22**(3), 179–84.

Wallersteiner, U., (1982), Evaluating a multi-user VDT work station in a working environment, *Proceedings of the Human Factors Society 26th Annual Meeting*, Seattle, 145–9.

Webb, R. D. G., Tack, D. and McIlroy, W. E., (1984), Assessment of musculo-skeletal discomfort in a large clerical office: A case study, *Proceedings of the International Conference on Occupational Ergonomics*, Toronto, 392–6.

Winkel, J. and Oxenburgh, M., (1990), Towards optimizing physical activity in VDT/office work, in Sauter, S. L., Dainoff, M., Smith, M. J. (eds) *Promoting Health and Productivity in the Computerized Office: Models of successful ergonomic interventions*, London: Taylor & Francis.

3

Three myths of ergonomic seating

Marvin J. Dainoff

Introduction

Three statements summarize certain current approaches to ergonomic seating:

1. Ergonomic seating requires a single postural orientation (90° upright) that is independent of operator task.
2. Ergonomic chair adjustments should be passive in the sense that chair surfaces 'float' with the operator's movements.
3. Users should not require training in how to sit in a chair.

Each of these propositions can be challenged. This paper discusses these challenges, and presents an alternative 'ecological' framework regarding seated work posture.

The 90° posture: Statement of the problem

The large-scale introduction of computer technology into the workplace during the 1970s was quickly followed by an 'epidemic' of visual, musculoskeletal, and stress-related complaints. Many of these complaints resulted from poor design of these new environments, and centred on the video display terminal (VDT).

Early efforts at redesigning such work-places focused on the viewing characteristics of the VDT. In retrospect, specification of work posture was considered a relatively straightforward anthropometric problem. It was generally believed that for the optimal seated work posture, the head is erect, the trunk is upright, the feet are flat on the floor, and major joint angles (elbow, hip, and knee) are fixed at 90°. Mark and Dainoff (1988) defined this as the 'cubist' approach to working posture, since it can be modelled as a series of cubes at right angles.

Within such a cubist framework, the designer's goal is to permit a proportion (typically 5th to 95th percentile) of a defined user population to attain a cubist working posture. Relevant anthropometric dimensions of that population (e.g., popliteal height, seated arm rest height) are then used to develop guidelines for ranges of vertical adjustability in seat pan, keyboard, and VDT screen height.

However, it soon became apparent that the cubist model was inadequate. With the advent of automation, it is also becoming increasingly important that seated postures be understood; the recognition has grown regarding serious health consequences that may result from working for prolonged periods of time at poorly designed work stations (Hettinger, 1985; Grieco, 1986; Aaras *et al.*, 1988; Winkel, 1988; Zacharkow, 1988).

Prolonged sitting without support increases stress on the spine and stresses ligaments and muscles of the lower and upper back and legs. Disc nutrition is adversely affected. Pressure gradients on discs and abdominal viscera increase; posture may contribute to recent reports of adverse pregnancy outcomes (Scalet, 1987, p. 91). These and other concerns regarding fixed work postures underscore the necessity of good ergonomic design.

The 90°/cubist model for seating has been attacked from two fronts. Mandal (1981) proposed that seat pans tilt forward, arguing that thighs and pelvis should incline forward while the trunk remains upright. He reasoned that such a posture would rotate the pelvis forward, thereby restoring the normal standing posture of the spine (lordosis) and relieving pressure on the spinal disc.

On the other hand, Grandjean *et al.* (1983) proposed that a high and tilted back rest should allow the trunk to recline. This posture also relieves pressure on the lower spine, since much of the weight of the trunk is transferred to the back rest. The backward leaning posture of Grandjean seems to be naturally preferred by most users, more closely resembling rest or relaxation. However, leaning backwards moves the eye away from the display. In addition, 90° arm angles cannot be maintained since fingers are now well above the keyboard.

Both of these postures can be of benefit according to task demands (Dainoff and Mark, 1987). The demands of the task, in turn, depend on the viewing characteristics of the screen and hard copy. In the US, text characters displayed on the screen presently tend to be larger than text characters printed on paper. For example, text characters on the IBM PC screen are almost twice as high as those printed on paper by the IBM Proprinter.

Thus, tasks requiring close attention to paper copy and/or high speed keying do not allow the user to lean back; in these cases, the forward tilt becomes preferable.

Research conducted in our laboratory indicates that users performing high speed data entry sat in the forward tilt posture and found it effective. Operators who performed screen-based editing tasks (with very low keying requirements) found the backward leaning posture effective (Dainoff and Mark, 1987).

In order to accommodate both postures, chair adjustment mechanisms must be more complex than those that simply vary the seat height (required to achieve the 90° cubist posture). Interestingly, both forward and backward postures require that work surface heights adjust or be higher than the conventional 71–79 cm.

The passive approach

These arguments suggest two basic design criteria:

1. chairs should afford a variety of seated postures, including both forward and backward inclinations; and
2. chairs should promote easy movement between positions.

By the mid-1980s, the office furniture industry responded to these pressures by offering products with a variety of adjustments. Many seat pans could tilt backward and, in some cases, forward as well. Back rests could frequently recline, either independently of the seat pan angle, or in some linked relationship. However, a new series of problems emerged as these more complicated chair mechanisms were introduced. The 'ordinary' act of sitting in a chair had to be mapped onto an unfamiliar 'system' with controls and adjustments. Instruction manuals were necessary to enable users to operate these chairs.

Consequently, anthropometric and biomechanical design criteria alone do not suffice. Seated posture cannot be understood without considering the user's means of control and co-ordination. Consider the operations entailed in moving from a fully forward to a fully rearward posture in an ergonomic chair with 'active' controls. With active controls, the operator must activate a button or lever to move the chair surface; once the control is released, the surface will fix in its new position.

In this case, the forward to backward movement requires three separate adjustments. First, the seat pan angle must change from forward to back. This change in angle will, for chairs with a centre pivot, result in excessively high feet, and seat pan height. Finally the back rest must recline from the vertical.

Are such operations too complicated for the average user? Some chair designers believe so. A popular alternative design is the so-called 'passive' (also called 'dynamic') chair. Such designs typically only require adjustment of seat pan height and (sometimes) tension for the chair to 'float'; seat pan and back rest then move with the operator. Seat pan and back rest angles are linked in some ratio (typically 1 : 2 or 1 : 3). Advantages of such passive adjustments include:

1. adjustment is simplified—only one active control (seat pan height) is required; and
2. movement between alternative work postures is enhanced.

However, this popular design introduces problems associated with stability and linkage. Mark and Dainoff (1988) and Moritz (1988) have analysed these concerns; their results are discussed below.

The Miami experiments

Method

A Fixtures Furniture Discovery Model DT chair was utilized, in which the seat pan and back rest adjusted independently. The seat pan adjusted from $-6°$ to $+8°$. The back rest adjusted from $-4°$ to $+20°$.

Twelve subjects alternated between keyboard intensive (data entry) and screen intensive (verification) manuscript typing tasks. Subjects worked for

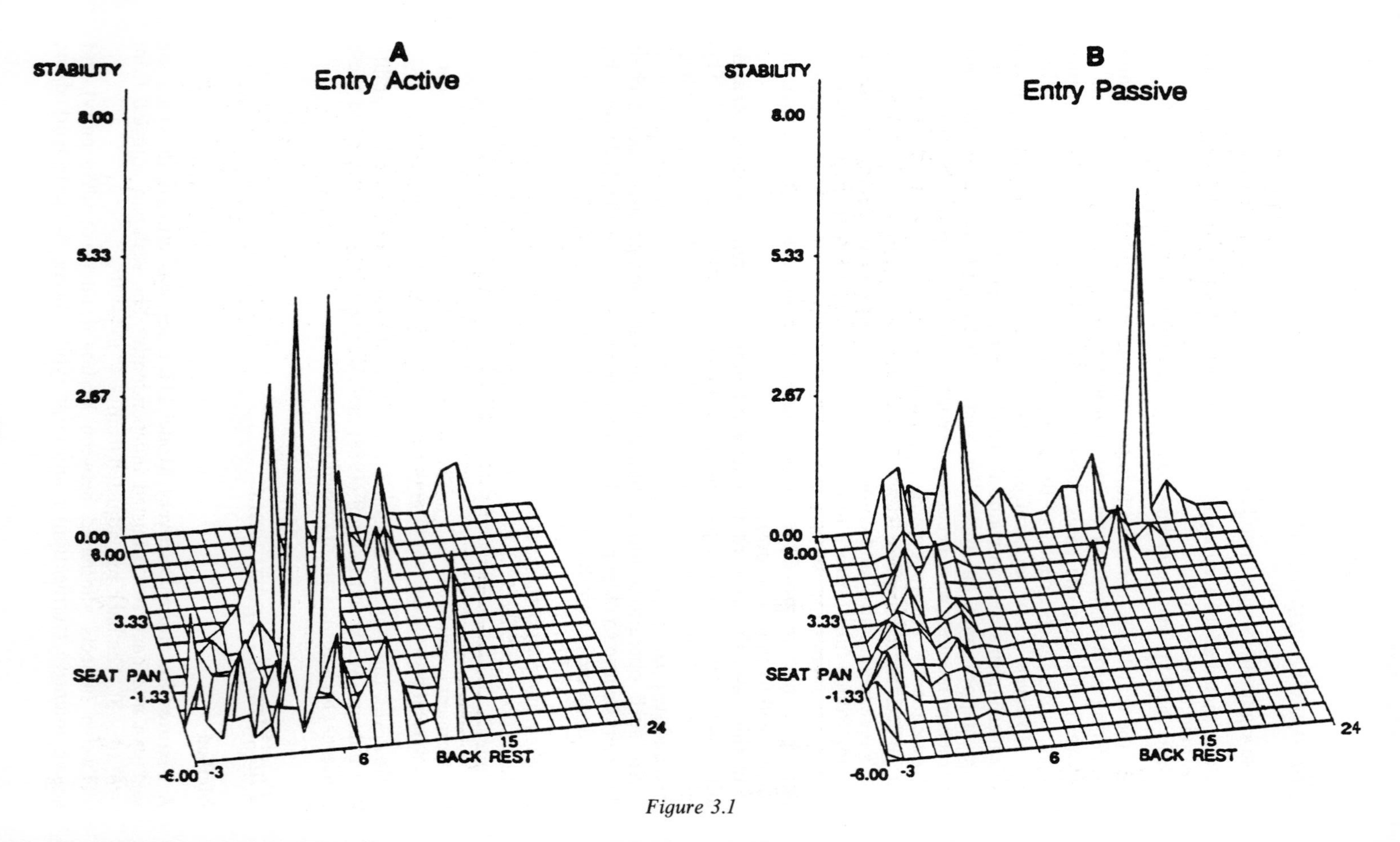
A
Entry Active
STABILITY
8.00
5.33
2.67
0.00
8.00
3.33
SEAT PAN
-1.33
-6.00
-3
6
15
BACK REST
24
B
Entry Passive
STABILITY
8.00
5.33
2.67
0.00
8.00
3.33
SEAT PAN
-1.33
-6.00
-3
6
15
BACK REST
24

Figure 3.1

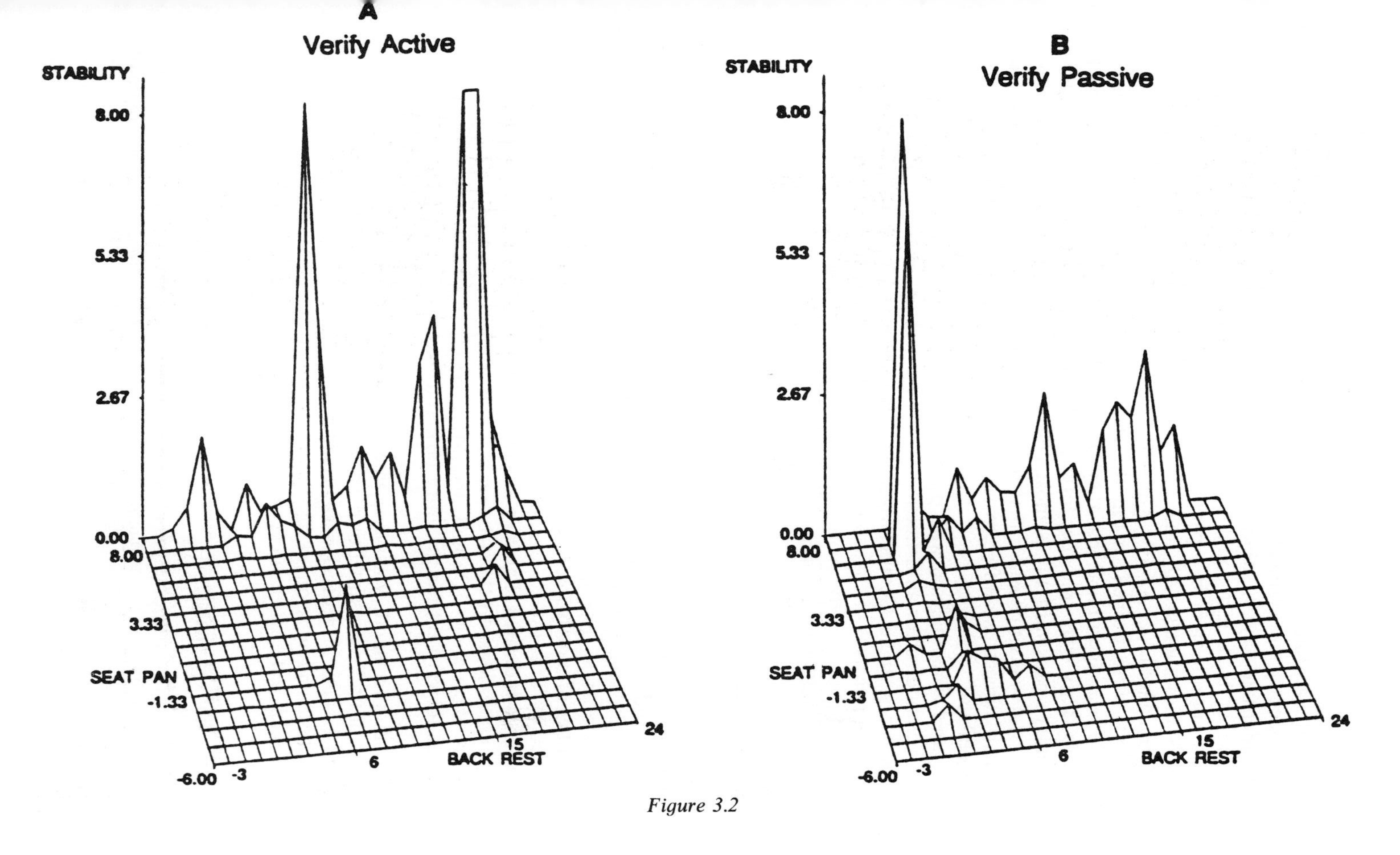
A
Verify Active
STABILITY
8.00
5.33
2.67
0.00
8.00
3.33
SEAT PAN
-1.33
-6.00
-3
6
15
BACK REST
24
B
Verify Passive
STABILITY
8.00
5.33
2.67
0.00
8.00
3.33
SEAT PAN
-1.33
-6.00
-3
6
15
BACK REST
24

Figure 3.2

180-minute sessions on two successive days, alternating between 90-minute periods of entry and verification tasks.

On alternate days, a subject's chair either adjusted in the active mode (the seat pan or back rest could be actively moved to a desired location, then locked in position), or in a passive mode (seat surfaces 'floated' passively with subject's movements). All four conditions (entry/verify; active/passive) were counterbalanced.

Subjects were trained to use the chair. Part of this training included a rationale bearing in mind that they might wish to use either forward or backward tilt. They were told that forward tilt is most effective for copy intensive data entry; backward tilt is most effective for screen intensive verification. However, it was emphasized that these principles were to serve only as guidelines and subjects were free to sit in whatever posture they considered most comfortable.

Postures were analysed with potentiometers that were connected to each of the three major chair adjustments; these were seat pan height, seat pan angle, and back rest angle. Computerized data capture allowed a continuous recording of all three axes of chair position throughout the course of the experiment. In addition, videotaped records of all sessions provided independent confirmation of potentiometer records.

Results—stability

A stability analysis was carried out across subjects within each of the four conditions. The analysis was represented in a three dimensional plot in which seat pan and back rest angles were plotted on the x and y axes and a derived measure of stability plotted on the z axis. Stability, in this case, was defined as the ratio of total time spent at a given (x, y) postural orientation divided by the number of times subjects moved into that orientation. Stable postures were identified by large ratios; unstable postures by small ratios.

This data indicated that very different postures were assumed in active and passive conditions. In the active conditions, subjects behaved as they had in previous experiments (Dainoff and Mark, 1987): they followed instructions and used forward tilt for the entry task and backward tilt for the verification task.

On the other hand, when the same subjects were in the passsive conditions, they did not follow instructions. They avoided the forward tilt position for the entry task, and, in general, assumed less stable postures (see Figures 3.1 and 3.2; in these and subsequent figures, seat pan angle is measured from horizontal and back rest angle from vertical).

Results—linkage analysis

In commercially available passive-linked chairs, seat pan and back rest angles vary in some fixed ratio; typically 1 : 2 or 1 : 3. That is, for every 1° of angle of rotation of the seat pan (backwards or forwards), the back rest moves 2° or 3°. Since subjects in the present study used a chair in which seat pan and back rest adjust *independently*, we were able to compare preferred ratios with industry practice.

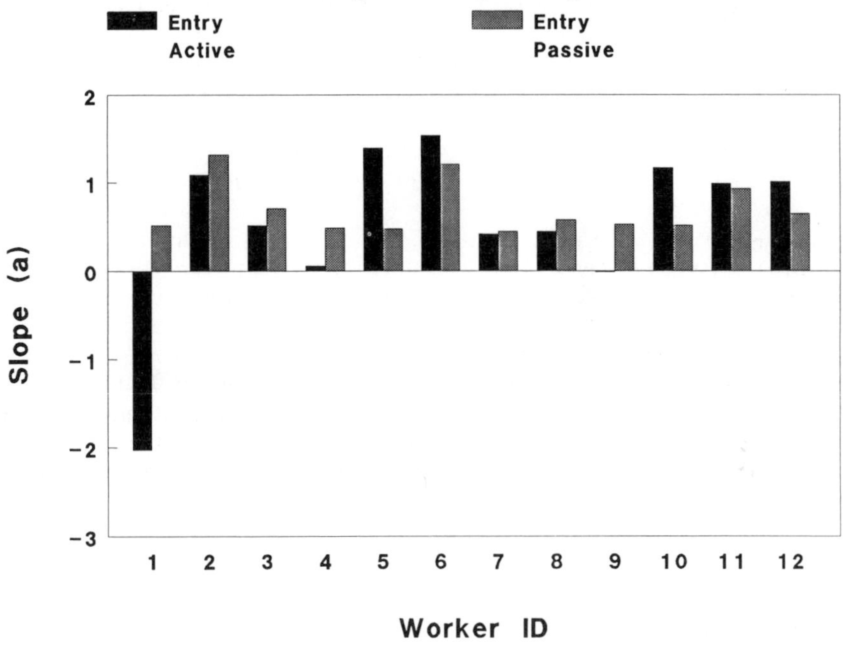

Figure 3.3

A scatter plot and regression equation was produced for each of the 12 subjects under each condition. In each case, back rest angle was regressed against seat pan angle. The slopes of these regression equations indicate the extent to which our subjects' preferred postures fell within the constraints imposed by industry practice. In particular, if a subject's preferences matched industry practice, the obtained slopes should fall close to region of 2.0–3.0.

In fact, these data indicated that there is a very poor match between industry practice and subjects' preferences for relative seat pan and back rest angles. The summary histograms for the entry task, show that none of the subjects even approached slopes in the range of 2.0–3.0. In the verification task there were, in fact, some subjects who entered this range, but the overall slope variability was very high. Furthermore, the individual scatter plots indicate that the back rest angle/seat pan angle relationships were highly non-linear; this was particularly true for the verify task. (See Figures 3.3 and 3.4.)

Sample individual scatter plots are included. These plots exemplify a typical pattern in which the increase in back angle against slope is very shallow over much of the range, but very steep as maximum seat pan angle is reached. Thus, the slope of the linear regression line did not accurately represent this non-linear data. (See Figures 3.5 and 3.6.) It is interesting that Yamaguchi (cited in Grandjean, 1973, pp. 104–106) also found that a non-linear relationship between seat and back rest angle was required to achieve minimum disc pressure.

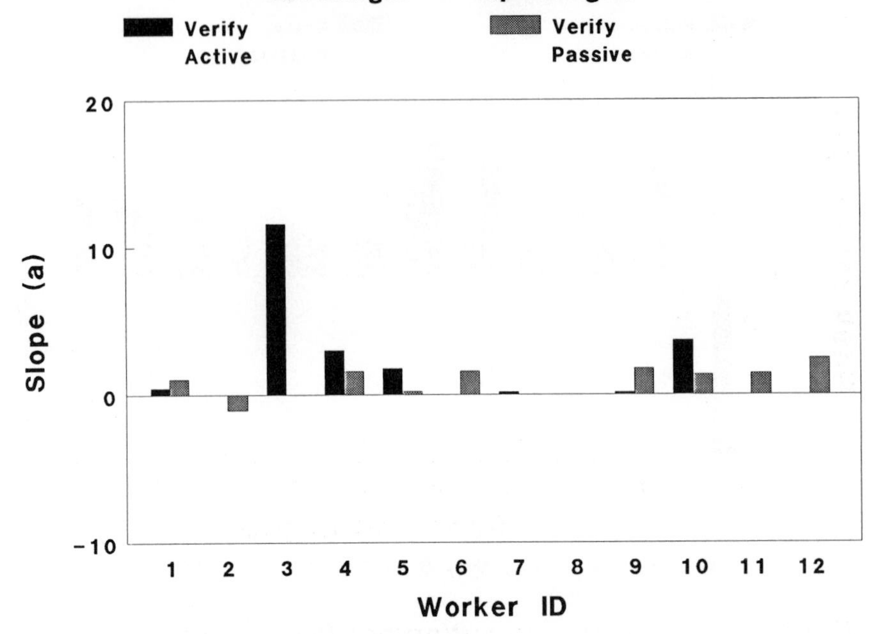

Figure 3.4

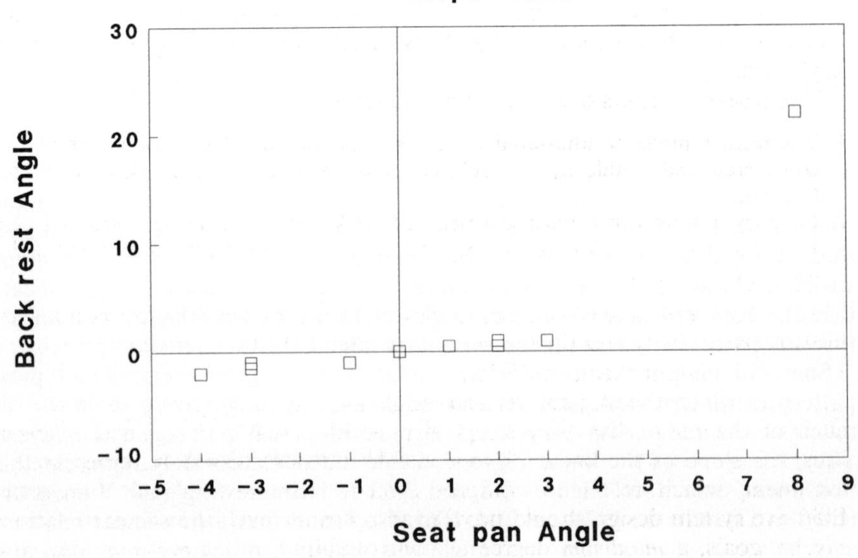

Figure 3.5

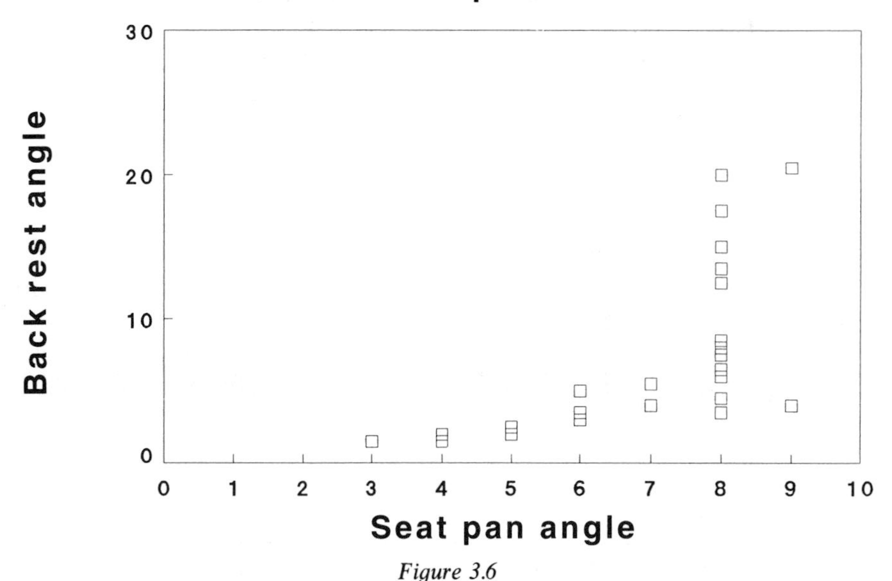

Figure 3.6

Teaching how to sit—summary and conclusions

The passive 'linked' chair design attempt to provide operators with a simplified, 'user-friendly' chair that facilitates postural changes with a minimum of active adjustments. As such (in our judgement) this chair represents a reaction against chair mechanisms which are so complicated that they require instruction manuals. However, this design approach is problematic when the operator must maintain a relatively fixed orientation with respect to the screen and keyboard.

The research results discussed above suggest:

1. The passive mode of adjustability provides less stability. This lack of stability was considered undesirable by our subjects, and could conceivably affect work performance.
2. Industry custom which links seat pan and back rest in a constant ratio of 1 : 2 or 1 : 3 constrains operators to postural orientations which they would avoid if they were able to choose.

This argument suggests a framework in which issues of operator control and user-interface are given their proper importance. In this context, the work of Frese (1987) might shed some light.

Frese differentiated between **control**, which refers to the possibilities for effective control of a system, **complexity**, which refers to the number of necessary decisions required to achieve goals through use of a system, and **complicatedness**, which refers to complex systems that are difficult to control. Effective system design should provide an *optimum* level of complexity to meet system goals, a *maximum* degree of controllability, but a *minimum* degree of complicatedness.

Thus, in Frese's terms, a chair with three independent degrees of freedom of control (seat pan height, angle, and back rest angle) may represent a system which is excessively complicated. Hence reducing the number of adjustment possibilities, by fixing linkages and passive adjustability, may reduce its complicatedness. However, in the process, the complexity level has also been reduced below the optimum level necessary to achieve postural control.

We prefer to increase the controllability of the chair. This can be accomplished in two interrelated ways:

1. increase the usability of chair control mechanisms by ensuring that the mapping between intention and action is clear cut; and
2. increase the effectiveness of user training.

These goals are interrelated. The motor skills required to move the chair surfaces from one configuration to another can be minimal with well-designed 'user-friendly' controls. At a higher level, however, these skills are irrelevant unless the users understand why they should alternate between these configurations. We recommend that the users be trained in both 'wellness' education and work efficiency, and present the chair controls as a means of exploring their workspace envelope, so that they can establish their own optimum postural orientations (for health and performance) that are specific to task demands and the operator.

References

Aaras, A., Westgaard, R. H., Stranden, E. and Larsen, S., (1988), *Postural load and the incidence of musculoskeletal illness.* Paper presented at NIOSH/Miami University Workshop: *Promoting Health and Productivity in the Computerized Office: Models of Successful Ergonomic Interventions*, Oxford, Ohio.

Dainoff, M. and Mark, L., (1987), Task and the adjustment of ergonomic furniture, in Knave, B. and Wideback, G. P. (eds) *Work with Display Units*, Amsterdam: North Holland.

Frese, M., (1987), A theory of control and complexity, in Frese, M., Ulich, E. and Dzida, W. (eds) *Human Computer Interaction in the Workplace*, Amsterdam: Elsevier.

Grandjean, E., (1973), *Ergonomics of the Home*, London: Taylor & Francis.

Grandjean, E., Hunting, W. and Pidermann, M., (1983), VDT workstation design: Preferred settings and their effects, *Human Factors*, **25**(2), 161–76.

Grieco, A., (1986), Sitting posture; An old problem and a new one, *Ergonomics*, **29**, 345–62.

Hettinger, T., (1985), Statistics on diseases in the Federal Republic of Germany with particular reference to diseases of the skeletal system, *Ergonomics*, **28**, 17–20.

Mandal, A. C., (1981), The seated man (Homo Sedans), the seated work position, theory and practice, *Applied Ergonomics*, **12**, 19–26.

Mark, L. and Dainoff, M., (1988), An ecological framework for ergonomic research, *Innovation*, **7**, 8–11.

Moritz, B., (1988), Stability of Seated Work Posture. Unpublished MA thesis, Miami University, Oxford, Ohio.

Scalet, E. A., (1987), *VDT Health and Safety*, Lawrence, Kansas: Ergosyst Associates.

Winkel, J., (1988), Towards optimizing physical activity in VDT/office work. Paper presented at NIOSH/Miami University Workshop: *Promoting Health and Productivity in the Computerized Office: Models of Successful Ergonomic Interventions*, Oxford, Ohio.

Zacharkow, D., (1988), *Posture: Sitting, Standing, Chair Design and Exercise*, Springfield, IL: Charles C. Thomas.

4

Use of adjustable VDT stands in an office setting

Paul Cornell and Doug Kokot

The introduction of VDTs in the office has presented many challenges for work station designers. Traditionally, desks and work surfaces had been set to accommodate reading and writing tasks, typically 29–30 in (73.6–76.2 cm). Typewriter surfaces were set slightly lower in order that the fingers, resting on the keys, were at approximately the same height. This design rule worked well for the tasks performed.

Office work has changed, however. Today nearly all workers – including executives – need to access electronic information, most often via a keyboard. This is a fundamental change in the nature of office work, both cognitively and physically. As in all jobs, the design of the work station needs to accommodate the task and tools used. When there is a mismatch, there is an increased risk of discomfort, fatigue and job dissatisfaction. The evolution in office work requires an evolution in design.

In terms of the physical aspect of office work, the increased use of keyboards is the most significant job change. This has raised concern about the potential health risks, particularly repetitive strain injury (RSI). Some attribute RSI complaints to improper work surface height. Of the four major contributors to RSI – force, repetition, posture, and vibration – posture may be adversely affected by work surface height.

Research linking RSI to posture, however, has been inconclusive. For example, studies examining moderate postural deviations have failed to establish a causal relationship (Armstrong, Fine, Goldstein, Lifshitz and Silverstein, 1987; Silverstein, Fine and Armstrong, 1987). Research specifically on office settings actually suggests that injuries are influenced more by the cognitive and social aspects of job design (Hadler, 1986; Bammer, 1989; Green and Briggs, 1989; Linton and Kamwendo, 1989). In spite of this, work surface adjustability is often suggested as a means of *injury* prevention e.g., Allison, 1990.

Recommendations for user adjustability do not always specify an adjustment range. In cases where a range is provided, it is usually based upon anthropometric models assuming an upright, rectilinear posture and summing body segment lengths (e.g., popliteal length plus torso length, etc.). The fifth percentile female typically provides one end point and the ninety-fifth percentile male provides the other (e.g., Rinalducci *et al.*, 1983; ANSI, 1988).

The reasoning behind this approach appears sound, but laboratory studies of preferred work surface settings have not been consistent with predictions (Brown and Schaum, 1980; Miller and Suther, 1981). In one of the more thorough studies on the subject, Grandjean, Hunting and Pidermann (1983) observed postures and settings 'distinctly different from those recommended in textbooks' (p. 161). The Grandjean *et al.* study was important because of its methodology: 68 subjects were observed over a five-day period *while at work.* Subsequent studies (Weber, Sancin and Grandjean, 1984; Cushman, 1984; Hozeski, 1986) corroborated the earlier findings that people prefer higher heights than predicted by the upright posture model. Despite the consistency of these findings, many hold fast to these theoretical, additive models – including recent American (ANSI/HFS 100-1988) and Canadian (CAN/CSA-Z412) guidelines.

Many assertions have been made about office worker preferences and needs, but there is little substantive data. A majority of the data that exists has been collected in the laboratory, with subjects performing cursory tasks over short time periods. Given the potential functional, aesthetic, financial, and health impact of adjustable work surfaces, it is critical to know more about their use. The purpose of the present study was to observe adjustable VDT stand usage in an actual work environment over a long period of time. We investigated three hypotheses. First, we expected the range of settings to be higher than the ANSI and CSA guidelines. Second, we did not expect that the settings would change much over time. And third, we predicted there would be a modest, but significant, correlation between anthropometric dimensions and preferred settings.

Methods

Subjects

It is often stated that those in greatest need of adjustability are the VDT intensive workers, i.e. those who spend six or more hours per day using a keyboard. A group of these workers was selected from an accounting department at Steelcase, in Grand Rapids, Michigan. These workers perform data entry, data inquiry, and account verification. Over the course of the study, 91 people were observed. Fifty-eight (64%) of the subjects were women. All participants had at least six months job experience. They had received some training on the use of the adjustable features of the work station and chair. Participation in the study was voluntary and not reimbursed.

Apparatus

The adjustable work station used in the study was the Steelcase 9051 VDT stand. It had separate keyboard and display surfaces and each surface could

Table 4.1 Work station measurements

	n	Mean	Std. dev.	Range
Seat height				
Measure 1	71	20.0	0.85	16.8–21.8 (inches)
		50.8	2.2	42.7–55.4 (cm)
Measure 2	70	19.9	0.82	17.0–21.8 (inches)
		50.5	2.1	43.2–55.4 (cm)
Measure 3	48	20.1	0.76	18.0–21.9 (inches)
		51.1	1.9	45.7–55.6 (cm)
Keyboard home row				
Measure 1	71	29.1	1.21	26.5–33.2 (inches)
		73.9	3.1	67.3–84.3 (cm)
Measure 2	70	29.1	1.22	26.8–33.2 (inches)
		73.9	3.1	68.1–84.3 (cm)
Measure 3	52	28.9	1.02	27.0–31.5 (inches)
		73.4	2.6	68.6–80.0 (cm)
Display surface height				
Measure 1	71	29.7	1.33	26.0–31.6 (inches)
		75.4	3.4	66.0–80.3 (cm)
Measure 2	70	29.6	1.13	27.1–31.6 (inches)
		75.2	2.9	68.8–80.3 (cm)
Measure 3	52	30.3	2.07	22.3–31.6 (inches)
		77.0	5.3	56.6–80.3 (cm)

tilt and adjust in height. The keyboard range was 22.8–32.25 in (57.9–81.9 cm) with tilt of 0 to 15 degrees. The surface was counterbalanced, allowing quick adjustment with a foot pedal. The display surface could adjust from 22.4 to 32.75 in (56.8–83.2 cm) and tilt ±7 degrees. A hand crank raised and lowered the surface. These free-standing units were adjacent to a fixed, 30-in (76.2 cm) high, 30 × 60 in (76.2 × 152.4 cm) work surface in an 'L' configuration. All work stations were surrounded on two or three sides by 65-in (165.1 cm) high panels. Adjustable task lights, document holders, and footrests were available.

Three kinds of chair were used. The Steelcase ConCentrx chair was the preferred alternative, and had a height adjustment range of 16.25–19.5 in (41.3–49.5 cm). Steelcase 454 and Task II chairs were also used. Their adjustment ranges were 17–21 in (43.2–53.3 cm). Most of the chairs had pneumatic height adjustment which could be operated while seated.

A variety of IBM terminals were used including: 3177, 3179, 3180, 3192, PC AT, and PS/2 Model 50. All of these had a detachable keyboard with variable tilt and a tilt/swivel monitor.

Procedures

In the first part of the study, data were collected in the evening without the knowledge of the workers. Three teams recorded the following information in each work station: keyboard surface height and angle, keyboard angle, display surface height and angle, display angle, seat height, and seat pitch. Two measurements were made a week apart on 71 work stations. Nine months later a

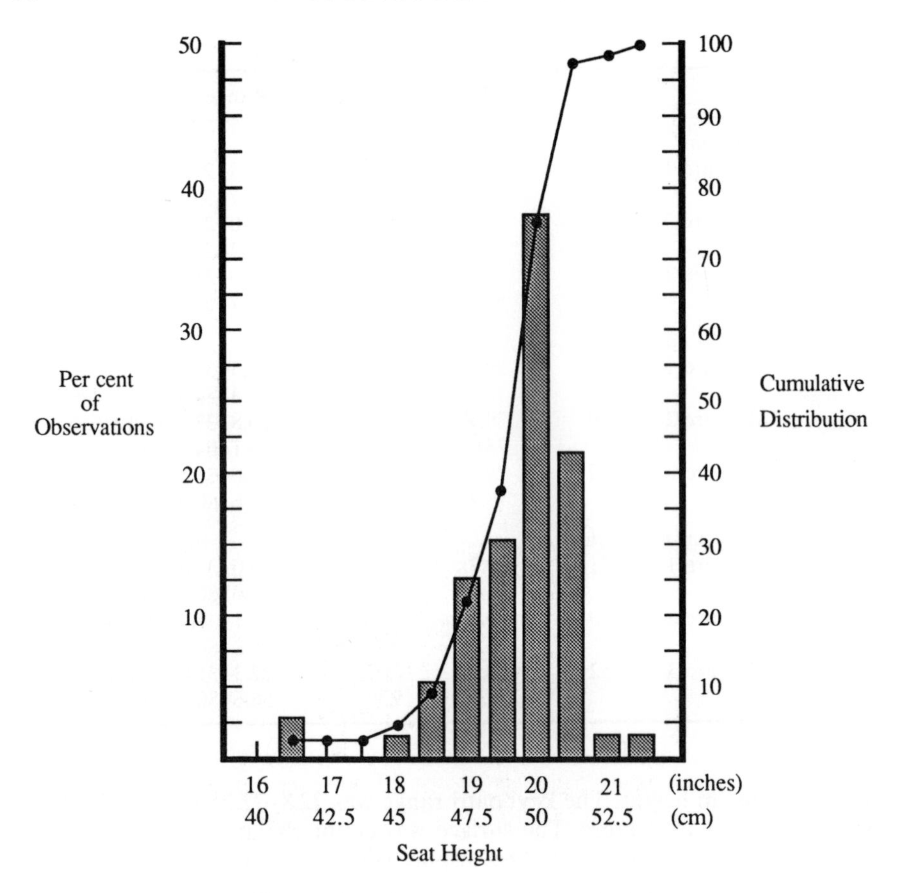

Figure 4.1

third set of measures were made on 50 work stations. The exact number of measures taken varied due to transfers, new hires, etc.

The measurements were taken with tape measures and anglometers. The dimensions were rounded off to the nearest 1/16th inch (0.16 cm). Seat height and angle were calculated using the technique specified in ANSI/HFS 100-1988. Measurement error was introduced by individual differences and difficulty in accessing certain areas. It is estimated that the dimensions were accurate to within 1/8th of an inch (0.31 cm).

After the third set of measures was taken, anthropometric data were collected from a group of volunteers. Standing height, eye height (from the seat), elbow height (from the seat) and popliteal length were measured. Of the 91 people in the total sample, 40 volunteered for the body measures.

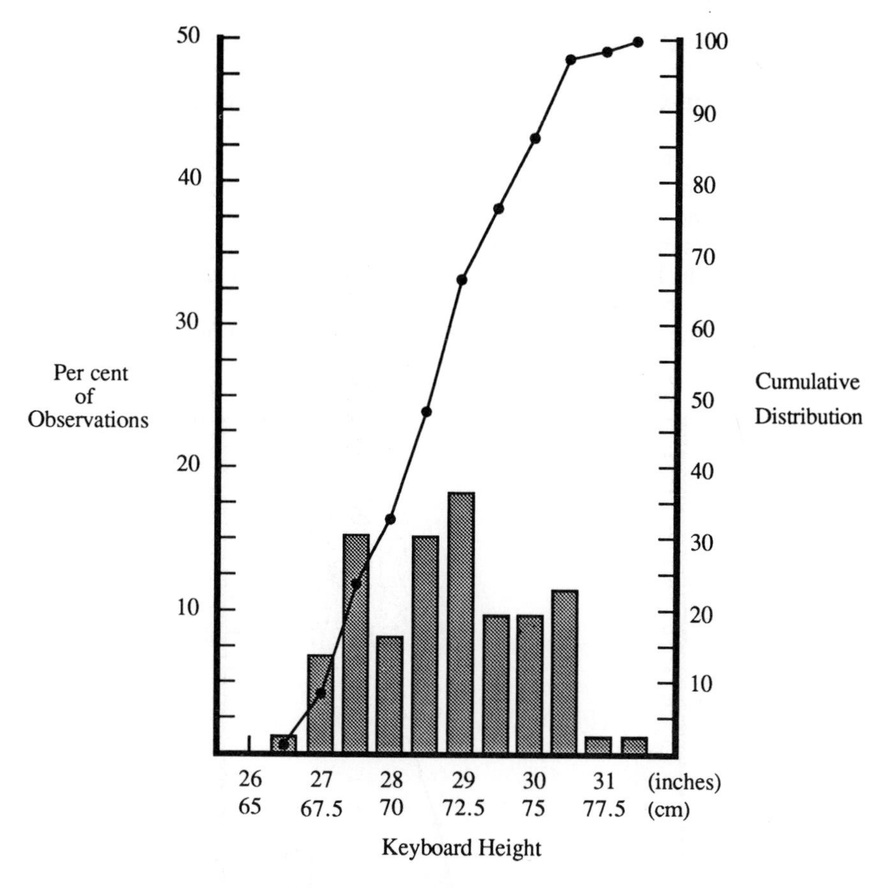

Figure 2. Distribution of Keyboard Home Row Height

Note: The histogram shows the percent of the total sample in each height interval (left-side axis). The line shows the cumulative distribution as a function of height (right-side axis).

Figure 4.2

Results

Some manipulations of the data occurred prior to statistical analysis. Seat height data were converted to reflect the ANSI measurement technique. To be consistent with other research, keyboard data were converted to reflect the home row of the keys. Surface height and keyboard design and angle were used to determine this dimension.

Preferred settings

The three observations of seat height, keyboard home row, and display surface height are shown in Table 4.1. Included are the sample size, mean, standard deviation, and range. The results are very consistent from one observation to the next.

The distribution of observed settings is shown in Figures 4.1, 4.2, and 4.3. These data are from the first measures taken ($n = 71$). The seat height data (Figure 4.1) clustered tightly around the mean of 20 in (50.8 cm). Approximately 74 per cent of the data were within a 1.5 in (3.8 cm) range. With two exceptions, no one used the lower end of the adjustment range. Several did have the seat at its maximum height.

Keyboard height had a flatter distribution (Figure 4.2). The keyboard home row could adjust from 24–33.5 in (61.0–85.1 cm), but there were only a few observations in the upper and lower three inches (7.6 cm) of this range. The

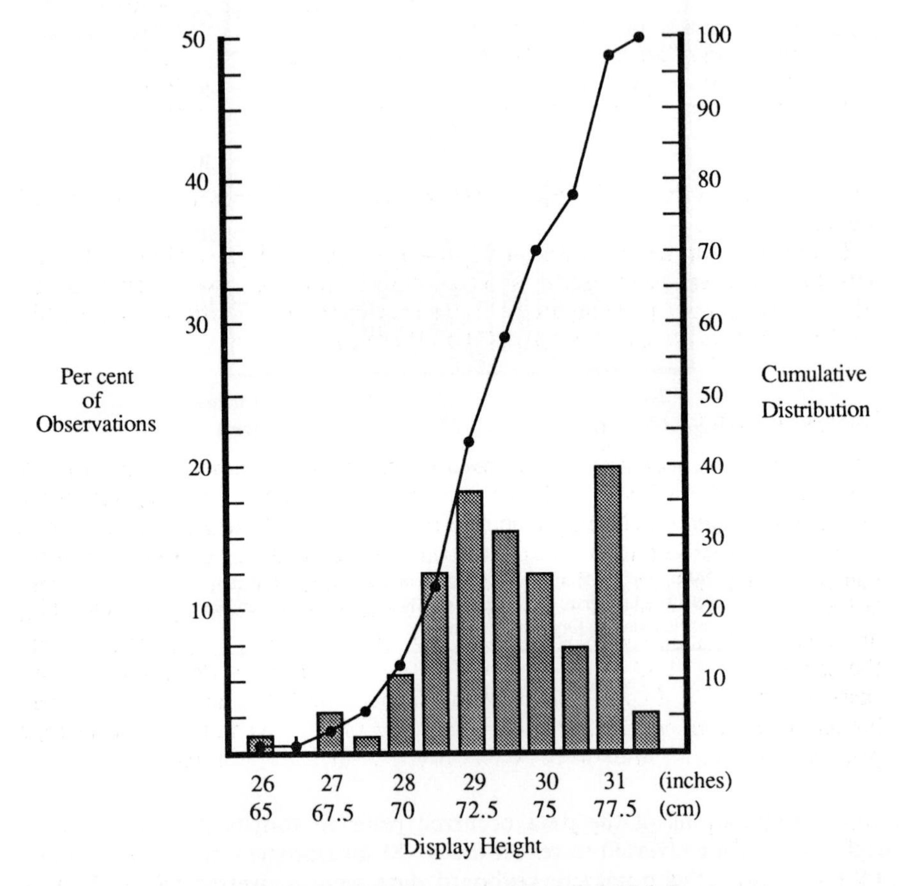

Figure 3. Distribution of Display Height

Note: The histogram shows the percent of the total sample in each height interval (left-side axis). The line shows the cumulative distribution as a function of height (right-side axis).

Figure 4.3

Table 4.2 Changes in preferred setting

Dimension	No change	Difference Between Observations <0.49 <1.3	 0.5 ≪ 0.75 1.3 ≪ 1.9	 >0.75 (inches) >1.9 (cm)
Seat height	62.8%	24.3%	8.6%	4.3%
Keyboard home row	57.1%	32.9%	10.0%	—
Display surface	91.4%	5.7%	—	2.9%
		Correlations		
		Seat height	Keyboard height	Display height
Measure 1 with measure 2 (n = 70)		0.89	0.96	0.86
Measure 1 with measure 3 (n = 29)		0.57	0.59	0.51
Measure 2 with measure 3 (n = 29)		0.68	0.61	0.52

four-inch range from 27–31 (68.6–78.7 cm) accounted for nearly 96 per cent of the data.

The display surface distribution approached a normal curve (Figure 4.3). As with keyboard and seat height, subjects did not use the lower portion of the adjustment range. Approximately 92 per cent of the observations were within the 3.5 in (8.9 cm) range of 28–31.5 (71.1–80.0 cm).

Changes in setting

There were 70 repeated measures between the first two observations, and 29 between the first and third and second and third. As shown in Table 4.2, in the majority of cases, no change in setting was observed. When a difference was noted, most were less than 0.5 in (1.2 cm). The correlations between measurements are also shown in Table 4.2. It is interesting to note that the correlations between the first and second observations were much higher. The data in Table 4.1 indicate the mean setting between observations remained roughly the same for seat and keyboard height, but the standard deviation in both cases was less. For display height, both the mean and standard deviation were higher. There was no change in equipment or task to account for this change over the nine month interval between observations two and three.

Settings and body dimensions

Of the 40 people who volunteered for anthropometric measurements, 26 were female (65 per cent). The data are shown in Table 4.3 The stature for a US ‘average adult’ (Diffrient, Tilley and Harman, 1981) is 67.5 in (171.5 cm). The sample is shorter than the average, probably due to the disproportionate number of males and females.

Correlations were determined between the four anthropometric dimensions and the observed work station heights. These are shown in Table 4.4. (The data are from the first observation.) None of the correlations between body

Table 4.3 Anthropometric dimensions

	n	Mean	Std. dev.	Range
Stature	40	66.6	3.97	60.0–73.0 (inches)
		169.1	10.1	152.4–185.4 (cm)
Eye height above seat	40	29.9	1.62	26.0–32.7 (inches)
		75.9	4.1	66.0–83.0 (cm)
Elbow height above seat	40	8.8	1.0	6.5–10.5 (inches)
		55.9	2.5	15.5–26.7 (cm)
Popliteal	40	17.8	1.52	14.5–21.0 (inches)
		45.2	3.9	36.8–53.5 (cm)

dimensions and settings were significant. Surprisingly, none of the correlations between work station settings were significant either.

Figure 4.4 shows the preferred setting of seat height, keyboard home row, and display as a function of stature. This plot graphically demonstrates that stature is not related to preferred work station settings.

Discussion

Our results were consistent with previous studies – the preferred settings were higher than predicted by models such as used in the ANSI guideline (see Table 4.5). Seat height was particularly high with 60 per cent of the data around 20 in (50.8 cm). This is surprising given the sample was 64 per cent female with a mean height an inch (2.5 cm) lower than an average adult. According to theory, one would expect much lower settings with such a sample. The data also tended to cluster around a central point, with a majority of the data contained within a narrow range.

Table 4.4 Correlation matrix

n	height	Seat height	Keyboard surface	Display stature	height	Eye height	Elbow Popliteal
Seat height	1.00	−0.01	−0.03	0.11	−0.05	−0.07	0.17
	(71)	(71)	(71)	(32)	(32)	(32)	(32)
Keyboard home	−0.01	1.00	0.18	0.19	0.32	0.22	0.09
	(71)	(71)	(71)	(32)	(32)	(32)	(32)
Display surface	−0.03	0.18	1.00	−0.06	0.06	−0.05	−0.06
	(71)	(71)	(71)	(32)	(32)	(32)	(32)
Stature	0.11	0.19	−0.06	1.00	0.73*	−0.16	0.77*
	(32)	(32)	(32)	(40)	(40)	(40)	(40)
Eye height	−0.05	0.32	0.06	0.73*	1.00	0.09	0.61*
	(32)	(32)	(32)	(40)	(40)	(40)	(40)
Elbow height	−0.07	0.22	−0.05	−0.16	0.09	1.00	−0.28
	(32)	(32)	(32)	(40)	(40)	(40)	(40)
Popliteal	0.17	0.09	−0.06	0.77*	0.61*	−0.28	1.00
	(32)	(32)	(32)	(40)	(40)	(40)	(40)

* Significant at $p < 0.05$.

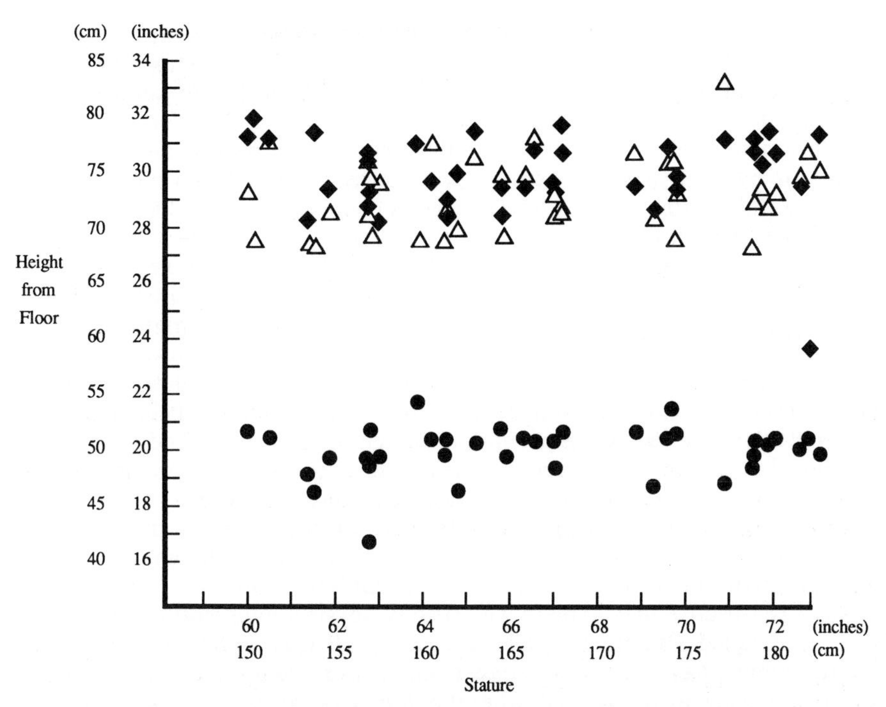

Figure 4. Workstation Settings as a Function of Stature

Figure 4.4

Table 4.5 Comparisons of research and standards

Source	Seat height	Keyboard home row	Display surface	
Grandjean *et al.*, 1983	16.9–22.4	28.0–43.2	26.4–35.8	(inches)
	43–57	71–87	67–91[1]	(cm)
Weber *et al.*, 1984	16.9–21.7	28.7–32.7	23.6–32.7	(inches)
	43–55	73–83	60–83	(cm)
Hozeski, 1986	16.9–22.0	26.0–29.9	25.6–33.9	(inches)
	43–56[2]	66–76	65–86	(cm)
Present Study	16.8–21.9	26.5–33.2	22.3–31.6	(inches)
	43–56	67–84	67–81	(cm)
ANSI[3]	16.0–20.5	24.2–29.2[2]	24.2–29.2[4]	(inches)
	41–52	61–74[2]	61–74	(cm)
CSA[3]	15.0–20.9	22.8–30.7	24.8–29.9	(inches)
	38–53	58–78[2]	63–76	(cm)

[1] 25 cm subtracted to approximate surface height.
[2] 3 cm added to approximate keyboard home row.
[3] Both documents permit fixed height surfaces as well.
[4] Recommended range when display and keyboard on same surface.

Only minor changes in setting were noted over time. Keyboard home row height changed the most, with 32.9 per cent of the observations differing by about one-half inch (1.3 cm). This suggests that a fixed height surface, set to accommodate specific users, may be sufficient for some applications. This form of adjustment – called maintenance adjustment – has been available on furniture for over 20 years. The correlation between the first two measures and the third (taken nine months later) suggests changes are more likely to occur over longer periods of time.

The lack of significant correlations between body size and preferred settings was surprising. The data in Figure 4.4 clearly show that stature does not predict user preference. It may explain why recommended ranges of adjustment do not match observations, as shown in Table 4.5. These results question the validity of the rectilinear, additive model so often used to determine ergonomic criteria. Existing models do not reflect operator behaviour and undoubtedly have led to inappropriate design specifications and recommendations. To prevent over- and under-design, predictive models need to be improved to reflect actual user preference, behaviour and need.

The failure of these models supports the notion that ergonomics must be viewed in a broader context, involving variables and factors in addition to the physical environment (e.g., Cornell, 1987; Sainfort, 1990). The absence of a broader perspective may explain why strictly physical interventions to prevent RSI have failed (Bammer, 1989; Green and Briggs, 1989; Hadler, 1990). Studies which have taken a broader view have found that work content, autonomy, pace, workload and other job design variables are highly related to musculoskeletal complaints and RSI (Linton and Kamwendo, 1989; Bammer, 1989). This is consistent with surveys of office workers which show that job design and management issues are of greater concern (Cornell, 1988). Ergonomics means more than just making things adjustable: the entire physical-psychological-social context must be considered in designing effective work places.

User adjustability is often suggested for VDT intensive tasks. Many office workers are not intensive keyboard users, but could benefit from a work-surface height designed to be more accommodating to VDT use. A single, fixed height surface is desirable for these workers, particularly if their work is multitasked, i.e., they switch from one task to another throughout the day. A recommended height can be derived from these data. The overall mean for keyboard *home row* was 29.1 inches (73.9 cm). Subtracting the approximate height of the keyboard yields a keyboard *work-surface* height of 27.9 in (70.9 cm). The overall mean for the display surface was 29.8 in (75.8 cm). For a single height surface, a reasonable solution would be the average of these two, or 28.8 in (73.2 cm).

There were some limitations to this study that should be examined in future research. First, it would be valuable to assess changes in settings throughout the day, not just at the end. Second, larger samples, balanced for height, age, and gender, should be studied. Third, a variety of tasks should be investigated systematically. Fourth, the impact of ergonomics awareness should be studied. And fifth, the influence of sedentary postures and rest breaks should be assessed. These studies would be beneficial in developing better models of user needs, behaviour and comfort, and in understanding the causes and prevention of RSI. Such knowledge would enhance worker performance as well as the quality of working life.

References

Allison, G., (1990), A view of the current state of office ergonomics, *The Work Station Report.*

American National Standards Institute, (1988), *Human Factors Engineering of Visual Display Terminal Workstations.* ANSI/HFS 100-1988.

Armstrong, T., Fine, L., Goldstein, S., Lifshitz, Y. and Silverstein, B., (1987), Ergonomics considerations in hand and wrist tendonitis, *Journal of Hand Surgery*, **12**, 830–7.

Bammer, G., (1989), Work-related neck and upper limb disorders associated with office work – Prevalence and causes. Lecture presented at the University of Western Australia.

Brown, C. and Schaum, L., (1980), User adjusted VDT parameters, in Grandjean, E. and Vigliani, E. (eds) *Ergonomic Aspects of Visual Display Terminals*, London: Taylor & Francis.

Canadian Standards Association, (1989), *Office Ergonomics*, CAN/CSA Z412.

Cornell, P., (1987), Performance and QWL in a data entry task, *Proceedings of the Human Factors Society 31st Annual Meeting*, 1350–4.

Cornell, P., (1988), Office environment index '88: A poll of workers, executives and facility managers on life and work in today's offices, *Proceedings of the Human Factors Society 32nd Annual Meeting*, 785–9.

Cushman, W., (1984), Data entry performance and operator preferences for various keyboard heights, in Grandjean, E. and Vigliani, E. (eds) *Ergonomic Aspects of Visual Display Terminals*, London: Taylor & Francis.

Diffrient, N., Tilley, A. and Harman, D., (1981), *Humanscale 7/8/9*, Cambridge: MIT Press.

Grandjean, E., Hunting, W. and Pidermann, M., (1983), VDT work station design: Preferred settings and their effects, *Human Factors*, **25**, 161–75.

Green, R. and Briggs, C., (1989), Effect of overuse injury and the importance of training on the use of adjustable workstations by keyboard operators, *Journal of Occupational Medicine*, **31**, 557–62.

Hadler, N., (1986), Industrial rheumatology, *Medical Journal of Australia*, **144**, 191–5.

Hadler, N., (1990), Cumulative trauma disorders, *Journal of Occupational Medicine*, **32**, 38–41.

Hozeski, K., (1986), Subjective preference and use of work station features, *Proceedings of the Human Factors Society 30th Annual Meeting*, 890–4.

Linton, S. and Kamwendo, K., (1989), Risk factors in the psychosocial work environment for neck and shoulder pain in secretaries, *Journal of Occupational Medicine*, **31**, 609–13.

Miller, I. and Suther, T., (1981), Preferred height and angle settings of CRT and keyboard for a display station input task, *Proceedings of the Human Factors Society 25th Annual Meeting*, 492–6.

Rinalducci, E. *et al.*, (1983), *Video Displays, Work and Vision.* National Research Council. Washington, D.C.: National Academy Press.

Sainfort, P., (1990), Job design predictors of stress in automated offices, *Behaviour and Information Technology*, **9**, 3–16.

Silverstein, B., Fine, L. and Armstrong, T., (1987), Occupational factors and carpal tunnel syndrome, *American Journal of Industrial Medicine*, **11**, 343–58.

Weber, A., Sancin, E. and Grandjean, E., (1984), The effects of various keyboard heights on EMG and physical discomfort, in Grandjean, E. (ed) *Ergonomics and Health in Modern Offices*, London: Taylor & Francis.

5

VDT chair and work-surface heights

Keiichi Ohno, Gonzaburo Sakazume, and Masayuki Iwasaki

Introduction

Ergonomics was introduced into Japan in the mid-1950s. This new-found knowledge was first applied to the design of school furniture (Obara, 1973). In 1971, ergonomics was extensively used to standardize office desk and chair dimensions (Yagi, 1981). Japanese Industrial Standards (JIS) S 1010 for Writing Desks for the Office (Japanese Industrial Standards, 1978), and S 1011, for Office Chairs (Japanese Standards Association, 1978), are widely adhered to in the design of office furniture in Japan. These JIS standards address writing tasks at the office.

The spread of office automation has placed new demands on office furniture and work environments to prevent discomforts associated with VDT operation. Guidelines that address the VDT work environment include the Occupational Health Guidelines for VDT Work (Ministry of Labour, 1985).

In recent years, adjustable chairs and desks have appeared on the Japanese market. However, the greater cost of such office furniture has hampered their widespread use. Consequently, fixed height chairs and desks are prevalent in many work places.

Recommended ranges of adjustability of office chairs and work-surface heights in Japan are provided by JIS standards (noted earlier). Established in 1976, JIS standards estimate that the average height of the Japanese people is 164 cm (Agency of Industrial Science and Technology, 1973).

The average height of Japan's young people has increased in recent years. The 1984 national nutrition survey conducted by the Ministry of Health and Welfare (1985) indicated that males aged between 20 and 24 years had a mean stature of 169.5 cm, as shown in Figure 5.1. This must be reflected in chair and desk height standards.

Two issues have come to the fore. First, can commonly recommended VDT chair and desk dimensions accommodate the spectrum of individual users? Second, how can we help the VDT worker understand the importance of

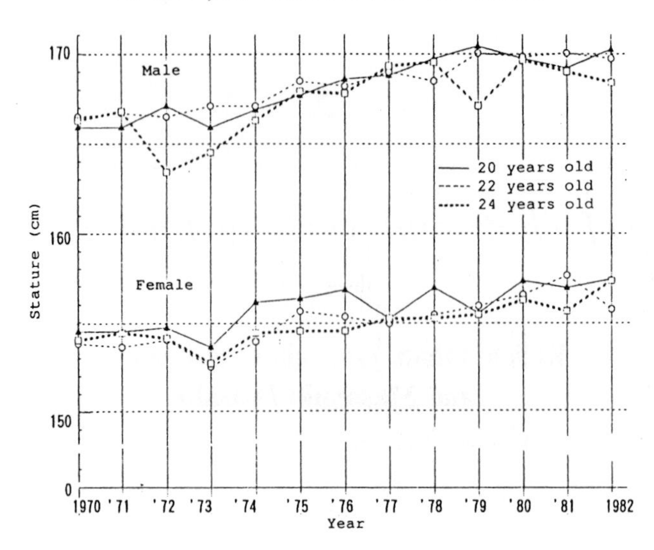

Figure 5.1 Increases of stature of young people in Japan

proper posture in VDT work and the correct adjustment of chair and desk heights?

Experiment to determine optimum chair and desk heights

Experimental system

An adjustable work station was developed to compare predicted chair and desk heights with those preferred by users. The objectives of this research were to collect anthropometric data of office workers for product design, and to assess how users adjust their work stations when they understand the basis for sitting correctly.

The experimental VDT work station (see Figure 5.2) consists of a motor-driven work-surface and chair, which is integrated with a personal computer (Fujitsu Model FM16-β). This computer monitors, records and analyses the chair and work station adjustments which are selected by users.

Experimental methods

Predicted user settings of chair and work-surface heights were calculated in accordance with subject stature and weight (reported by subjects in a questionnaire). Some of the subjects sat at the work station and selected preferred chair and work-surface heights.

Calculation of theoretical chair and work-surface heights

All subjects recorded on a questionnaire their name, sex, stature, weight, and age (see Table 5.1).

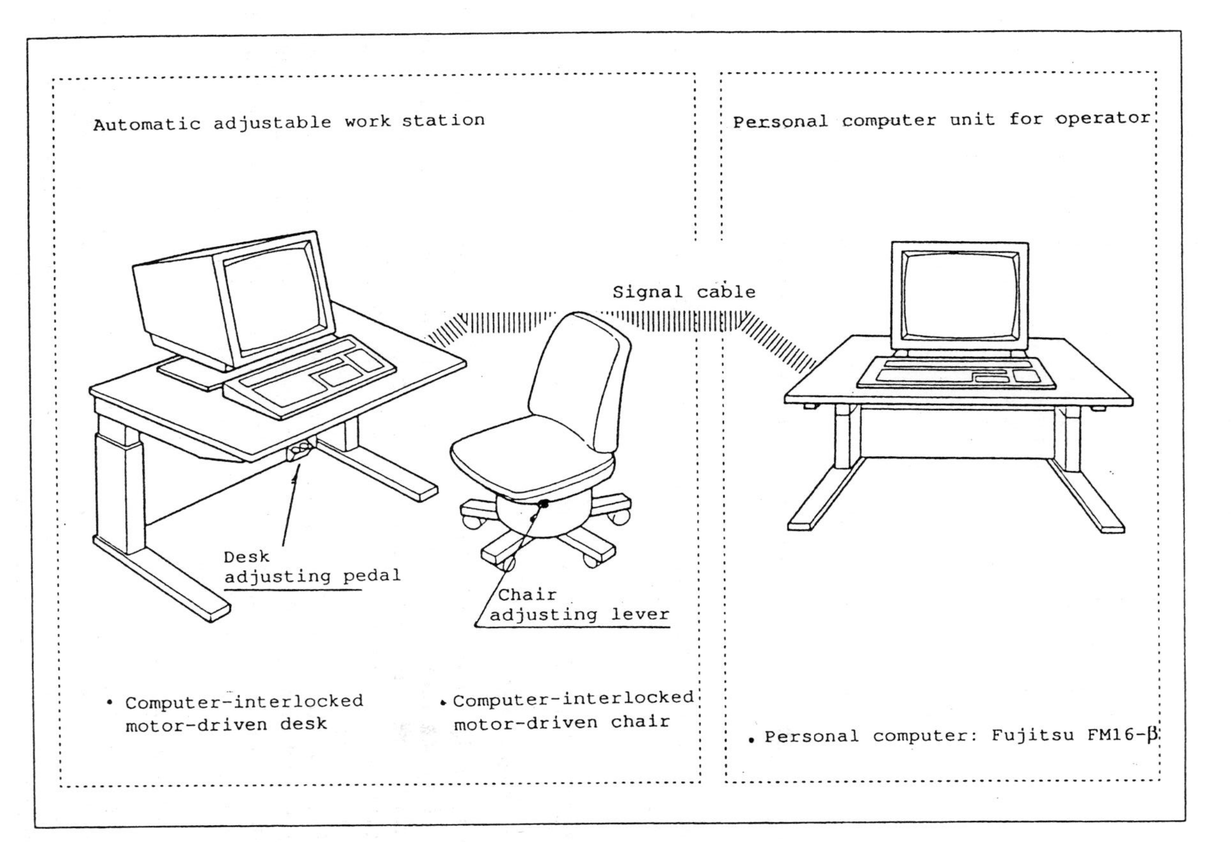

Figure 5.2 Schematic illustration of automatic adjustable work station system

Table 5.1 Questionnaire form.
Compatibility with your chair and desk

Name:		Sex: Male/Female		Stature:	cm	Weight:		kg
Age:	15 ≤	16–20	21–25	26–30	31–35	36–40	41–45	51 ≥

In a preliminary experiment, 50 adults were asked to set the chair and desk at their preferred height in the experimental system shown in Figure 5.1. Data on these selected settings were then compared with their stature. The slopes of the lines connecting the plotted data were calculated to derive (5.1) and (5.2) above.

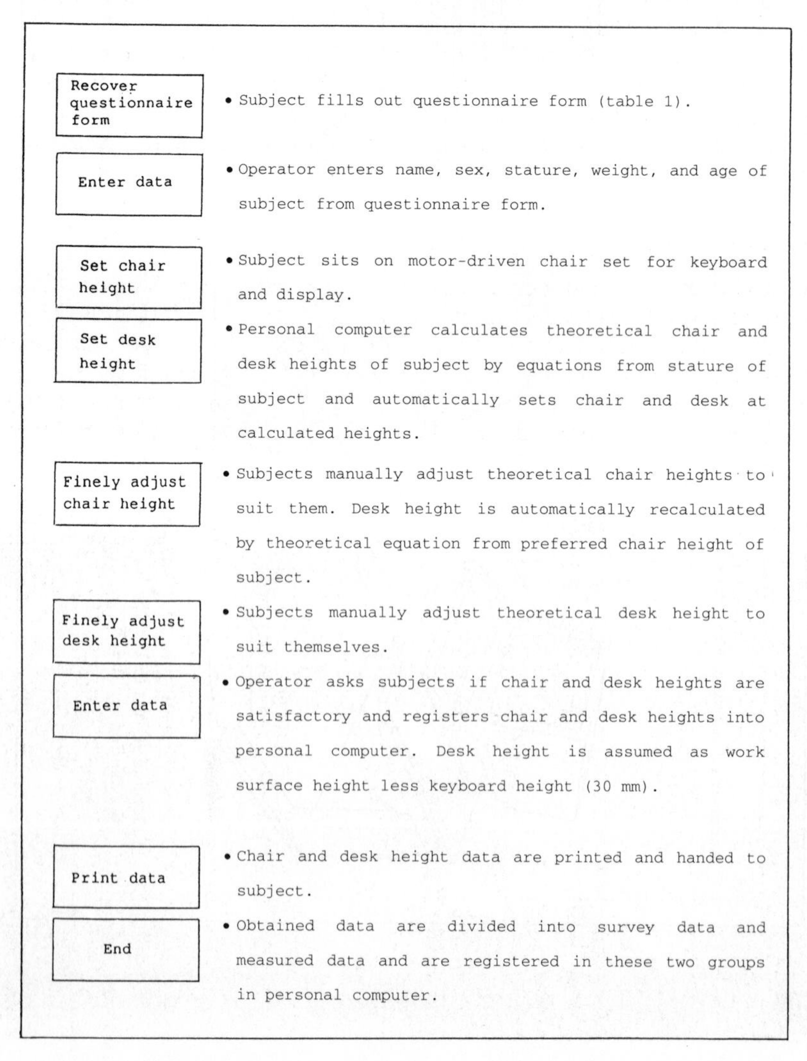

Figure 5.3 Procedure for measuring subjects' preferred chair and desk heights

Theoretical chair and work-surface heights were derived from (5.1) and (5.2) for subjects based on their stature, and supplied to subjects.

$$\text{Chair height (in mm)} = (\text{Stature} - 1500) \times 0.15 + 380 \quad (5.1)$$

$$\text{Desk height (in mm)} = \text{Chair height} + (\text{Stature} \times 0.04 + 200) \quad (5.2)$$

Measurement of preferred chair and desk heights

Of the subjects who filled out the questionnaire, those who wanted to do so set the chair and desk at their preferred heights.

The predicted chair and desk heights of each subject were calculated by (5.1) and (5.2), and based on their stature. After the chair and desk of the automatic adjustable work station were set at these predicted levels, subjects then selected their preferred chair and desk heights.

The questionnaire data will be subsequently referred to as survey data, and the preferred chair and desk heights as measured data. Both survey and measured data were printed and recorded by computer.

The procedure for measuring the preferred chair and desk heights of the subjects is shown in Figure 5.3.

Experimental subjects, period, and place

The experiment was conducted in Japan's major cities (e.g., Sapporo, Sendai, Tokyo, Nagoya, Osaka, and Fukuoka) from May 1986 to March 1990 (3 years and 10 months total). Subjects were office workers who visited office-related exhibitions, including employees of the organizer, Uchida Yoko Co., Ltd.

An experimental view is shown in Figure 5.4. The man at centre is a subject at the experimental work station. The woman on the left is an operator who

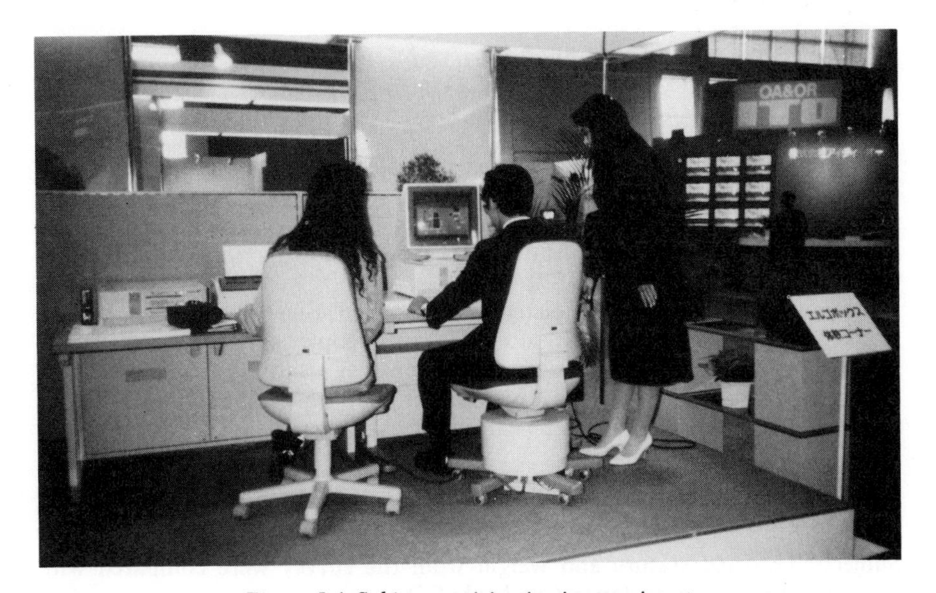

Figure 5.4 Subject participating in experiment

Figure 5.5 Measured data presented on display screen

encoded the data. The woman on the right guided the subject through the experiment.

Figure 5.5 shows the data displayed on the computer screen which is placed on the desk of the experimental work station. Subjects can determine their predicted and preferred chair and desk heights from this screen.

Experimental results and analysis

Subject characteristics

Three thousand, five hundred and sixty people participated in the questionnaire and experimental research. There were many more male subjects than female subjects because the experiment was limited to volunteers at male-dominated office exhibitions. About 2000 survey and experimental data items were each collected (see Table 5.2).

Subject age and sex are depicted in Figure 5.6. Since these subjects were selected from office exhibition attendees few people were 20 years old or younger. The age distribution of male subjects was relatively even between 21 and 51+ years. However, age distributions of female subjects were largely concentrated between 21 and 25 years, as is characteristic of the female office work force in Japan.

Subject stature and weight

Subjects' reported stature and weight from the survey were compared with those of other studies in Japan (see Table 5.3).

Table 5.2 Total sample size

	Subjects who responded to the questionnaire (Survey data*)	Subjects who participated in experiment† (Measured data*)	Total
Male	1848	1672	3520
Female	337	271	608
Total	2185	1943	4428

* Data from subjects who responded to the questionnaire form are referred to as survey data, and chair and desk heights selected by subjects are referred to as measured data.
† Subjects who participated in experiment also responded to the questionnaire.

Data on stature appear to be reliable, and both stature and weight increase with age. Subjects were found to be 4 cm taller and 4 kg heavier than those described by the Anthropometric Data Editorial Committee (Japan Interior Designers Association, 1984), which performed measurements with a highly accurate Martin anthropometer.

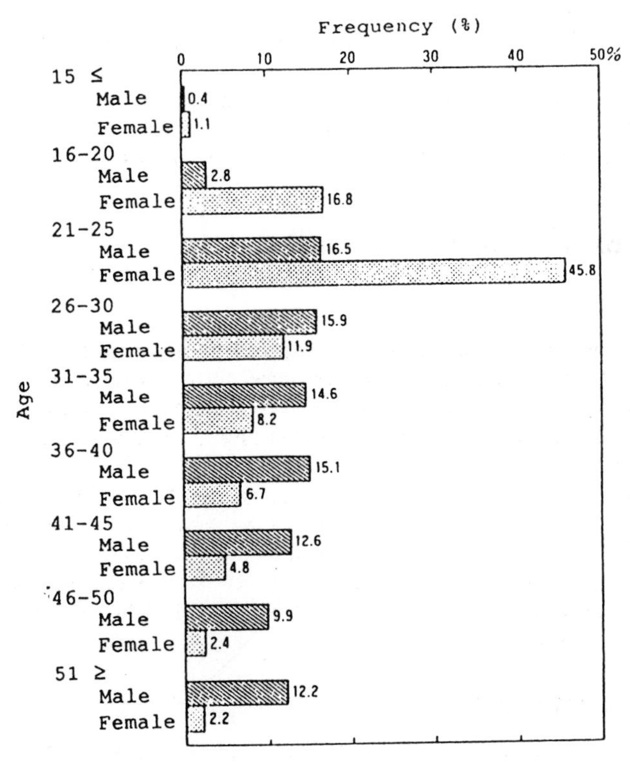

Figure 5.6 Age distributions of male and female subjects

Table 5.3 Body dimensions of Japanese adults from four sources

	Male		Female	
	Stature (cm)	Weight (kg)	Stature (cm)	Weight (kg)
Anthropometric Data Editorial Committee (1970)*	165.1 SD 5.24	58.8 SD 6.78	154.4 SD 4.95	48.7 SD 5.00
Body dimensions of Japanese Air Self-Defence personnel (Aramaki and Tagami, 1980)	167.8 SD 5.90	63.2 SD 7.18	— —	— —
F&F Design Research Institute (1980–1984)*	168.8 SD 6.48	62.8 SD 6.90	157.3 SD 4.97	48.8 SD 5.80
This study (1986–1990)	169.1 SD 5.79	62.8 SD 8.05	157.9 SD 5.32	48.8 SD 5.25

* Source: Japan Interior Designers Association and MITI Design Project Group, (1984).

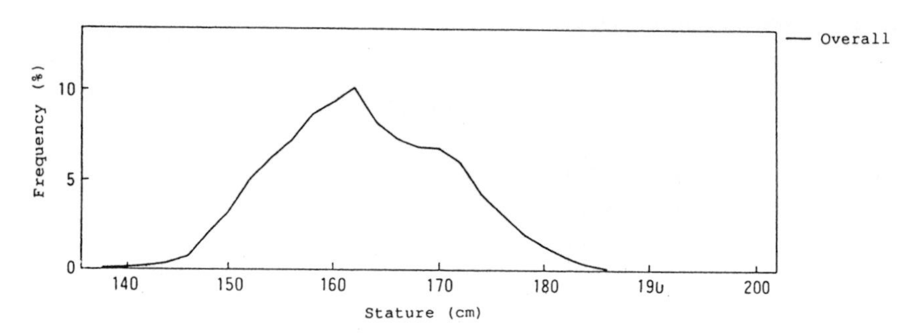

Figure 5.7 Overall stature distribution of subjects

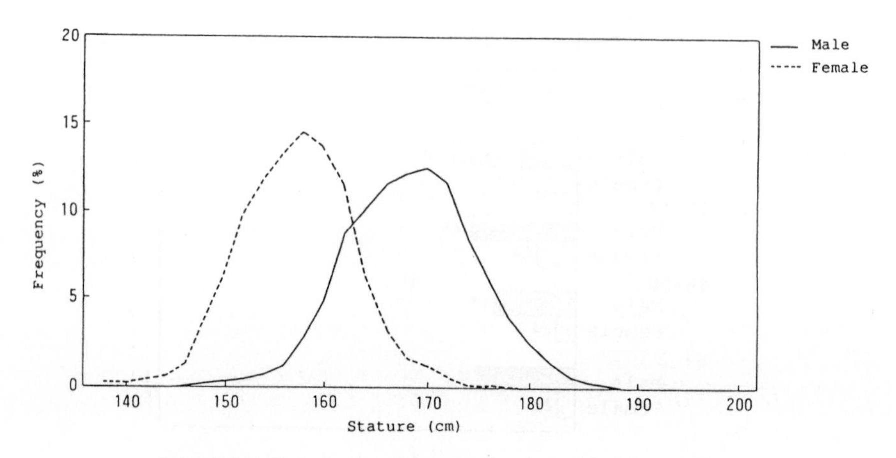

Figure 5.8 Stature distributions of male and female subjects

The distribution of subject statures was with a mean of 167.5 cm and a range from 153.7–181.3 cm with ±2 SD, (see Figure 5.7). Statures of male and female subjects are normally distributed, with a mean of 170 and 158 cm, respectively (see Figure 5.8).

The distribution of subject weights ranged from 43.2–81.5 kg with ±2 SD, with a mean of 62.3 kg (Figure 5.9). Male and female subjects' weights were normally distributed, with a mean of 64.6 and 48.4 kg, respectively (Figure 5.10).

Measurement of preferred chair and desk heights

The keyboard and CRT display of the experimental system were placed on the desk to simulate VDT work. The keyboard was 30 cm higher than the desk. The subjects wore shoes during the experiment.

The subjects' preferred desk heights are shown in Figure 5.11. Males and females preferred desk heights were almost normally distributed, with a mean of 67.5 cm and 66.0 cm, respectively (see Figure. 5.12). The height of the writing surface was obtained by adding the keyboard thickness of 30 cm to the

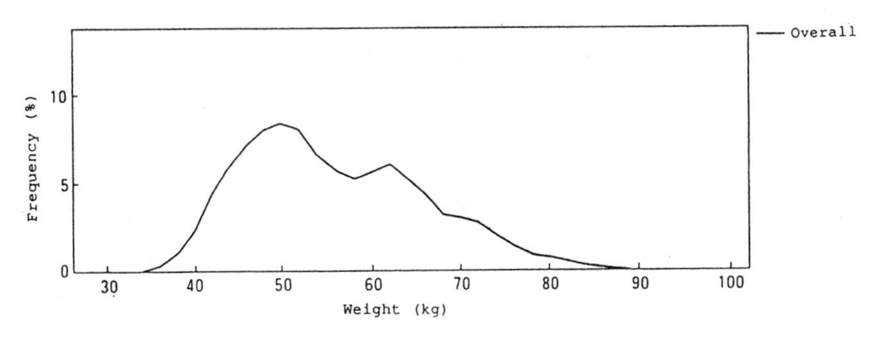

Figure 5.9 Overall weight distribution of subjects

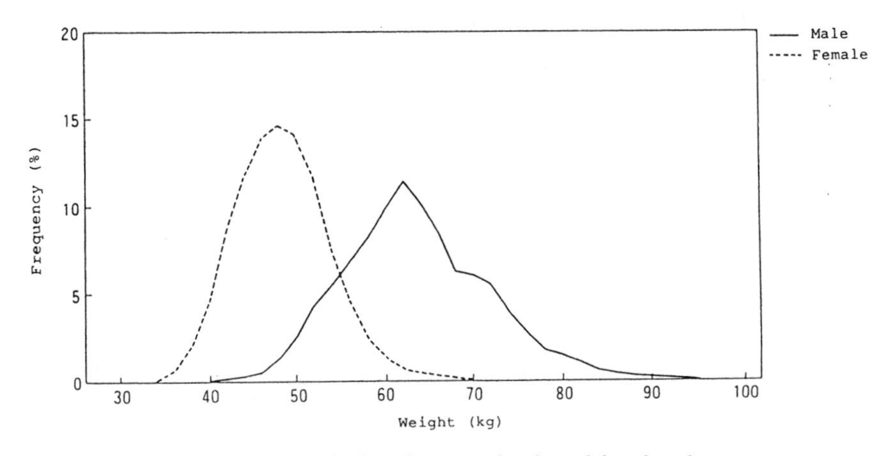

Figure 5.10 Weight distributions of male and female subjects

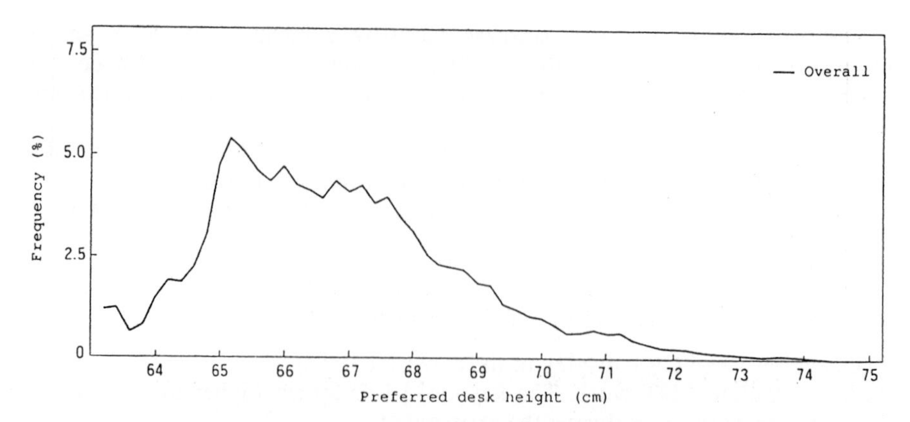

Figure 5.11 Overall preferred desk height distribution of subjects

preferred desk height. This height averaged about 70 cm and corresponds closely with the 70 cm desk height specified in JIS S 1010 (Japanese Standards Association, 1978). The desk height range of 60–75 cm recommended in the Occupational Health Guidelines for VDT Work (Ministry of Labour, 1985) is equivalent to the 95th percentile of preferred desk height of this study.

The distribution of preferred chair heights is depicted in Figure 5.13. The mean male and mean female preferred chair heights were 41 cm and 39.8 cm, respectively (see Figure 5.14). The shape of these distributions is narrower for females than males. This is probably because the experiment was conducted outside the office, and because short female subjects wore high-heeled shoes, while tall females wore low heels. The chair height range of 35–45 cm recommended in the Occupational Health Guidelines for VDT Work (Ministry of Labour, 1985) is equal to the 95th percentile of preferred chair height in this study.

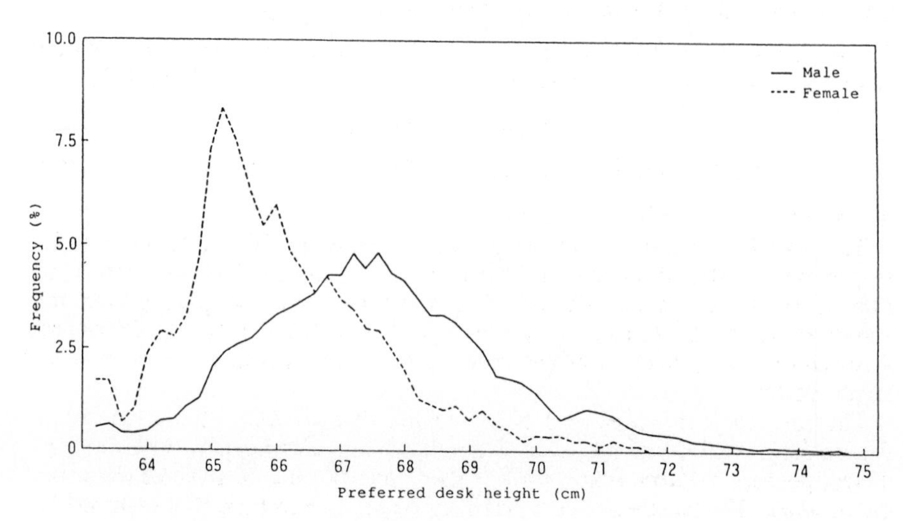

Figure 5.12 Preferred desk height distribution of male and female subjects

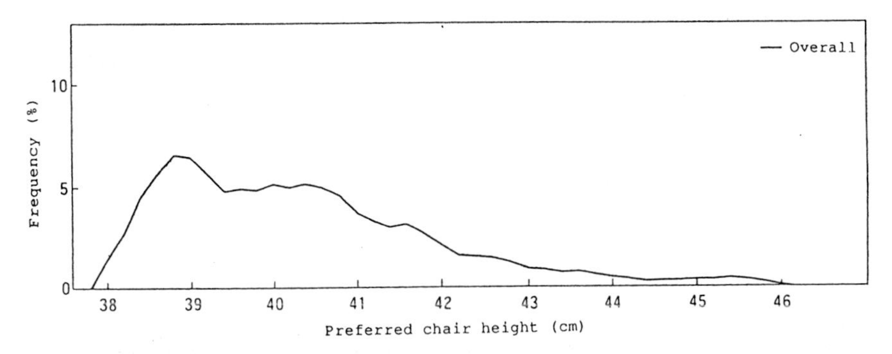

Figure 5.13 Overall preferred chair height distribution of subjects

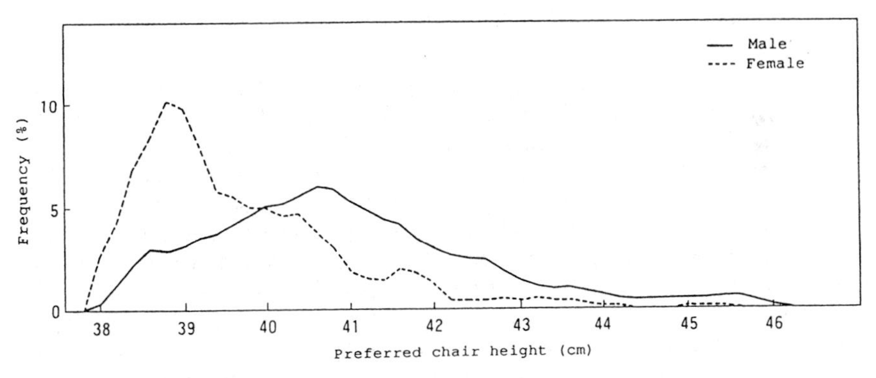

Figure 5.14 Preferred chair and desk height distribution of male and female subjects

Comparison of predicted and preferred chair and desk heights

The preferred chair and desk heights of subjects were compared with predicted chair and desk heights calculated by the following equations (Obara, 1973):

$$\text{Chair height} = \tfrac{1}{4} \times \text{Stature} - 1 \qquad (5.3)$$

$$\text{Desk height} = \text{Chair height} + (\text{Sitting height} \times \tfrac{1}{3} - 1) \qquad (5.4)$$

where sitting height = 0.55 × stature.

The actual equations derived from data collected in this study had smaller slopes than predicted equations described above (Figures 5.15 and 5.16). This finding may have resulted because office desks in Japan are typically fixed at a standard height of 70 cm, as specified in JIS S 1010 (Japanese Standards Association, 1978). Most subjects work at, and are accustomed to, fixed 70 cm height desks.

The predicted and preferred chair heights of male subjects are shown in Figure 5.17. Again, females selected a smaller range of heights than was predicted, perhaps because many wore higher heels at such exhibitions than they did at work. The female subjects preferred higher chairs than was expected for their stature (Figure 5.18), presumably as a result of the shoes they wore.

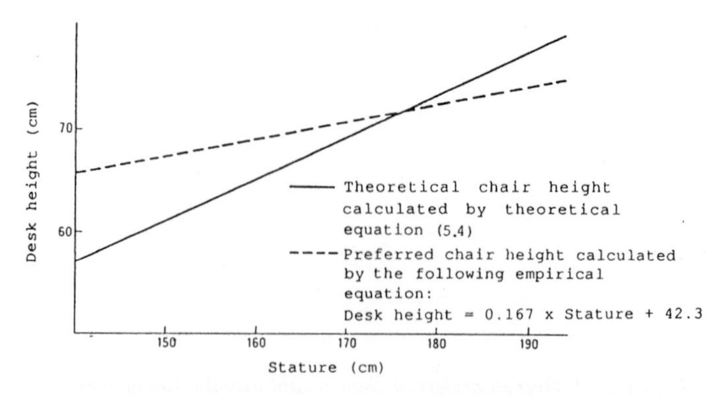

Figure 5.15 Comparison of theoretical and preferred desk heights of male subjects

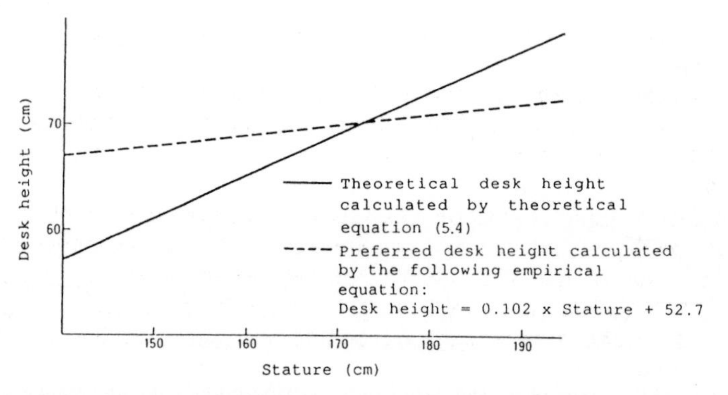

Figure 5.16 Comparison of theoretical and preferred desk heights of female subjects

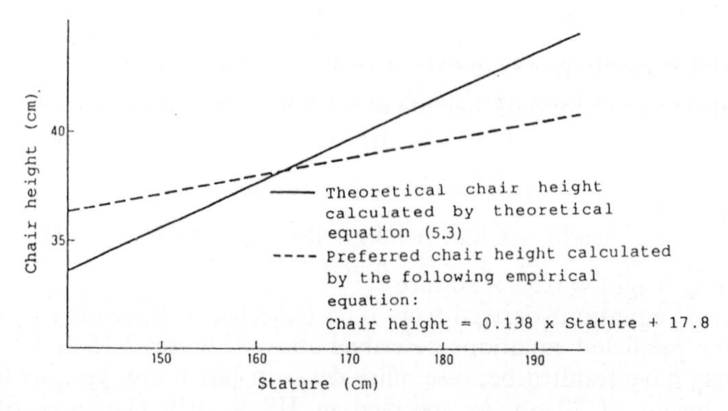

Figure 5.17 Comparison of theoretical and preferred chair heights of male subjects

Factors governing desk and chair heights

When data were analyzed, stature was found to be closely correlated with the desk height (see Tables 5.4–5.7) as expected, for males, stature was the most highly correlated with selected desk and chair heights (see Table 5.4). Weight

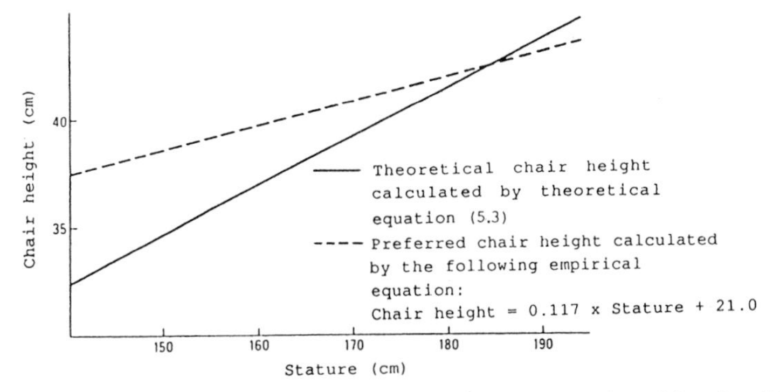

Figure 5.18 Comparison of theoretical and preferred chair heights of female subjects

and age factors are also significantly correlated with their preferred chair and desk heights.

Since weight and age are more closely correlated with stature than with preferred desk and chair heights, the experimental data of male subjects were analyzed by multiple regression analysis. Some of these results are shown in Table 5.5.

When the multiple regression analysis was conducted with preferred chair height as the dependent variable and stature and weight as independent variables, there was a significant correlation between preferred chair height and weight. When the age data were included with the independent variables of stature and weight, a significant correlation was observed between preferred chair height and age.

Preferred desk height was significantly correlated with age, but not weight. Stature of females as a lone variable correlated with the preferred desk and chair heights (see Table 5.6). Multiple regression analysis of the female data resulted in a significant correlation between preferred chair height and weight. However, the correlation found between the preferred desk height and age of male subjects was not observed in female subjects (see Table 5.7).

Conclusion

Data on both stature and weight (obtained by questionnaire and preferred chair and desk heights measured in the experiment) underscore an increase of mean stature compared to previous studies in Japan, and suggest that sex differences exist in selected chair and desk heights. Such data have implications for office furniture design. These findings include the following.

1. The mean stature and weight of Japanese people has increased by about 4 cm and 4 kg, respectively, in the past 20 years.
2. The mean preferred chair and desk heights of the subjects was 40 and 70 cm, respectively.
3. Comparisons of preferred and predicted chair and desk heights (calculated from subject stature) suggest that the former changed less with stature than the latter. That is, variability was less than expected as a function of

Table 5.4 Results of simple correlation analysis for male subjects

	1. Age	2. Stature	3. Weight	4. Obesity	5. Preferred chair height	6. Preferred desk height
1. Age	1.00000**	−0.3307**	0.0538*	0.3723**	−0.1758**	−0.1013**
2. Stature	−0.3307**	1.0000**	0.4750**	−0.4955**	0.4899**	0.5057**
3. Weight	0.0538*	0.4750**	1.0000**	0.5286**	0.1883**	0.2549**
4. Obesity	0.3723**	−0.4955**	0.5286**	1.0000**	−0.2866**	−0.2358**
5. Preferred chair height	−0.1758**	0.4899**	0.1883**	−0.2866**	1.0000**	0.7489**
6. Preferred desk height	−0.1013**	0.5057**	0.2549**	−0.2358**	0.7489**	1.0000**

Table 5.5 Results of multiple regression analysis for male subjects

	Desk height		Chair height	
Independent variable	Two independent variables	Three independent variables	Two independent variables	Three independent variables
Stature	0.14526**	0.14504**	0.16428**	0.17563**
Weight	−0.01164*	−0.01154*	0.00446	−0.00038
Age		−0.00136		0.07001**
Constant term	17.24053**	17.27948**	39.50835**	37.50047**

Table 5.6 Results of simple correlation analysis for female subjects

	1. Age	2. Stature	3. Weight	4. Obesity	5. Preferred chair height	6. Preferred desk height
1. Age	1.00000**	−0.1923**	0.2614*	0.4355**	−0.0417	−0.0452
2. Stature	−0.1923**	1.0000**	0.4560**	−0.5097**	0.2941**	0.3741**
3. Weight	0.2614**	0.4560**	1.0000**	0.5332**	−0.0319	0.0806
4. Obesity	0.4355**	−0.5097**	0.5332**	1.0000**	−0.2489**	−0.2778**
5. Preferred chair height	−0.0417	0.2941**	0.0319	−0.2489**	1.0000**	0.6935**
6. Preferred desk height	−0.0452	0.3741**	0.0906	−0.2778**	0.6935**	1.0000**

Table 5.7 Results of multiple regression analysis for female subjects

	Desk height		Chair height	
Independent variable	Two independent variables	Three independent variables	Two independent variables	Three independent variables
Stature	0.07962**	0.08619**	0.11892**	0.12794**
Weight	−0.02950*	−0.03697*	−0.03217	−0.04241*
Age		−0.05214		0.07160
Constant term	28.64048**	27.76699**	48.77765**	47.57806**

increased height. There were marked differences between male and female subjects in preferred chair and desk heights.
4. Selected chair and desk heights were also correlated with weight, age, and stature of subjects.
5. The popliteal length and seat pan heights of subjects were not compared. Consequently, further study is needed to interpret these findings, and establish the basis for the correlation of subject age and weight with chair and desk heights.

Findings (1) and (2) about stature and weight are considered reliable enough to be applied to product development because of the large sample size.

Male and female subjects selected very different chair and desk heights. Since office furniture has historically been developed for male workers, female employees may well need special design provisions to accommodate them in the office.

The mean preferred desk height was 70 cm. Because it is easier to accommodate short workers at a high work station than the converse, the desk height (if fixed) will need to be designed for tall office workers.

As indicated in finding (3), the preferred chair and desk heights of subjects measured in this study increase less than was predicted with increases in stature. This probably resulted because subjects are accustomed to the 70 cm fixed height desks. The continuing growth of mean stature has further implications for future designs. Uchida Yoko will develop its own equations for calculating the optimum chair and desk heights of specific office workers and will train them in how to use their chairs and desks.

Finding (4) has implications for chair and desk heights. If these results are confirmed in other research, furniture may need to be developed for different weight and age groups of office workers. However, the correlation of weight and age with selected chair and desk heights requires further study to understand the relationship between anthropometric dimensions of office workers and chair and desk heights.

The experimental work station was demonstrated and tested at exhibitions such as the Business Show (organized by the Nippon Administrative Management Association) and the Office Environment Exhibition (organized by the Japan Management Association), whose attendees are office workers. Exhibition attendees who participated in the experiment expressed an understanding of the potential for incurring fatigue due to improper chair and desk heights when they compared their preferred chair and desk heights with those predicted and displayed by computer. Many office workers volunteered for the experiment as a result of a heightened recognition and interest in the design of the VDT work environment.

The automatic adjustable work station system was exhibited at the 1986 Annual Meeting of the Kanto Chapter of the Japan Ergonomics Research Society (JERS) and at the 1987 Nationl Annual Meeting of the JERS (Kondo *et al.*, 1986). At presentations at the two meetings, many people provided valuable comments on the work station and research. These include:

- Product development must incorporate research on characteristics of office workers more effectively, such as relating to age.
- Users need training in how to sit properly; instructions on how to perform the adjustments should be attached to the chair.

- Traditional equations used for calculating chair and desk heights require revision.

Findings of this research agreed with recommendations provided in the Occupational Health Guidelines for VDT Work (Ministry of Labour, 1985). The experimental work station system generated much interest at the Ministry of Labour. Further, the associated software program was adopted as a teaching aid for the VDT Work Health Education Instructor Course; this was organized by the Japan Industrial Safety and Health Association under the supervision and guidance of the Ministry of Labour. This program had been used as a course aid for three years, from July 1987 to July 1990.

References

Agency of Industrial Science and Technology, (1973), *How to Use New Office Chairs and Desks* (Metal Furniture Manufacturers Association of Japan, Tokyo), 11, (in Japanese).

Aramaki, T. and Tagami, K. (eds), (1980), *Anthropometric Measurements of Air Self-Defense Force Personnel* (Aviation Medicine Experimental Team, Tokyo), 21–4, (in Japanese).

Japan Interior Designers Association and MITI Design Project Group, (1984), *Interior Design of Government Offices* (Japan Interior Designers Association, Tokyo), 122–4, (in Japanese).

JIS S 1010, (1978), *Standard Sizes of Office Writing Desks* (Japanese Standards Association, Tokyo), 2, (in Japanese).

JIS S 1011, (1978), *Standard Size of Chairs for Office* (Japanese Standards Association, Tokyo), 4, (in Japanese).

Kondo, A., Iwasaki, M. and Sakazume, G., (1986), Research on VDT usage: Report of furniture height simulation demonstration and data analysis results in VDT work. in, *Proceedings of Annual Meeting of the Kanto Chapter*, Japan Ergonomics Research Society, 5–12, (in Japanese).

Manabe, S. and Nagamachi, M. (eds), (1968), *Introduction to Ergonomics* (Asakura Shoten, Tokyo), 174–82, (in Japanese).

Ministry of Health and Welfare (ed), (1985), *Fiscal 1986 Edition of National Nutrition* (Results of 1983 National Nutrition Survey) (Daiichi Shuppan, Tokyo), 155–6, (in Japanese).

Ministry of Labour, (1985), *Occupational Health Guidelines for VDT Work* (Japan Industrial Safety and Health Association, Tokyo), 17–18, (in Japanese).

Nakao, Y., Togawa, T., Furukawa, T., Saito, S. and Tsukahara, S., (1981), Measurement of anthropometric dimensions, *Japanese Journal of Ergonomics*, **17**, 199–249, (in Japanese).

New Office Promotion Council (ed), (1988), *What Will Become of the Japanese Office?* (International Trade and Industry Research Institute, Tokyo), 24–41, (in Japanese).

Noro, K. (ed), (1990), *Illustrated Ergonomics* (Japan Standards Association, Tokyo), 421–9, (in Japanese).

Obara, J. (ed), (1973), *Interior Design 2* (Kajima Shuppankai, Tokyo), 21–4, (in Japanese).

Yagi, A. (ed), (1981), *History of Steel Furniture Industry* (Kindai Kagu, Tokyo), 243, (in Japanese).

Yamaguchi, M., (1985), Chair and sitting comfort, *Japanese Journal of Ergonomics*, **21**, 233, (in Japanese).

Yokomizo, K. and Komatsubara, A., (1987), *Ergonomics for Engineers* (Nihon Shuppan Service, Tokyo), 65–72, (in Japanese).

PART III

Anthropometrics

6

Overcoming anthropometry barriers to computer–human modelling for seat design and evaluation

John A. Roebuck, Jr.

Introduction

Many aspects of seat design may be evaluated in early stages by use of computer graphics mathematical models of human forms and their articulations. Computer-based, interactive mathematical models of humans are being developed as versatile and powerful tools that can potentially facilitate engineering analysis of seating in many ways, long before mock-ups and scale models are constructed and simulators made operational.

Computer aided design (CAD) definitions of work space geometry and environmental constraints can be evaluated in a preliminary manner with regard to body fit and clearance to the seat and adjoining work space, access clearance, vision obstructions, seat adjustments, capability of humans to exert force and reach to manual controls. Such controls may be on the seat itself or part of other human-machine interfaces such as aircraft cockpits, office equipment, computer work stations and factory assembly stations.

However, at this time it must be recognized that graphic computer human models, and much of computer aided design are undergoing rapid growth. Neither is fully mature in regard to processes or hardware capabilities. Effective usage of these modern design media calls for a thoroughly developed and extensive set of compatible anthropometric data, design approaches to define body segment articulations and effective methods to utilize the data and the mathematical models in evaluation of seats and work/occupancy spaces.

Yet, in today's technological state-of-the-art there are still specific barriers lurking among the available data and processes. Mathematical human models are still relatively limited in scope, and not as accurate as desired for realistic evaluation. One reason for this condition is that there are many complex anthropometric and biomechanical relationships, supporting data and statistical interactions yet to be discovered and applied for accurate model formulation. These barriers must be overcome in order to effectively utilize computer human models.

The remainder of this chapter describes several such barriers and offers examples of approaches to overcome many of them, either immediately or over a period of future development in the sciences of anthropometry and biomechanics. These advances may help the seating industry and human factors/ergonomics specialists in general to progress more quickly toward goals of enhanced quality and productivity in design development. The advantages of computer human modelling will increase as such models become more versatile, easy to use and accurate.

Overcoming inadequate information and confusion

Through accidents of history and neglect of the proper study of man, many pervasively irritating, information-related barriers have been created in the path of progress. Some of these are legacies embedded in the literature and historical practice of anthropology. Like brick walls or boulders strewn on a path, they must be bypassed or broken down. However, most of the barriers discussed here are more appropriately visualized as various sizes of potholes, cracks, crevices and canyons reflecting inadequacy of information, especially of the type and quantity needed for computer human modelling. The process of 'overcoming' these is often to 'fill in the crack' with new knowledge or to 'build a bridge over the canyon' by a combination of clever analysis and new data.

Seen from outside the field, it appears that the many reports and thick volumes written on these subjects should contain all the knowledge needed to formulate adequate models. However, many readers of this paper well understand how audacious is the concept of modelling the tremendous variety of human sizes, shapes, performance capabilities and limitations. They know about the extremely uneven state of development in many critical areas, and the utterly fantastic complexity of the human body. So, what can be done about it?

General long-range approach

Briefly, the long-range view indicates a real need for a new and better, more comprehensive cycle of anthropometric and biomechanical surveys and detailed research. Past surveys actually need to be replicated on current populations using the recent knowledge gained from struggling with the the problems. This time, researchers must 'do it right' for the new purposes of obtaining integrated, comprehensive and specific data needed *for human modelling* by graphic computer analysis. Most designs of surveys in the past have not considered even adequately the poor draughtsman trying to draw a human form in a work site geometry. Current computer models are even more demanding, and not much good at fudging, or 'winging it'. Work space evaluation needs are increasingly different from the original anthropological goals of comparing racial groups or even later needs for design of clothing. Of course, before embarking on large-scale surveys one should perform small-scale measurement studies of selected samples from the expected populations (or, to be more accurate, their antecedents that can be actually measured today). Such preliminary 'mapping of the territory' is both prudent and necessary, before exploring it in fine detail.

Barrier: Availability of population data

One of the first questions to be asked is: Are *any* of the currently available anthropometric data appropriate for today's or tomorrow's maintenance problems? This question is raised because for the past 80–100 years there has been a remarkable secular (historical) growth in Stature and related length dimensions of many of the world's populations (NASA, 1978a), and in some US military populations as shown in Figure 6.1. The chart indicates that many years often pass between one anthropometric survey and the next. Experience in the recent past has shown that aircraft and aerospace projects have required a five to ten year development cycle from design to deployment (and some much longer). Thus, it is more likely that old data will be used than current data. By developing the practices of concurrent engineering it is hoped that American industry can shorten such development cycles significantly. Still, by the time an air vehicle is in operation for a few years the anthropometric data base which set requirements for aircraft projects may be two or more decades old.

Historical Growth Estimation Technique

- **TREND DATA ESTABLISHED FROM USAF PILOT EXPERIENCE (SURVEYS)**
- **DATA — SHOWN FOR EARTH GRAVITY ENVIRONMENT — AVERAGE MALE**

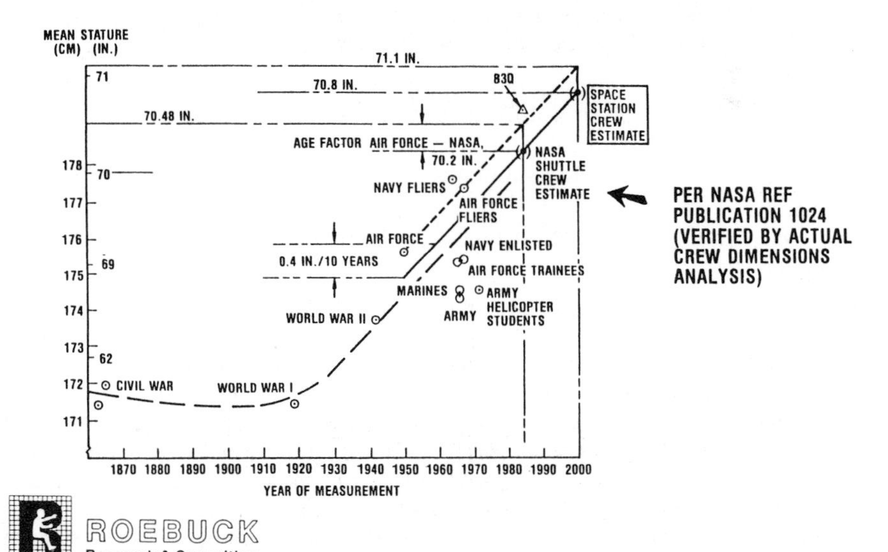

Figure 6.1 Average stature versus calendar year of US military men and NASA crew, showing projected growth trends

Long-range considerations: Forecasting secular trends

One method to offset the above-described problem is to prepare *forecasts* of anthropometric secular trends and use these estimations for design requirements. Examples of such processes performed for the space shuttle and the Freedom space station (and some of the resulting pitfalls) have been described in a paper by Roebuck *et al.* (1988). For the long-range view, it is necessary to continue developing the art and science of forecasting, accounting for known trends in population immigration, ageing, attrition, socio-economic and other factors (Roebuck *et al.*, 1975; NASA, 1978; Kroemer *et al.*, 1986). Such new approaches should be *specified* by the USAF as NASA has for the space station. Also, current specifications, such as MIL-STD-1472, ought to be revised to account for expected secular trends.

Short-term considerations: Available forecasts

Fortunately, we now have available the anthropometric forecast made during the early 1970s to predict changes in anthropometry circa 1985 for male NASA pilots (NASA, 1978). Also, during 1986 a forecast was performed to extend the predictions to the year 2000 (NASA, 1987). These forecasts were performed with the goal of predicting anthropometry of flying personnel for aerospace vehicles. However, until forecasts are developed specifically for maintenance personnel, these available forecasts might be used with reasonable adjustments for differences in age, education and other known factors. For example, non-pilots have in the past been generally shorter and of wider variability than flying personnel.

Unfortunately, the last decade has been one of some uncertainty in regard to the problem of forecasting. Some populations in the Free World have apparently begun to level out or stop their growth. Examples are those of Norway (NASA, 1987a), British civilians (Pheasant, 1986), and perhaps civilians in the USA. Early returns from the 1988 survey of US Army males (Gordon *et al.*, 1989) seem to indicate that growth of Army males has slowed. However, this may reflect a difference in socio-economic conditions for Army personnel. Caution is urged, and further near-term surveys and trend studies should be done to help understand and gauge the real trends more accurately.

Barriers: Missing and unsatisfactory external anthropometric length dimensions and interrelations

'*Potholes*': Another typical barrier related to external anthropometry is the mismatch between model requirements and available data from anthropometric and biomechanics surveys. For example, the author recently compared available data from several major surveys to the needs for only 14 dimensions for a simple model, the Crewstation Assessment of Reach, Version IV (CAR IV) (Harris and Iavecchia, 1984). It was found that at least two dimensions were missing from each survey (Roebuck, 1989). Most of these 'potholes' were lengths and heights. These do not necessarily cause traffic gridlock, but they do add the work of estimating dimensions that were not measured in a particular population of interest.

'*Confusing street signs*': Sometimes the question of dimension availability is confused by the many differences in titles for anthropometric measurements.

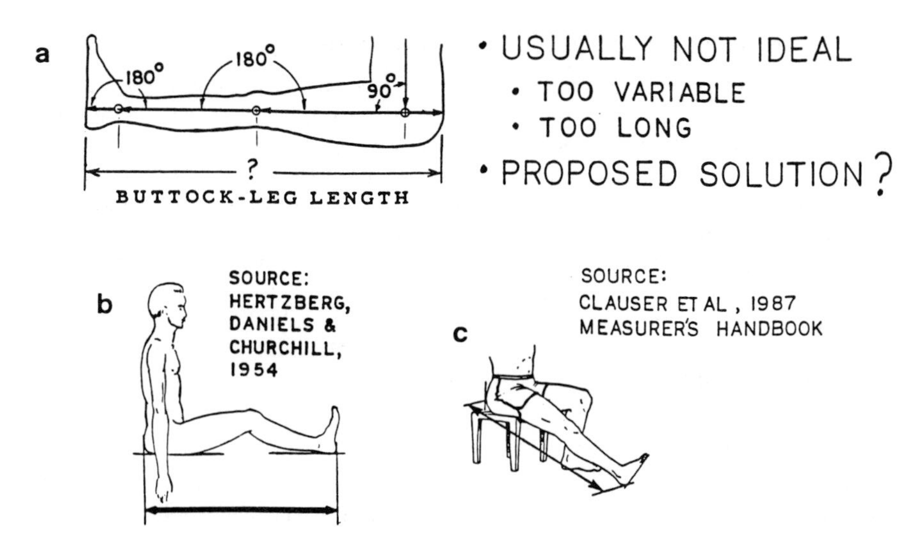

Figure 6.2 Ideal posture for Buttock–Leg Length measurement versus actual postures for 1950 USAF survey and 1988 Army survey

This is an unsatisfactory condition which needs constant attention and upgrading of comparative documents to overcome. Also, it is one that could be minimized by periodic meetings of specialists in the field to set standards. One such meeting, hosted by the Air Force, was reported by Hertzberg (1968). It is long past time to convene such a meeting again and expand its scope in light of modern human modelling needs.

'*Wrong turns*': Problems can arise from inappropriate choices of measurements, as exemplified by the case of Buttock–Leg Length. The ideal definition of the posture for obtaining this measurement is depicted in Figure 6.2a. All of the leg link lengths (distances between effective joint centres of rotation) are aligned perfectly, and oriented at 90 degrees to a theoretical trunk back plane. However, Figure 6.2b, from Hertzberg, Daniels and Churchill (1954) reminds us that in actual practice many men cannot achieve the theoretical ideal posture. At best, such men can achieve a less satisfactory posture as shown in Figure 6.2c, where no links are perpendicular to the trunk. As a result, use of data based on either measurement 6.2b or 6.2c is subject to uncertainty about accuracy of the mean and the variance. In summary, all single-measurement attempts to define Buttock–Leg Length are unsatisfactory for all foreseeable purposes in design and in modelling. Following are some alternatives which make manageable many of these problems of missing lengths.

Long-range solutions: Alternative measurements, deriving missing dimensions

As a future simple answer for the Buttock-Leg Length problem, a more constructive recommendation is offered: Instead of one unsatisfactory measurement, make two easier and more accurate, related measurements and *derive* the result (Roebuck, 1989). One example of possible approaches is described in

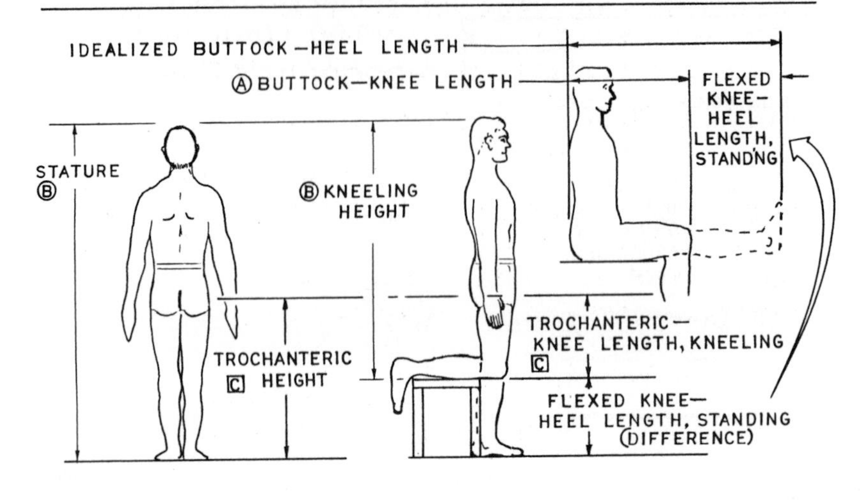

Figure 6.3

Figure 6.3, which suggests use of a measurement called Kneeling Height. By subtracting Kneeling Height from Stature, the difference obtained can be used to determine a length to be added to Buttock–Knee Length in order to derive Buttock-Leg Length.

Short-term solutions: Summing two statistical distributions

An example problem is solved in Figure 6.4 to illustrate an approach for combining two known distributions which have normal or near-normal distributions. The calculation process shown applies to adding Hand Length and Elbow–Wrist Length for male USAF Flying Personnel measured in 1967 (NASA, 1978*b*). The result is called Forearm–Hand Length, and is one of the 14 dimensions mentioned above for the CAR IV reach model (Harris and Iavecchia, 1984). A similar set of formulae (only two signs are changed) is used for subtraction of two known dimensions. As an example, the distance from top of knee in sitting position to the approximate joint centre in the 1988 U.S. Army populations (Gordon *et al.*, 1989) could be found by subtracting Lateral Femoral Epicondyle Height from Knee Height, Sitting. Note that useful anthropometric data for even one dimension in the context of human modelling of populations typically must consider many numbers: As a minimum, each dimension must be considered first as a statistical distribution and second as *a set of relationships to other dimensions* in the model. Thus, each dimension carries implications for an entire system of numbers such as:

- a measure of central tendency (e.g., the arithmetic mean);

SUM OF 2 KNOWN DIMENSIONS

PROBLEM: FIND MEAN AND STANDARD DEVIATION FOR USAF 1967 FLYING PERSONNEL (MALES) FOREARM HAND LENGTH.

GIVEN/APPROACH: FOLLOWING DATA AND CALCULATION TABLE FORMAT:

DIMENSION	MEAN	S.D.	CORRELATION COEFF. 1	2	3
1. HAND LENGTH	7.52	.32		.643	
2. ELBOW WRIST L.	11.81	.56	.643		
3. FOREARM-HAND L.	19.33	.80			

•ADD MEANS: $M_3 = M_1 + M_2$

•CALCULATE S.D.: $S_3 = \sqrt{S_1^2 + S_2^2 + 2R_{12} S_1 S_2}$

Figure 6.4

- a measure of variability (e.g., the standard deviation);
- measures of inter-relations to other dimensions in the population (e.g., coefficients of correlation and/or regression equations); and
- measures of deviation from a normal (Gaussian) distribution.

Without all of the above data in some form, it is difficult to create accurate mathematical models which describe the multiple combinations which can occur within the individuals of a population. (Parenthetically, it is only fair to note that the first two measures can be mathematically derived from other equivalent data, such as two widely separated percentiles, assuming that the statistical distribution is essentially normal.)

Many length dimensions meet the criterion of normality reasonably well. Breadths and depths often are not so satisfactory, but are less likely to be added or subtracted. Unfortunately, not all survey reports include coefficients of correlation, or they are separately reported in less obvious places at a later time. However, alternative approaches are given later for estimating standard deviations when coefficients of correlation are missing.

Estimating coefficient of correlation for new distributions

Another step in preparing data for use in the CAR IV model is inputting coefficients of correlation between *all* the 14 variable input dimensions used in the model. (Such correlations are needed for the Monte Carlo synthesizing process described later.) Therefore, if one of the necessary dimensions must be estimated, that new dimension must also have known or derived correlation coefficients between it and the remaining variable input dimensions used in the model. Solutions for this type of problem are not found in current anthropometry methods books. One sample problem solution is illustrated in Figure 6.5, which deals with the correlation between a newly derived dimension distribution and each of the addends used to derive it.

Barrier: Insufficient multivariate regressions

In adding to bivariate correlations, human modelling typically requires multivariate distributions which can help to accurately define dimensions of depth, breadth and circumference of whole human forms. For most past large-scale studies, such data are not readily available or not available at all. In some cases the problem is mainly one of economics and priority of efforts. In other cases, the data were simply never analysed.

Long-range solutions
Future anthropometric surveys should include many multiple regression relationships and provide the original data in computer files for reconstitution as needed to develop more of such relationships as needed. Such data could be valuable for many other populations in the data banks that are used for human modelling, particularly those which apply to the maintainer population in the military forces and to the pitifully few data applicable to the US civilian work force that builds military and civilian commercial equipment.

CORRELATION: SUM VS. ADDEND

PROBLEM: FIND CORRELATION COEFFICIENTS:
- FOREARM-HAND LENGTH VS. HAND LENGTH
- FOREARM-HAND L. VS. ELBOW-WRIST LENGTH

GIVEN/APPROACH: FOLLOWING DATA AND CALCULATION TABLE FORMAT:

DIMENSION	MEAN	S.D.	CORRELATION COEFF. 1	2	3
1. HAND LENGTH	7.52	.32		.643	.846
2. ELBOW WRIST L.	11.81	.56	.643		.952
3. FOREARM-HAND L.	19.33	.80	.846	.952	

- CORRELATION, 3 VS. 1: $R_{13} = \dfrac{S_1 + R_{12}S_2}{\sqrt{S_1^2 + S_2^2 + 2R_{12}S_1S_2}}$

- CORRELATION, 3 VS 2: $R_{23} = \dfrac{S_2 + R_{12}S_1}{\sqrt{S_1^2 + S_2^2 + 2R_{12}S_1S_2}}$

Figure 6.5

Short-term solutions
For a small number of military populations, multivariate regressions have been calculated and published for many combinations of useful dimensions. Examples where this has been done are reported analyses of 1968 Air Force Women (Clauser *et al.*, 1978), 1967 USAF Flying Personnel (Churchill and McConville, 1976) and 1988 US Army (Cheverud *et al.*, 1990). The Air Force has made available certain types of access to its extensive computerized data banks. Working with these data, it is possible to develop additional multiple regression relationships.

Barrier: Missing correlations hinder standard deviation estimation

In many cases adequate correlation data are not available in the literature, and the sources do not offer means by which they can be calculated. As a result, the foregoing simple formulae for combining dimensions, such as shown in Figure 6.4, cannot be applied.

Long-range solutions
Future large-scale anthropometric studies should include analysis charts that show the regressions of standard deviation as a function of the mean of the dimension distributions. These should be shown separately for lengths, breadths, depths and circumferences. This approach is offered as a general, long-

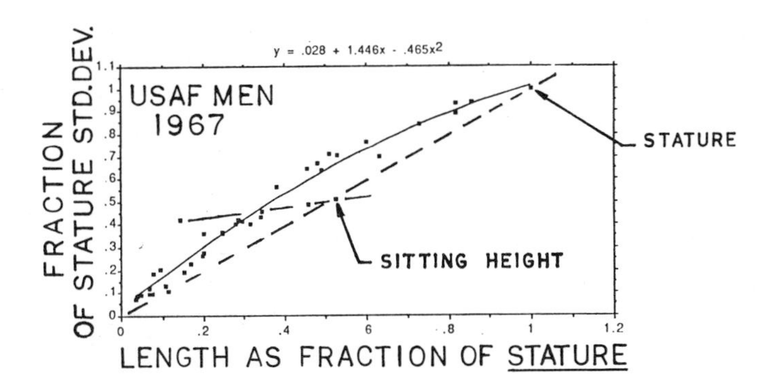

Figure 6.6

range solution goal because it has been shown helpful in the past and is a currently available method of estimation, as explained below. However, extensive research still needs to be done to more fully exploit and fine-tune this promising method, using data from past surveys.

Short-term solutions: Analysing relative variability trends

Lacking data on bivariate correlations or other information on variability, the estimation of variability of dimension distributions can be aided by analysis of trends of standard deviation as a function of mean of each distribution, using graphic presentations (Roebuck *et al.*, 1975; Pheasant, 1986). The initial formats may use a direct plot of standard deviation *vs.* mean. However, in Figure 6.6 a new, more generally useful approach is shown. The data have been normalized in terms of their relation to stature. That is, all mean values have been divided by the mean of stature and all standard deviations have been divided by the standard deviation of stature. Such regression lines for lengths and heights display a convex upward trend, starting near the origin of the graph and passing very close to the co-ordinates for stature (1, 1).

Now, consider the same type of data for a quite different population, that of Air Force Women in 1968 (Clauser *et al.*, 1968). We know that the absolute values of the means and standard deviations of such a population will be different from those of the males. However, if the data are shown in the normalized format, and the two graphs are superimposed as shown in Figure 6.7, the result displays clearly that the pattern of *relative* variability is almost iden-

SUPERIMPOSED DATA - M & F
NORMALIZED BY STATURE

- NEAR-IDENTICAL REGRESSIONS
- SIMILAR SECONDARY PATTERNS

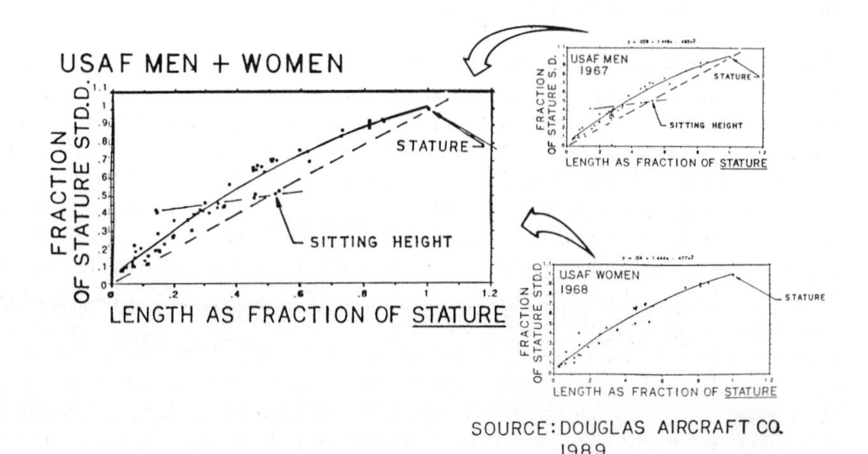

Figure 6.7

tical. Readers are invited to participate in further exploration of these variability patterns with the goal of improving estimation accuracy.

Barrier: Lack of data on breadth, depth and contours of limbs and clothing

While missing data on lengths are annoying, even greater gaps arise when trying to develop a graphic model that mathematically describes enfleshment, that is, external contours of the nude human body and of external surfaces of clothing and hair styles. Although the general consideration of contours is related to specific questions of *breadths and depths*, the latter are first considered as a separate topic in this paper. Graphic models require anthropometric data on depths and breadths of the limbs, especially the legs, to complete orthographic views of body outlines. However, only circumferences are generally measured at specific stations along the legs. Sometimes a few breadths at the elbow and knee epicondyles and at the ankles are included. Among the few exceptions to this situation are the photometric data on 250 young men reported by McConville *et al.* (1963). Other reports from which such data can be extracted are the small number of stereophotometric studies. Even the latter reported data need considerable analysis to pull out the needed information. Further, the number of sample subjects are relatively small compared to the USAF large-scale surveys that number in the thousands of subjects.

Long range solutions: Measure more breadths and depths
The obvious solution in the long run is to plan for the needed dimensions in future large-scale anthropometric surveys. Again, this approach needs further detailed definition and agreement of international groups of interested professionals, followed by publicity and educational efforts for the community of anthropologists, biomechanics specialists, and others who have a stake in the new results. The conference on standardization reported by Hertzberg (1968) is an example of what could be done again with a modest level of support by the Air Force or other government agencies.

Short-term solutions: Estimating depths and breadths from circumferences
In the meantime, there are some useful and simple approximation methods available for the specific problem of breadths and depths. Recent studies by the author have discovered some nearly linear relationships between measured circumferences and depths and breadths at the same stations of the limbs. The main source for such data has been the study of three-dimensional data for clothing manikins reported by McConville, Alexander and Velsey (1963). Figure 6.8 is one example of depth and breadth dimensions recently derived from data for Air Force males. Lacking any other data, one could use such relationships to estimate depths and breadths for Air Force maintainers and even for civilian males and females. However, it clearly would be better to have actual confirming data on diverse populations. In the example graph there is also shown a line for the function of the knee circumference divided by π. This curve can be considered the diameter of a circle with the same circumference as the measured body dimension. Not surprisingly, this reference curve nearly parallels those for depth and breadth.

BREADTH & DEPTH ESTIMATION

- FROM CIRCUMFERENCES
- NEEDED FOR LIMBS, JOINT LOCATIONS

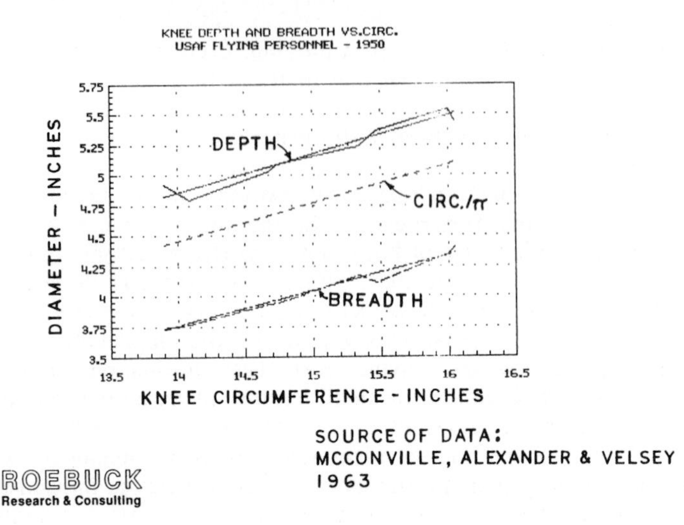

Figure 6.8

As discussed later, such methods to estimate depths and breadths from circumferences can also be used to estimate the sagittal and lateral locations of centres of rotation for joints.

Barrier: Insufficient data on contours and limited capability to summarize the data

Theoretically, many of the above problems could be solved by really complete and accurate 'contour maps' of the human body, based on larger anthropometric surveys, using stereo video, stereo photography or laser scan techniques. Then, the needed lengths, breadths, depths, circumferences and even volumes, areas and mass properties could be calculated as needed by later analysis. In fact, whole-body contour data on living persons are only available on modest sample sizes (one of the largest has $n = 46$). Even these small sample studies have been limited to a small number of populations. Further, most of these studies use only one body posture (standing), so that sitting dimensions are not derivable.

Admittedly, these approaches are expensive and complex. Even to describe a minimal number of useful breadths, depths, and offsets from internal link axes at key stations (where contours display local maxima and minima) requires between 100 and 200 dimensions. When one considers the amount of data and processing time associated with stereometric measurements of contours, the problem becomes even greater. The available detail stereometric data typically provided is about 5000 data points. Future studies using the laser technology that is currently under development is likely to have more comprehensive outputs and even larger data sets. The situation presents a difficult dilemma: without the extensive data there is a barrier of insufficiency that may be considered a yawning chasm. On the other hand, when contour data do become plentiful there will be a glut of numbers which will require creative mathematics for analysis and major advances in data handling procedures for storage and retrieval. The long-range problems are already seen in the stereo data currently available.

Long-range solutions: Selection of contours and development of analytical formulations

The number of data points for extensive contour measurements can be somewhat mitigated by careful selection of key stations for measurement of contours. Key stations along the limbs include those at which maxima or minima occur, such as at ankle height, calf height, knee, gluteal furrow, etc. In the areas of the limbs between these key stations, where contours are changing slowly, fewer data are needed to provide needed engineering information. Rather than obtain a measurement every 2 cm, one could obtain measurements every 4–5 cm apart. Much better would be measurements at specified *percentages* of distances between joint centres and key landmarks.

A more technically intriguing approach is to seek mathematical formulations that closely approximate the curves of the human body in areas important to work space design and clothing design. At least one mathematical technique to develop formulae of adequate accuracy seems to exist today (Carrier, 1989). However, the length of the formulae and the number of coefficients required may prove too extensive for practical applications. Further work in this area is still needed.

Near term solutions: Use of contour data

The small number of available studies can be used to derive several useful types of information, considering them as generally applicable patterns of external shapes, and scaling the results to conform with other data, such as average body dimensions in principal planes. Analyses of depths and breadths *vs.* circumference can be performed on the contour data to develop some generalized regression data.

Barrier: Contour data on clothing exteriors

Consideration of the practical aspects of modelling in the electronic media raises questions of how to portray dimensions of clothing and the resultant impacts on work place design. These questions are particularly important to design for maintainability, since the wearing of bulky cold weather clothing can seriously degrade dexterity and impair access to restricted spaces. Currently limited data on clothing additions to the typical external anthropometric dimensions are insufficient when modelling in three dimensions and attempting to depict the entire outer surface of garments. In fact, data on outer contours of items such as shoes appear to be completely ignored in the human factors and biomechanics literature. If we cannot model such additional bulk adequately, the value of human models may be rather minimal for just the cases where they should be of most advantage.

Long-range solutions: External clothing dimensions

Clearly, attention must be directed toward a systematic and thorough, three-dimensional measurement study of typical maintainer and equipment operator clothing external contours and how they relate to the body parts they cover. Anthropometric and biomechanical specialists need to expand their horizons and apply their expertise to produce extensive measurements on clothed persons. These clothing data need to be correlated with the nude data and developed into predictive relationships for future use.

Beyond the static measurements and relatively constant-shaped safety glasses and hard hats, there is a need for software concepts to model the folds, shifts and protrusions of pants, skirts, gloves, jackets and safety harness as work is performed. There is need to examine issues of safety in clothing, to predict possible snagging of clothing during ingress or egress of tight spaces, or to avoid safety hazards around rotating machinery or robots.

Further serious considerations apply to the restrictions of angular motion range and reduction of forces that can be applied when encumbered by clothing.

In the more general sense, hair styles are also a consideration of bulk, snagging hazards and clothing fit that may be of concern for maintenance personnel. While not classified as clothing, they are a part of the general problem of external contour modelling. These factors offer many opportunities and challenges for future research.

Short-term solutions

For the present this author can offer only the suggestions that careful study be made of the individual articles of clothing and the available data on bulk additions related to specific external dimensions. Creative and adaptive approaches are needed to model key clothing dimensions.

Barrier: Lack of data on locations of joint centres of rotation and link lengths relative to external skin landmarks

Most mathematical geometry models of humans for work space design evaluations depend explicitly or implicitly on the concept of a stick figure, or a set of links connecting average centres of joint rotation. The usual assumption is that such links have fixed lengths. This popular fiction is a highly useful approximation for many types of model used for fit and function analysis. However, a basic problem is that data from standard, large-scale anthropometric surveys rarely report locations of such joint centres and link lengths.

Importance of connectability, a long-range concern

The foregoing example of summing two known distributions illustrates an important need to be considered in the planning of anthropometric surveys and of mathematical models: Many anthropometric length dimensions must be measured and selected so as to be *connectable.* For example, Hand Length and Elbow–Wrist Length both must be measured to the same landmark on the wrist in order to be accurately and simply summed by the procedure described. Coincidently, these dimensions also closely define one aspect of the location of the wrist centre of rotation, a valuable aid to modelling the lower arm/hand relationship. In contrast, some valuable data for work space design such as Buttock–Knee Length and Knee Height, Sitting cannot be so combined. Thus, the many surveys that do not provide needed locations of the approximate knee joint centre effectively create barriers which make difficult the determination of many dimensions such as Buttock–Leg Length (also called Buttock–Heel Length, or Functional Leg Length).

Long-range solutions: Changes in attitudes and goals

Future anthropometric surveys should focus on collecting functionally useful data on locations of effective, or average locations of joint centres of rotation as well as overall, external dimensions. In planning such measurements there needs to be a change in attitude from the academic scientist toward that of the design engineer. Rather than concentrate on the details of the joint motion at each angle, an overview concept needs to be derived in which the attempt is to minimize total error in predicting reaches and clearances. The following criteria are offered as a start.

- The effective joint centre locations within the body and the selected constant lengths of the interconnecting links should provide for accurately defining the major, standard external dimensions in standing, sitting and reaching postures.
- The extended limb links and Stature should be correct if they are derived using the link lengths.
- End-to-end distances along body segments while they are being held with major joints at right angle postures should be correct. Thus, the use of the joint centres and link lengths should create no surprises and should be consistent with the standard anthropometric dimensions, such as Shoulder–Elbow Length, Sitting Height and Knee Height, Sitting.
- Effective joint centres and link lengths should provide for minimal error over the full range of motion of the joints during reaching activity. Measurements at the mid-range of the joints is particularly valuable to model comfortable postures accurately. Otherwise, and possibly in addition, measurement at 45° angles is desirable for joints with a wide range of travel, such as the elbow, knee, hip and shoulder complex.

With these guidelines, progress can be made toward a new set of simple, practical, standard measurements which can utilize the methods of Reuleaux (1875). These methods are well known in the biomechanics literature, and need not be explained here again. Such modest changes in measurements can have a major benefit for modelling the human body for engineering purposes.

Short-term solutions: Flesh links

While the above ideal measurements are not available, there are ways to derive new and useful approximations of the desired joint centre locations. In the process of devising these techniques it has been found helpful to adopt a new general approach and define new concepts in order to organize the data and perform the estimation calculations. The basic concepts of *internal* link lengths described by Dempster (1955) have been extended to cover a new set of entities called 'flesh links'. These links are defined as distances or vectors between defined points on the internal link/joint set and specified *points on the surface of the human body*. Thus, they measure the depth of flesh (including bone) from the basic axes of angular position definition.

The points of origin within the body may be either joint centres or 'stations' along the links between the joint centres. The distances to the skin surfaces (or clothing surfaces as the case may be) may be measured perpendicular to links or at some specified angle convenient for the purpose. For example, in deriving flesh lengths for a man standing in a natural, balanced posture, the internal links for the upper and lower limb typically lean forward 2–6 degrees. Horizontal distances from a vertical reference axis to posterior and anterior points on the skin are conveniently defined in the horizontal plane, rather than perpendicularly from the link axes. In contrast to internal links, flesh links may be considered as changing length as the flesh is compressed or displaced by sitting and standing, etc.

The power of this new concept is twofold: it draws attention to the many missing data which define key points on contours of the external shape of the body; and it offers a formal method of defining and organizing the needed depths and breadths as they are derived. Data tables can be set up to define sets of links that functionally describe the important protrusions and valleys of the human form. Such links are generally directly related to standard segment length measurements rather than rough approximations based only on Stature, as defined by Dempster (1955) or by Dempster, Sherr and Priest (1964).

The basic concepts for these links have appeared previously in various forms. Even in the Crewstation Assessment of Reach (CAR IV) non-graphical model there are certain links and offset distances defined that exactly fit the above definition of flesh links. For example, the so-called 'ankle link' can be considered a flesh link from the ankle joint to the surface of the foot at the heel. In the book by Roebuck, Kroemer and Thomson (1975) they are called terminal links. Having broken the conceptual barriers around internal links, it is also useful to define other types, such as the following:

- Surface links: Distances between two points on the surface of the body or clothing.
- Pseudo links: Distances between a part of the body (usually a joint) and another key, non-body point such as the centre of a tool handle or a control knob.

A need for brevity precludes further details about applying this concept. However, Figure 6.9 illustrates some examples of flesh links and their suggested titles for the leg and foot.

Barrier: Lack of data on combined joint motion ranges and globographics

After the link lengths and joint centres are established relative to the external dimension landmarks there is a need to model their relative angular posture and ranges of motion. As a general rule it is assumed that the preferred comfort orientation of a joint is near the middle of its range of motion. However, we have little data on joint range of motion as regards the following aspects:

- Limits of motion determined by adjacent and distal joint orientation (two-joint and three-joint limits).
- Range of motion in other than principal planes of the body, particularly combinations of joint motions which include *axial* rotation of the limb.

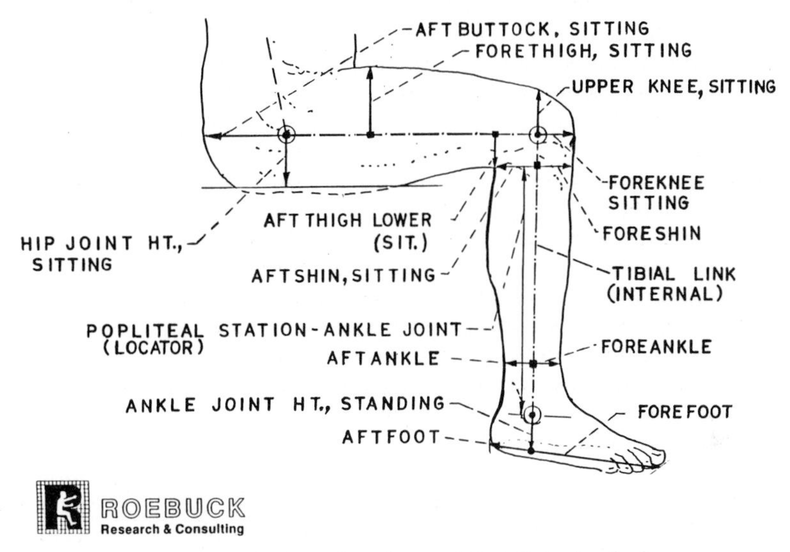

Figure 6.9 Examples of several flesh links for the leg

- Range of motion data relative to the specific population that is being modelled (data are usually from college students, not USAF maintainers, for example).
- Restrictions caused by various types of clothing (e.g., gloves, parkas, chemical protection suits).
- *Starting orientation for the measurement* is missing too often in the definition of the measurement. In particular, this poses a serious problem for mathematical models that are based on the internal link postures as indicators of angular position. Many researchers use external surfaces (such as the lower or upper surface of the arm) as starting orientations.

Some of the most glaring lacks of angular motion data involve the shoulder and hip joints. The common practice of measuring range of motion only in principal planes defines a very tiny percentage of the actual sinus cones of motion in these 'ball and socket' types of joint. The only recourse of the designer is to consider motion range in each principal plane as acting independently. Joint motions surely are not independent at all the extremes of combined up-down, left-right and axial motion range. As a result, models will likely indicate more range of travel than can actually be achieved at orientations other than in principal planes.

Long-range solutions

There is a need to perform research on much larger samples of humans which are representative of seated operator populations and the conditions under which they work. To do so requires new, more easily used tools and procedures, and a new kind of thinking that accepts the need for such added measurements as part of anthropometry. Ideal data concepts are illustrated in Figure 6.10. Near the top of this figure is a 'globographic' presentation of range of motion for shoulder flexion-extension and adduction-abduction, following the examples of Dempster (1955). Thus, the basic globographic concept widely expands the amount of coverage considered in most surveys. However, there is still something lacking. The globographic figures of Dempster (1955) and even the recent globographic depictions of shoulder joint motions of Engin and Peindl (1987) do not indicate concurrent effects of *axial rotation* of the humerus.

Lower on the figure is one special case example of a general method proposed by this author for depicting axial rotation capability by using two vectors oriented along the axis of the limb (the lower leg in this case). The length of these vectors extending radially from the surface of the globe define surfaces outboard of the globe that indicate the amount of available inward and outward axial rotation for each posture combination of flexion-extension and adduction-abduction. (Note that for the knee there is practically no adduction-abduction, so the 'surfaces' are reduced to lines on a plane.)

No short-term solutions

Unfortunately, there are probably no comprehensive, immediate solutions for this type of complex problem. Although incomplete, the data of Engin and Dempster could be incorporated as general improvements for most current, near-term modelling effects. For many models this would represent a major

JOINT RANGES DEPICTED BY LINKS
AND GLOBOGRAPHIC REPRESENTATION

- BONES CONSIDERED AS STRAIGHT LINES (LINKS) BETWEEN JOINT CENTERS
- CONCEPTUAL GLOBE FIXED TO ONE LINK
- INTERSECTION OF MOVING LINK WITH GLOBE TRACES RELATIVE MOTION RANGE (SWING)
- AXIAL ROTATIONS SHOWN AS SURFACES OUTBOARD OF GLOBE GENERATED BY RADIAL VECTORS

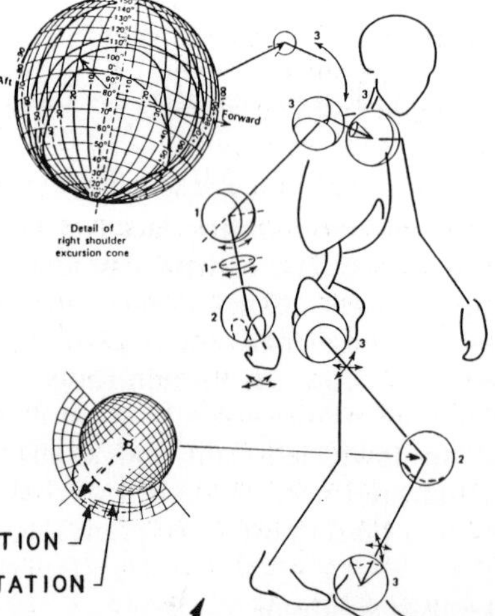

LATERAL ROTATION
MEDIAL ROTATION

SOURCE: DEMPSTER, 1955

Figure 6.10 Conceptual globographic representations of joint motion range, including axial rotation advance in thinking about joint range modelling. Some creative extrapolation is needed, since adequate data on large populations do not exist. Also, small-scale experiments may be performed to offer additional interim data until large-scale studies are performed.

Advancing the state of the art: Percentage accommodation versus percentiles

Sometimes the barriers one faces in a project involve decisions on which is the better and more useful of two good approaches. For years anthropometry specialists and human factors generalists have tried to overcome the 'average man syndrome' (Daniels, 1952). They tried to convert the thinking of engineers toward specifying *percentiles* for dimensions. Out of this concept and background grew the practice of combining many high or low percentiles into sets of common-percentile, articulated drafting templates or manikin designs for convenience in early design layouts and for evaluating layout drawings. Used and interpreted properly, with replaceable parts of larger and smaller percentiles, such manikins have real value in avoiding the many errors that went with the average man concept.

Yet, many human factors and ergonomics specialists knew perfectly well that such manikins generally do not accurately characterize the chance of successful accommodation if two or more body dimensions, angles, etc., are critical to a design. Moroney and Smith (1972) and others have pointed out this concern clearly and forcefully. Several alternative and generally better approaches have been developed that recognize the importance of worst-case combinations of dimensions for different body segments. For example, a set of six specified combinations of body dimensions have been derived for Navy cockpit design, whereas eight sets of specified body dimensions have been defined for USAF cockpit design. Bittner (1987) has derived 17 sets of body dimensions, called CADRE. The design and evaluation process is somewhat more extended and complicated by this approach, in that more manikins must be defined and constructed, either as articulated templates or as electronic models. Yet, even these generally more comprehensive concepts may not be adequate as design criteria for maintainability.

Another more general approach is to seek a 'goodness-of-fit' number, which may be called *per cent accommodation.* This approach is currently in use in the automotive field for passenger car design evaluation. As embodied in a mathematical modelling technique, this approach attempts to generate several hundred statistically valid sets of body dimension combinations, possibly using a random selection process. Each of these combinations (synthetic individuals) is then tested for fit, reach, etc. appropriate to the evaluation in question. This approach more closely approximates the process used in a large-scale evaluation of a mock-up or prototype by use of human subjects.

As a general concept, some form of this random generation approach appears preferable for the wide range of unusual body postures and access evaluations in complex work spaces that are characteristic of maintenance activity. The CAR IV reach accommodation model is one example of a model that has incorporated the concept with Monte Carlo generation of synthetic individuals. As it stands now, the CAR IV model is *not* an 'off-the-shelf' candidate to solve many of the vision, clearance, strength and other concerns of maintainability evaluations. However, its Monte Carlo routines could provide the basis for useful routines for adaptation to generalized human modelling for human factors evaluations.

Summary and conclusions

This paper has highlighted several technical difficulties and organizational/procedural needs related to anthropometry, biomechanics and their usage in computer-based human models. Needs for fundamental changes are identified, including a new understanding of potential uses of data on anthropometry and biomechanics for the burgeoning era of computer-based graphic simulation. Three-dimensional definition of landmarks and form are needed, and all measurements must relate to common axes, thus tying them together, end-to-end.

Methods to overcome specific difficulties have been offered, considering both long-range and immediate solutions. Pleas were made for forecasting of anthropometric trends and for more extensive and better surveys to better know our people as regards statistical changes brought about by immigrations, population shifts, ageing and other demographic concerns. New concepts described include the use of 'flesh links', empirical approaches to estimating standard deviations, approaches for estimating missing breadths and depths of limbs by using circumferences, and methods to estimate locations of joint centres from external surfaces of the body. Also recommended are the convening of congresses and committees to standardize on the new measurement methods, to define new terminology, and to update the data bases and collations in conformance with the new era in computer aided design methodology.

A preferred statistical randomization approach for manikin design and evaluation simulations was recommended. It is based on using many different combinations of possible body dimensions to determine a percentage accommodated instead of relying on a set of common percentiles in a few manikin models.

A modestly revolutionary thought was also offered: Government encouragement of advancements in manufacturing and design technology through funding of automation and promoting simultaneous engineering for maintainability improvements, should also include *funding and organizing large-scale measurement surveys of the civilian work force*, the people whose hands and heads and muscles actually build the tools and assemble the aircraft used by the military services. Such an investment would provide to the USA a type of government support that is already benefiting some of our most competitive trading partners (e.g., Japan, Taiwan). It could have significant, wide-spread benefits for the US economy, as well as aid to design of more readily maintainable airplanes.

If the lessons described in this paper are applied and the new programs advocated are funded and promoted, then we can deal with the barriers identified, and we *shall overcome*!

References

Bittner, A., (1987), CADRE: A Family of Manikins for Workspace Design, in Asfor, S. S. (ed), *Trends in Ergonomics*, Amsterdam: North-Holland, pp. 733–740.

Carrier, R., (1989), Krigeage Applications. Lecture delivered to technical meeting of Human Factors Society, Los Angeles Chapter at University of Southern California, for Genicom Consultants, Montreal, Canada.

Cheverud, J., Gordon, C. C., Walker, R. A., Jacquish, C., Kohn, L., Moore, A. and Yamashita, N., (1990), *1988 Anthropometric Survey of US Army Personnel: Bivariate Frequency Tables*, Technical Report Natick/TR-90/031, United States Army Natick Research, Development and Engineering Center, Natick, Mass. 017600–5000. (Also see Cheverud *et al.*, Natick/TR-90/032,-/033,-035 for other statistical analyses.)

Churchill, E. and McConville, J. T., (1976), Sampling and Data Gathering Strategies for Future USAF Anthropometry. AMRL-TR-74-102, Webb Associates, for Aerospace Medical Research Laboratory, AMD, AFSC, WPAFB, Ohio.

Clauser, C. E., McConville, J. T., Tucker, P. E., Churchill, E., Reardon, J. A. and Laubach, L. L., (1968), Anthropometry of Air Force Women, AMRL-TR-70-5, Aerospace Medical Research Laboratory, AMD, AFSC, WPAFB, Ohio.

Clauser, C., Tebbetts, I., Bradtmiller, B., Ervin, C., Annis, J., McConville, J. T. and Gordon, C., (1987), Measurers Handbook: US Army Anthropometric Survey, 1987–1988 (draft), Anthropology Research Project, Inc., Yellow Springs, Ohio.

Daniels, G. S., (1952), The 'Average Man?', Technical Note WCRD 53-7, WPAFB, Ohio.

Dempster, W. T., (1955*a*), Space Requirements of the Seated Operator. WADC Technical Report 55–159, Wright Air Development Center, Air Research and Development Command, USAF, WPAFB, Ohio.

Dempster, W. T., Sherr, L. A. and Priest, J., (1964), Conversion Scales for Estimating Humeral and Femoral Lengths and the Lengths of Functional Segments in the Limbs of American Caucasoid Males. *Human Biology*, **36**, 246–62.

Engin, A. E. and Peindl, R. D., (1987), On the Biomechanics of Human Shoulder Complex—I. Kinematics for Determination of the Shoulder Complex Sinus. *J. Biomechanics*, **20**(2), 103–17.

Garrett, J. W. and Kennedy, K. W., (1971), *A Collation of Anthropometry, Volumes I and II*, AMRL-TR-68-1, Aerospace Medical Division, AFSC, WPAFB, Ohio.

Gordon, C. C., Churchill, T., Clauser, C. E., Bradtmiller, B., McConville, J. T., Tebbetts, I. and Walker, R. A., (1989), *1988 Anthropometric Survey of U.S. Army Personnel: Summary Statistics Interim Report*. United States Army Natick Research Development and Engineering Center, Natick, Mass. 01760-5000.

Harris, R. and Iavecchia, H., (1984), Crewstation Assessment of Reach Revision IV (CAR IV) User's Guide. Technical Report 1800.10a, Analytics, Willow Grove, PA 19090 for: HFE Technology Development Branch, Naval Air Development Center, Warminster, PA. Task Report Contract No. N62269-82-D-0131, Del. Order 10, Task 1.

Hertzberg, H. T. E., (1968), The Conference on Standardization of Anthropometric Techniques and Terminology, *Amer. J. Anthropol.*, **28**(1), 1–16 (N.S.).

Hertzberg, H. T. E., Daniels, G. S. and Churchill, E., (1954), Anthropometry of Flying Personnel—1950. WADC TR 52-321, Wright Air Development Center, WPAFB, Ohio (AD 110-573).

Kroemer, K. H. E., Kroemer, H. J. and Kroemer-Elbert, K. E., (1986), *Engineering Physiology – Physiologic Bases of Human Factors/Ergonomics*, New York: Elsevier.

McConville, J. T., Alexander, M. and Velsey, S. M., (1963), Anthropometric Data in Three-Dimensional Form: Development and Fabrication of USAF Height-Weight Manikins. Technical Documentary Report AMRL-TDR-63-55, Behavioral Sciences Laboratory, 6570th Aerospace Medical Research, Aerospace Medical Div., AFSC, WPAFB, Ohio.

Moroney, W. F. and Smith, M. J., (1972), Empirical Reduction in Potential User Populations as the Result of Imposed Multivariate Anthropometric Limits. NAMRL-1164. Pensacola: Naval Aerospace Medical Res. Lab.

NASA, (1978a), *Anthropometric Source Book, Volume I: Anthropometry for Designers*, NASA Reference Publication 1024, National Aeronautics and Space Administration, Scientific and Technical Information Office, Washington, D.C.

NASA, (1978b), *Anthropometric Source Book, Volume II: A Handbook of Anthropometric Data*, NASA Reference Publication 1024, National Aeronautics and Space Administration, Scientific and Technical Information Office, Washington, D.C.

NASA, (1987), *Man-System Integration Standards*: NASA-STD-3000, Vol. 1, NASA Johnson Space Center, Houston, Texas.

Pheasant, S., (1986), *Bodyspace: Anthropometry, Ergonomics and Design*, London: Taylor & Francis.

Reuleaux, F., (1875), *Theoretische Kinematik: Grundzuge einer Theorie des Maschinenwesens*, Braunschweig, Germany: F. Vieweg und sohn. (Also translated by A. B. W. Kennedy, 1876, *The Kinematics of Machinery: Outline of a theory of machines*, London: Macmillan.

Roebuck, J. A., (1989), Anthropometry Applications to Industrial Ergonomics, Workshop briefing presented to Industrial Ergonomics and Safety Conference 1989, Cincinnati, Ohio.

Roebuck, J. A., Kroemer, K. H. E. and Thomson, W. G., (1975), *Engineering Anthropometry Methods*, New York: Wiley.

Roebuck, J. A., Smith, K. E. and Raggio, L. J., (1988), Forecasting Crew Anthropometry for Shuttle and Space Station, *Proceedings of the Human Factors Society 32nd Annual Meeting*, Human Factors Society, Santa Monica, CA. (Also: Rockwell International paper STS 88-0717, Space Transportation Systems Division, Downey, CA.

7

Anthropometry and advanced ergonomic chairs

Marvin J. Dainoff, James Balliett, and Phillip Goernert

Introduction

During the late 1970s, widely distributed computer technology began to enter the office. As the Video Display Terminal (VDT) began to appear on office desks, the realization emerged that office work environments would have to be redesigned. Accordingly, many national and international organizations began to issue guidelines and standards for the VDT workplace.

Virtually all such recommendations for proper design of chair and work surface configurations were based on a 90° upright seated work posture. Mark and Dainoff (1988) called this the cubist approach; since major body angles (legs, hips, trunk, arms) are all at right angles, seated posture can be modelled with a series of cubes.

The cubist approach simplifies the communication of design recommendations. Once a user population is defined, only a few anthropometric dimensions are required to specify ranges of seat adjustments which fit some proportion of that population. However, as new research evidence increases our understanding of seated posture, the cubist simplification becomes increasingly untenable (Grandjean *et al.*, 1983; Mandal, 1981; Kroemer, 1987).

It has been argued (e.g., Dainoff and Dainoff, 1986; Dainoff *et al.*, 1987; Dainoff and Mark, 1989; Mark and Dainoff, 1988; and others) that effective chair and work station design should support *alternative* work postures. These include both the forward leaning posture emphasized by Mandal (1981) [seat pan tilted forward, trunk erect], and the backward leaning posture proposed by Grandjean *et al.* (1983) [seat pan and back rest tilted back].

Advanced ergonomic chair and workstation designs which support these alternative postures are becoming increasingly available in the market. While these designs can benefit end users, they also complicate the job of writing standards. It is much less straightforward to prescribe ranges of chair and work surface adjustability that include forward and backward leaning postures.

The following discussion explores these issues in order to modify and expand specifications contained within the recent US Standard (ANSI/HFS 100-1988). In so doing, we will develop ranges of keyboard support heights (desk heights) which afford comfortable working postures in both forward and backward leaning postures. These keyboard support heights must be defined in relation to *seat pan* height and angle. To accomplish this goal, we define a new composite anthropometric dimension: *seated fingertip height* (SFH). SFH depends upon whether the operator is in a forward or backward leaning posture. This range will be compared with the original values published in ANSI/HFS 100. Finally we recommend that the obtained range of the anthropometric dimension SFH match the corresponding range of adjustability of the keyboard support height.

The ANSI approach

Rationale and scope

Although major sections of the ANSI/HFS standard deal with work environments (luminance, acoustics, thermal), visual display characteristics, keyboard charcteristics, and measurement techniques, this paper only addresses the section devoted to chairs and work surfaces.

ANSI presented a parallel set of recommendations within this furniture section, the specific solution and the general solution. The ANSI specific solution assumes the operator sits upright or near upright with the feet resting on the floor or footrest, and accommodates CRT-based VDT operators who range in size from the 5th percentile female through the 95th percentile male of the US population. The general solution, on the other hand, represents a technology-independent, posture-independent, population-independent set of general principles for workstation design which can be applied to a wide variety of populations and work postures. The general solution is described in terms of body positions and angles, and embodies general principles of postural orientation, the specific solution is described as furniture dimensions, representing a particular implementation of the general solution for a certain circumstance.

Selected components

We used selected components of the ANSI/HFS standard to derive the composite anthropometric dimension, seated fingertip height. We assume that the optimal height of a work surface should allow the fingertips to be placed close to the keyboard with a minimum of effort. In determining this height, the relationships described below between chair and worksurface dimensions must be carefully considered.

Seat height (ANSI/HFS 8.7.1)

The general solution for seat height requires that the seat pan height should allow users to place their feet firmly on a support surface.

In the specific solution, the lower limit for seat height is determined by the popliteal height of the 5th percentile female (35.5 cm) plus a correction for heel height (5 cm), leading to a recommendation of 40.5 cm. This value is increased to 40.64 cm to correspond to 16 in.

The upper limit for seat height is determined by the 95th percentile male popliteal height (48.77 cm). No heel correction is allowed.

Keyboard support height (ANSI/HFS 8.4.1)

In the ANSI/HFS general solution, the height of the keyboard support surface is based on two factors:

1. There should be sufficient knee clearance under work surface;
2. The resulting postures should be within the following set of angular relationships:
   (a) The forearm should be within $70 + Y/2$ degrees and $90 + Y/2$ degrees from the superior frontal plane, where Y is the seat back angle from vertical in degrees.
   (b) The angle between the upper arm and the forearm should be greater than 70° and less than 135°.

The corresponding specific solution defines a range of adjustable keyboard support surfaces from 58.5 to 71.0 cm. The lower limit for keyboard support height is derived by adding the 5th percentile female seated elbow height (18.03 cm) to 5th percentile female corrected seat height (40.64; see above). The resulting computed recommendation for the lower limit of an adjustable keyboard support surface is 58.67 cm; this is rounded to 58.5, or 23 inches.

The upper limit for keyboard support height is derived by adding the 95th percentile male seated elbow height (29.4 cm) to 95th percentile male seat height uncorrected for heel height (48.77; see above). The resulting recommendation for the upper limit of an adjustable keyboard support surface should be, by simple addition, 78.17 cm. In point of fact, ANSI's actual recommendation for the upper limit is 71.12 cm; this was a compromise value that recognized that large males may not be accommodated.

Seatpan angle (ANSI/HFS 8.7.4)

The ANSI general solution stipulates that the thigh and lower leg angle should be between 60° and 100° with the lower leg perpendicular to the floor.

The specific solution requires that if the seat pan angle adjusts, it should range between 0° and 10° back.

Chairs with forward-tilting seat pan angles (negative values) are explicitly permitted in this standard, but not as part of the specific solution, since a range of forward angles is not specified.

Angle between seatpan and seat back (ANSI/HFS 8.7.5)

The general solution requires that the torso-to-thigh angle range between 90° and 100°. Further, for forward tilting postures, the upper torso should not be constrained forward of the vertical (90°).

The specific solution requires that for adjustable seatpan back rest systems, the pan-to-back angle must include some part of the range of 90° to 105°.

Generalization to alternative postures

Introduction and rationale

The ANSI/HFS standard differentiates between a general solution, in which requirements are stated in terms of relationships between body parts, and a specific solution, which presents requirements in terms of furniture dimensions. The latter requirements are presumably derived from the principles and relationships of the former, but restricted to near-upright posture. As mentioned above, forward-leaning seat pans are addressed in the general, but not specific solutions. Hence, it must be inferred that forward-leaning postures are not considered near-upright.

At the same time, the specific solution allows for a backward inclined seat pan of 10° and a seat-to-back angle of 105°. This combination allows a back rest angle of 25° from vertical, and corresponds almost exactly to the backward leaning posture described by Grandjean *et al.* (1983).

If, therefore, a combination of seat pan and back rest leading to a 25° backward inclination is still considered near upright within the definition of the specific solution, it seems reasonable to also incorporate a forward seat pan inclination of −10°. In so doing, ANSI/HFS also requires that the back rest not constrain the trunk forward of 90°.

Finally, all of the ANSI/HFS recommendations assume that the upper arm always remains aligned with the trunk in the superior frontal plane. This seems unrealistic in practice, even in a 90° upright posture. As standard reference work on occupational biomechanics indicates that 25° of flexion of the upper arm from the superior frontal plane is acceptable for seated work (Chaffin and Andersson, 1984, p. 311).

Goals and approach

This chapter attempts to expand the ANSI/HFS specifications for keyboard support surface height by incorporating a series of alternative postural configurations within the constraints discussed in the previous paragraph. Our approach resembles that of ANSI/HFS 8.4.1, discussed above, for the 90° posture in that proper working height is derived by adding two anthropometric dimensions: popliteal and seated elbow height. According to Pheasant (1986, p. 38) and Roebuck *et al.* (1975, p. 160), the mean of any new composite normal distribution can be formed by adding a set of means of constituent normal distributions. Thus, the 50th percentile of a new composite anthropometric dimension can be defined as the sum of the 50th percentile points of a set of tabled (empirical) anthropometric dimensions.

In the present model, because of the complexity introduced by components which depart from simple right angles, the new composite dimension, seated fingertip height (SFH), is the sum of six separate components. Moreover, these components must be (mostly) defined in terms of centres of rotation at the joints and body segment links. (Link segments are defined in terms of distances between joint centres of rotation.) Pheasant (1986, Table 7.1) presents average major body links for males and females expressed as a percentage of stature.

Thus, each link-based component in the overall composite model can be expressed as a percentage of the stature of the 50th percentile male or female in the user population of interest. The central assumption in our overall approach is that adding these components is equivalent to adding 50th percentile points for individual anthropometric dimensions.

Upper and lower limits for the user population are computed from the 50th percentile male and female values for SFH. This is accomplished by estimating the standard deviations of the two new distributions (SFH-male, SFH-female) for each postural configuration. These estimates are then multiplied by the appropriate z score from tables of the normal distribution. For 5th/95th percentile values, these are ± 1.64. Our estimation method is suggested by Pheasant (1986, p. 40), and based on the empirical regression of mean (50th percentile) against standard deviation of tabled anthropometric data.

While the model, as described below, can be applied to any population, we have utilized the anthropometric tables for the US population prepared by Kroemer (1983); these serve as a basis for the ANSI/HFS standard. Values from Kroemer's tables were also used to produce a pair of regression equations (one for each gender) to estimate composite standard deviations. The actual equations appear in Appendix A.

Description of model

Definition of terms (see Figures 7.1 and 7.2)

Independent Variables:

Asp: Seat Pan Angle From Horizontal
At: Trunk Angle From Vertical
As: Shoulder Angle From Trunk
Ae: Elbow Angle From Upper Arm

Constants:

SPD: Seat Pan Depth
SRP: Seat Reference Point
HH: Heel Height

Links (L) and Anthropometric Dimensions (A):

PH: Popliteal Height (A)
HfSRP: Hip in front of SRP (L)
BPL: Buttock-Popliteal Length (L)
HaSRP: Hip above SRP (L)
HSL: Hip-Shoulder Length (L)
SEL: Shoulder-Elbow Length (L)
EFL: Elbow-Fingertip Length (L)

Auxiliary Computations:

$\alpha = \arctan (\text{HaSRP}/(0.5\ \text{SPD-HfSRP})$

Hip to Seat Center [r] = HaSRP/sin α

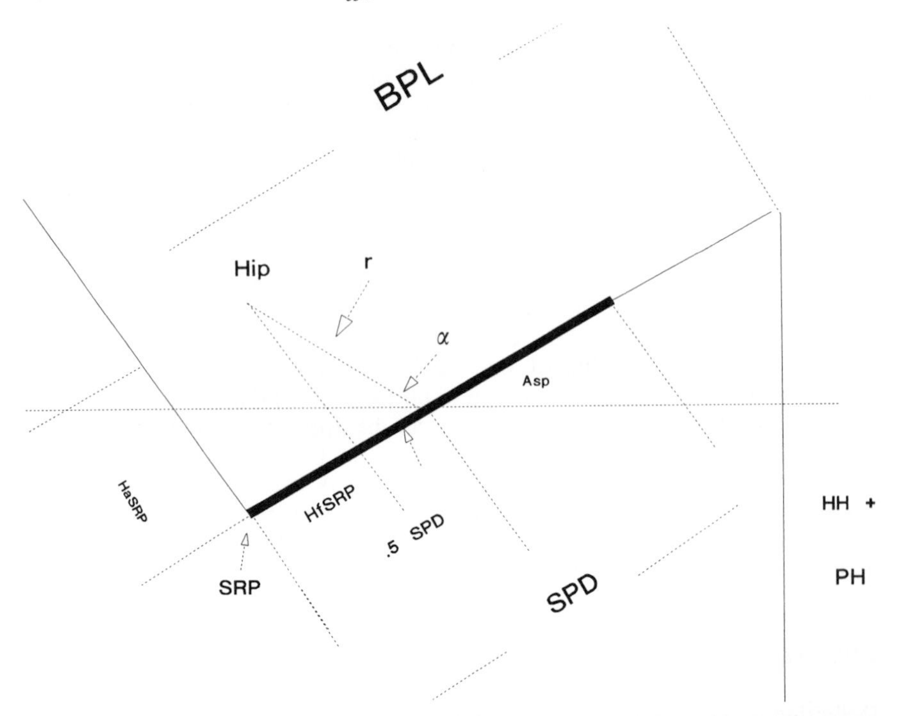

Figure 7.1 Components of lower section of the model. HH + PH: heel height plus popliteal height; Asp: seat pan angle from horizontal; SPD: seat pan depth; Alpha: angle between hip-to-seat centre and seat pan; r: hip-to-seat centre; BPL: buttock-popliteal length; HfSRP: Hip in front of seat reference point; SRP: seat reference point; HaSRP: hip above seat reference point

Components of the model (*see discussion below*)

Component 1: + Heel Height + Popliteal Height
Component 2: − sin Asp (BPL − 0.5 SPD)
Component 3: + Hip to Seat Centre (sin [α-Asp])
Component 4: + cos At (Hip Shoulder Length)
Component 5: − cos (As + At)(Shoulder Elbow Length)
Component 6: + cos [Ae − (As + At)] (Elbow Finger Tip Length)

Seated Finger Tip Height (SFH) = Sum(Components 1, ..., 6)

Discussion

Component 1 of the model specifies the height of the seat pan when in the 90° posture in terms of popliteal height plus heel correction (as in ANSI/HFS). Assuming that the centre of rotation is in the middle of the seat pan (0.5 SPD), the seatpan must be raised when tilted forward and lowered when tilted back in order to maintain a 90° angle between the lower leg and the floor. Component 2 accounts for this correction. Tilting the seat pan will also affect the position of the hip and trunk, raising the hip and trunk with a forward tilt and lowering the hip and trunk with a backward tilt. Component 3 locates the

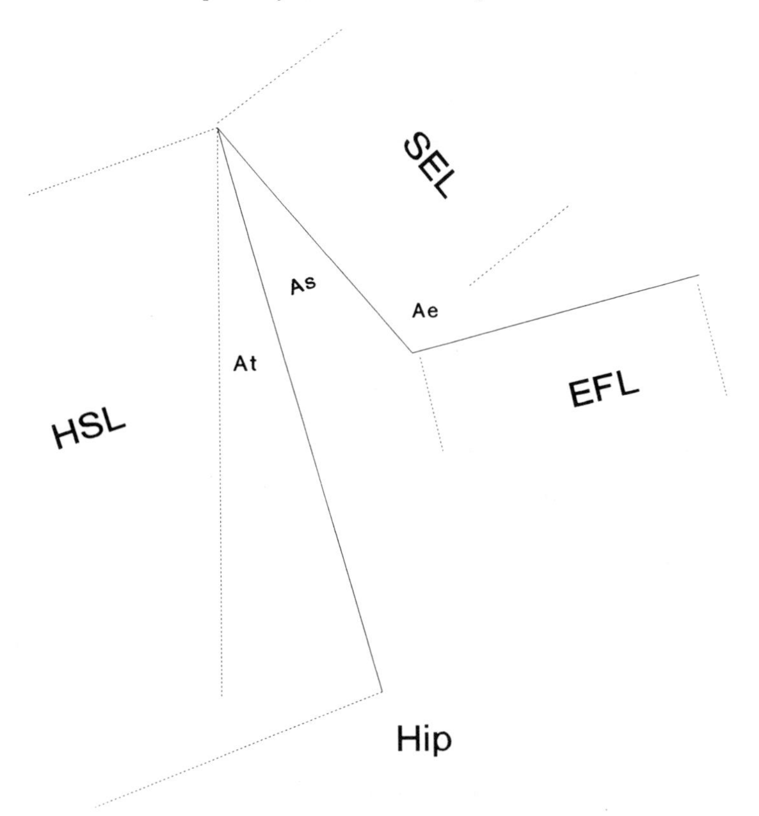

Figure 7.2 Upper section of model. HSL: hip-shoulder length; At: trunk angle from vertical; As: shoulder angle from trunk; Ae: elbow angle from upper arm; SEL: shoulder–elbow length; EFL: elbow–fingertip length

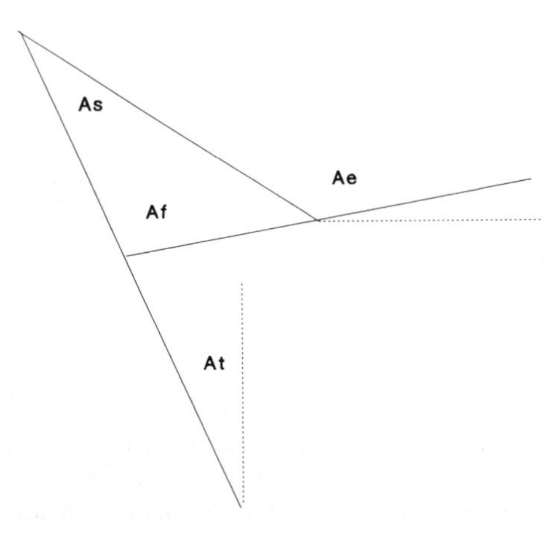

Figure 7.3 Relationships among elbow, forearm and shoulder angles. As: shoulder angle from trunk; At: trunk angle from vertical; Af: forearm angle from trunk (superior frontal plane); Ae: elbow angle from upper arm

vertical position of the hip above the seat pan as it rotates around the seat centre of rotation. These computations rely on Seat Reference Point, which is defined as the intersection between the planes of the seat pan and back rest.

From the hip, the vertical distance to the shoulder is determined by component 4. As with the hip, the shoulder is at its highest position when the seat pan is tilted forward and the trunk upright. The shoulder is at its lowest position when the seat pan is tilted all the way backwards and forms a 105° angle with the back rest. The line from the hip to shoulder rotates around the hip joint, defining trunk angle from the vertical.

Component 5 locates the point of the elbow, and component 6 locates the fingertips. The deviation of the upper arm from vertical will increase the vertical height of the elbow. In addition, decreasing the angle between the upper arm and forearm will increase the height of the wrist and hand. ANSI/HFS makes no recommendations concerning appropriate hand angles; we assume that there is no angle between the hand and forearm.

Computational analysis

Constraints

In exercising the model, limiting values were defined in terms of the four independent variables described above. These constraints follow from the specifications found in ANSI/HFS along with our own modifications.

Seat pan angle (Asp)

$$-10^\circ \leq \text{Asp} \leq +10^\circ$$

Comment: ANSI/HFS 8.7.4 specifies +10° of seat angle from horizontal; we have added −10° for forward tilt. This is implied (but not specified) by the proviso that the thigh can be at a 100° angle from the perpendicular lower leg.

Trunk angle (At)

$$0^\circ \Leftarrow \text{At} \Leftarrow 25^\circ$$

Comment: ANSI/HFS 8.7.5 specifies the included angle between seat pan and back rest to be between 90°–105° and that the upper torso not be forward of the vertical in forward tilt. For backward tilt, the maximum trunk angle of 25° from vertical is achieved by combining the allowable +10° of seat pan angle with a 105° included angle.

Shoulder angle (As)

$$\text{At} \leq \text{As} \leq (25^\circ - \text{At})$$

Comment: ANSI/HFS does not provide criteria for shoulder angles. Chaffin and Andersson (1984, p. 311) indicate that 25° of shoulder flexion is allowable; however their discussion implies that the trunk is vertical. Our approach maintains a maximum of 25° of flexion from vertical. When trunk angle is 25°, allowable shoulder angle (As) is zero; when trunk angle is zero, allowable As is 25°.

Elbow angle (Ae)

$$[(70^\circ + 0.5\ \text{At}) + \text{As}] \leq \text{Ae} \leq [(90^\circ + 0.5\ \text{At}) + \text{As}]$$

Comment: As discussed earlier, ANSI/HFS 8.4.1 discusses the location of the forearm in terms of three angles; these are forearm to superior frontal plane, forearm to upper arm (elbow angle), and seat back to vertical. If the seat back to vertical angle is considered equivalent to trunk angle (At), and Af is defined as forearm angle, these criteria can be redefined as follows:

ANSI/HFS Criterion 1: $(70 + 0.5\ \text{At}) \leq \text{Af} \leq (90 + 0.5\ \text{At})$
ANSI/HFS Criterion 2: $70 \leq \text{Ae} \leq 135$

When the shoulder angle (As) is not 0°, the (projected) value of forearm angle (AF) is equal to Ae–As since, by geometrical analysis, Ae = Af + As. Thus, ANSI/HFS criterion 1 can be expressed in terms of elbow angle Ae as seen above. Furthermore, when As is at its maximum value of 25°, the resulting values for Ae (83.5 – 102.5) fall within the range specified by Criterion 2. (See Figure 7.3.)

Model execution

Six postural configurations are defined by utilizing the four constraints described above. The six-part equation of the model is then solved for each configuration. Seated finger tip height (SFH) for 50th percentile males and females, 5th percentile females, and 95th percentile males are listed in Table 7.1. Appendix A lists all parameters employed in computing the solutions, along with sources for parametric values.

Table 7.1 also includes three additional configurations. For purposes of calibration, we have added configurations C and A. Configuration C indicates the results of solving the model for a 90° (cubist) posture in which Ae is 90° and all other angles are 0°. Configuration A indicates the values specified by ANSI/HFS 8.4.1 for keyboard support surfaces. Finally, configuration L represents the results of a systematic exploration of intermediate values of parameters within the configuration space defined by the limiting values presented earlier (Dainoff *et al.*, 1989).

Results

Table 7.1 contains estimated values of the composite anthropometric dimension, seated fingertip height, for a set of postural configurations which define limiting conditions for near-upright forward and backward leaning postures. These limiting conditions are prescribed by ANSI/HFS 100-1988 with the addition of 10° of forward seat pan tilt and 25° of shoulder flexion.

The intention of this exercise is to estimate required ranges of keyboard support surface height for forward and backward leaning postures. Since fingertips rest on a keyboard, rather than a keyboard support surface, the thickness of the keyboard itself must be subtracted from fingertip height to obtain a final estimate of keyboard support surface height. In this paper, we use a key-

Table 7.1 Derived values of seated fingertip height for postural configurations defined by body angles (At, As, Ae) and seat pan angle (Asp) for 50th and 95th percentile U.S. females and males

Configuration	1	2	3	4	5	6	L	C	A
Angle:									
Asp	−10	−10	−10	−10	10	10	10	0	0
At	0	0	0	0	25	25	10	0	0
As	0	0	25	25	0	0	0	0	0
Ae	90	70	115	95	90	70	95	90	90
Percentile:									
Female 50th	77.92	92.03	80.50	94.61	81.10	92.84	68.93	71.89	
Female 5th	72.93	86.14	75.34	88.57	75.90	86.90	64.49	67.27	61.50
Male 50th	80.19	95.92	83.02	98.76	84.82	97.90	70.94	73.82	
Male 95th	85.25	101.95	88.25	104.95	90.16	104.05	75.42	78.49	74.00

Definitions of angles:

Asp: Seat Pan Angle
At: Trunk Angle
As: Shoulder Angle
Ae: Elbow Angle

Configuration:
L: Lower bound-empirical
C: Cubist, computed by model
A: Cubist, from ANSI

board thickness of 3 cm. Therefore, within the constraints of the limiting values used in the equations, values of SFH in Table 7.1 should provide an estimate of keyboard support surface height plus 3 cm.

Calibration: Comparison of models at 90° configuration

Examining Table 7.1, it is instructive to compare configuration C, which is the 90° cubist version of our model, with configuration A, which is the specification from ANSI/HFS 8.4.1 plus 3 cm of keyboard thickness. Our model yields a fingertip height estimate which is 5.8 cm greater than ANSI/HFS for the 5th percentile female, and 4.5 cm greater for the 95th percentile male. However, as discussed above, the actual published ANSI/HFS recommendation for the 95th percentile male is actually 7.2 cm lower than it should have been by strictly anthropometric computations. Therefore, we argue that our procedures actually underestimate the ANSI/HFS values for 95th percentile male by 2.7 cm.

What are the sources of these discrepancies? There are, at the 90° configuration, three differences between the models:

- the use of a set of link segments *vs.* anthropometric dimensions;
- the method of estimating 5th/95th percentile values; and
- the size of the heel correction.

We can estimate the potential difference between link *vs.* anthropometric dimensions by doing a pure comparison at the 50th percentile with the same heel and keyboard correction for both methods. If we add popliteal and seated elbow heights at the 50th percentile with appropriate heel and keyboard correction, the new ANSI/HFS values are 70.1 cm for the female and 73.8 cm for the male. Comparing these values with those of configuration C in Table 7.1, we see that the discrepancy is now only 1.8 cm for the female and virtually nil for the male. Thus, it is reasonable to presume that most of the obtained discrepancy is due to differences in heel correction and method of estimating 5th/95th percentile values. (We used the estimated standard deviation procedure; ANSI/HFS simply added two 5th or 95th percentile values.) More importantly, we feel the link segment method can be generalized to other postures.

Major differences from ANSI/HFS

The magnitudes of the differences between 90° cubist models are small; however, allowing alternative postures produces values of seated fingertip height (SFH) which far exceed the published ANSI/HFS recommendations. As can be seen in Table 7.1, configurations 2, 4 and 6 have values of SFH over 100 cm; these exceed the 74 cm ANSI/HFS upper limit by over 25 cm. It is interesting that, even for the 5th percentile female, configurations 2–6 exceed the ANSI/HFS upper limit. On the other hand, configuration L, 5th percentile woman, defines the minimum value (64.49 cm) of the configuration space as determined by systematic exploration of intermediate parametric values. While this falls within the published ANSI/HFS limits, it falls below the 5th percentile female value for the cubist posture using our model (configuration C).

Discussion

Alternative postures require higher work surfaces

What is the outcome of this exercise? Specialists in workplace ergonomics (Grandjean *et al.*, 1983; Mandal, 1981; Kroemer, 1987; and others) have argued that the 90° cubist posture does not describe how people sit at work. To the extent that work station standards and recommendations are based on the cubist model, the chair designed from such standards is not likely to support work postures which are most comfortable and/or biomechanically optimum.

Given this argument, our goals have been to explore a method for expanding work station standards so as to accommodate alternative postural configurations. For our current model, we initially adopted existing limiting values of seat pan (0°–10°) and back rest angles (0°–25°) from the ANSI/HFS specific solution. These values represent, in ANSI/HFS terms, a near-upright posture. Our modification of the model then consists of: (a) allowing an amount of forward tilt in the seat pan (−10°) equal to that already prescribed for backward tilt, and (b) allowing for 25° of shoulder flexion. It might be argued that allowing this degree of shoulder flexion is excessive. However, Grandjean *et al.* (1983) observed a mean value of 23° of shoulder flexion in his operators, and we have seen similar results in our laboratory.

The outcome has been a derived set of values which locates a range of fingertip working heights characterizing the US population from the 5th percentile female to the 95th percentile male for the set of postural configurations defined by the above limiting values. To describe a work station as ergonomic, it must include keyboard support surfaces which adjust to accommodate such obtained fingertip ranges. Unfortunately, the upper limit values from our model are considerably higher (by 25–30 cm) than the ANSI/HFS recommendations for ergonomic work stations. Thus, we agree with Grandjean, Mandal, Kroemer and others that much higher work surfaces are needed to support optimal postures.

Oversimplifications and other problems

Can our model be used as a basis for a revised set of standards? We think this would be premature. There are several oversimplifications and other concerns with the model as presently constructed, which need to be addressed.

Hand posture

The current model simply assumes that the hand is a linear extension of the arm, with an assumed wrist angle of 0°. However, when the forearm is not horizontal, but the work surface is, some degree of wrist angle will be needed. Further, the model does not address the bending downward of fingers during keying. At the same time, Grandjean *et al.* (1983) argued that the keyboard support surface should angle down to conform with the arm angle. Given current concerns over wrist disorders in keyboard operators, it seems prudent to reduce the wrist angle as much as possible. Hence, our original oversimplification may be reasonable.

Centre of rotation of seat pan

The current model assumes, for simplicity, that the centre of rotation of seat pan is in the plane of the seat pan surface. In fact, the seat pan pivot is usually below the seat pan, and, in some chairs, is not even at the seat pan centre.

Body and chair dimensions

To use link dimensions, the hip joint must be referenced to the Seat Reference Point. A standard chair was specified with a given seat pan depth and location of centre of rotation to solve the model. We are currently modifying our model to generate a solution with reference only to body dimensions.

Conclusions

This paper presents a model of seated postures which examines the changes in seated fingertip height for a set of alternative postural configurations, and argue that these changes require a much greater range of adjustability of keyboard support height than is customarily prescribed by standards and recommendations. While these results are based on ANSI/HFS and the US population, the model may easily be generalized to other standards and populations.

However, the intent of this model was to examine varying postures on only one workstation dimension. The impact of these changes on other crucial attributes such as screen viewing angle and seat–work-surface clearance (Noro, 1989) have not been examined. Ultimately, our approach with models such as those of Noro, to guide the development of standards which take into account for both operator preferences and biomechanical efficiency.

Appendix

Model Parameters (cm)

	Male	Female	Source
Heel Height:	2.3	4.0	a
Popliteal Height:	0.254	0.249	b
Seat Pan Depth (SPD):	40.6		c
Hip in Front of Seat Reference Point (SRP):	0.07	0.082	d
Buttock Popliteal Length:	0.285	0.302	d
Hip Above SRP:	0.043	0.043	d
Hip Shoulder Length:	0.288	0.304	d
Shoulder Elbow Length:	0.174	0.172	d
Elbow Finger Tip Length:	0.241	0.233	e

Asp: Seat Pan Angle From Horizontal
At: Trunk Angle From Vertical
As: Shoulder Angle From Trunk
Ae: Elbow Angle From Upper Arm

Source:
a. Pheasant (1986, p. 70)
b. Kroemer (1983), expressed as a percentage of stature
c. ANSI/HFS 100-1988, Section 8.7.2.
d. Pheasant (1986, Table 1), links expressed as a percentage of stature
e. Pheasant (1986, Table 1), sum of links for forearm and hand, expressed as a percentage of stature

Regression equations

These equations provide for estimates of standard deviations of composite anthropometric dimensions in terms of means of such dimensions. They were obtained by regressing 50th percentile points (X) against standard deviations for each linear anthropometric dimension in Kroemer's (1983) tables.

Female: Est SD $= 0.0375X + 0.754$

Male: Est SD $= 0.038X + 0.893$

References

ANSI/HFS, (1988), American National Standard for Human Factors Engineering of Visual Display Terminal Work Stations: ANSI/HFS 100-1988. (Santa Monica, California: Human Factors Society).

Chaffin, D. B. and Andersson, G. B. J., (1984), *Occupational Biomechanics.* (New York: John Wiley & Sons, Inc.).

Dainoff, M. J. and Mark, L. S., (1989), Analysis of seated posture as a basis for ergonomic design, in *Work with Computers: Organizational, Management, Stress and Health Aspects*, (eds M. J. Smith and G. Salvendy), (Amsterdam: Elsevier Science Publishers), 348–54.

Dainoff, M. J., Balliett, J., Goernert, P. N., McCarthy, J. and Mark, L. S., (1989), Adjustability ranges for VDT work stations. Paper presented at the Second International Scientific Conference: Work with Display Units, Montreal.

Dainoff, M. J., Mark, L. S., Moritz, R. and Vogele, D., (1987), Task, seat adjustability and postural change, in *Proceedings of the Human Factors Society 31st Annual Meeting* (Santa Monica, California: Human Factors Society), 879–83.

Dainoff, M. J. and Dainoff, M. H., (1986), *People & Productivity: A Manager's Guide to Ergonomics in the Electronic Office.* (Toronto: Carswell Publishing).

Grandjean, E., Hunting, W. and Pidermann, M., (1983), VDT work station design: preferred settings and their effects, *Human Factors*, **25**, 161–76.

Kroemer, K. H. E., (1983), Design parameters for video display terminal work stations, *Journal of Safety Research*, **14**, 131–6.

Kroemer, K. H. E., (1987), Computer work stations: preferred posture and line of sight, in *Proceedings of the Human Factors Society 31st Annual Meeting* (Santa Monica, California: Human Factors Society), 1005–8.

Mandal, A. C., (1981), The seated man (*homo sedans*), the seated work position, theory and practice, *Applied Ergonomics*, **12**, 19–26.

Mark, L. S. and Dainoff, M. J., (1988), An ecological framework for ergonomic research, *Innovation*, **7**, 8–11.

Noro, K. and Hattari, T., (1989), Application for simulation model for designing human-computer interface. Paper presented at the 3rd International Conference on Human–Computer Interaction, Boston.

Pheasant, S., (1986), *Bodyspace*. (London: Taylor & Francis).
Roebuck, J. A., Jr., Kroemer, K. H. E. and Thomson, W. G., (1975), *Engineering Anthropometry Methods*. (New York: John Wiley and Sons).

PART IV

Posture

8

Effects of posture on mental performance: we think faster on our feet than on our seat

Max Vercruyssen and Kevin Simonton

'. . . everyone knows that people do not usually sleep standing up or sitting upright, and that they "nod off" when seated comfortably (Branton, 1987, p. 4).'

Introduction

Have you ever considered which posture is best for mental performance? This chapter addresses this question by reporting some experiments which have received international attention by the news media but which have yet to be reported in scientific journals.

In 1988, a science writer released a media bulletin entitled *We Think Better Standing Up* (Davidson, 1988; *see also* Kazanjian, 1989) describing posture (lying *vs* sitting *vs* standing) research conducted at the University of Southern California. This release was picked up by major national and international wire services which resulted in an overwhelming onslaught of requests for interviews. During the first wave of attention in the months that followed, this research was a topic of talk shows as well as public radio and television science broadcasts, internationally. Excerpts from the release appeared in newspapers and magazines around the world, in multiple languages, using such English titles as

'Stand Up and Think' (*Entrepreneur*, 1989)
'Stay on Your Toes' (*Reader's Digest*, 1989)
'Rise to the Occasion: Executive Couch Potatoes Should Stand and Deliver' (Schroepfer, *American Health*, 1989)
'We're Tops on our Toes' (Elias, *USA Today*, 1988)
'How to Think Faster' (*Science Impact*, 1988)
'An Upright Posture Can Tone Up Your Gray Matter' (Bozzi, *Longevity*, 1989)

'Stand Up for Clearer Thoughts' (*USA Today*, 1989)
'Researchers Take a Stand on Mental Quickness' (*Los Angeles Daily News*, 1988)
'Keep on Your Toes' (*Science Digest*, 1989*a*, *b*)
'Thinking on Your Feet' (*Health & You*, 1989).

The writers of these articles appeared to be interested in confirming their personal hypotheses that people work better while standing than sitting. We then received a second wave of reprint requests from fellow scientists and inquiries from junior scholars in search of science fair projects. Throughout this period there were also interview requests from clever freelance and staff writers from tabloids or scandal sheets. Of course, we avoided all contact with these rascals to keep our research from being exaggerated in content and then used to create an era of scientific authenticity for articles such as 'Woman gives birth to two-headed baboon'. At the time of this writing, two years after release of the 1988 media bulletin, we have had over 400 requests for interviews, reprints, ergonomic assessments, and help for students on school science projects. The unexpected popularity of this subject caught everyone off-guard, to say the least.

Why so much interest in how posture affects mental performance? We think the main reason is that the findings seem generally expected, and it is always easy to talk or write about cases in which science confirms common knowledge. Despite the immense popularity of this topic, what matters is the reliability and meaningfulness of these findings.

Why are you reading this chapter? If it is because you want to know the relationship between working posture and mental performance, you may be disappointed that we don't know the answer, yet. However, we do have some new facts that are helping us to piece together a tentative answer. On the other hand, if you are looking for scientific evidence that will confirm your long-felt beliefs, or support products you design/manufacture/sell, we might have something for you, but please don't stretch the science too much.

The science of seating involves many facets, one of which is the impact of posture on central nervous system integrity and mental performance. The purpose of this chapter is to review some of the posture research conducted during the past decade, and to advance some preliminary conclusions on how posture (lying, sitting, and standing) might influence the speed with which we can process information. Research contained herein reports postural effects on reaction time as the measure of central nervous system integrity, the rate of information processing, and general speed of behaviour. Reaction time (RT) is a popular dependent measure because it is easily controlled in the laboratory, has a rich history of previous exploration against which to compare findings, is easily translated to real-world applications, and provides a metric which is time-based, allowing actual quantification of latencies in human operation of human-machine-environment systems. Eight sections follow: background, six experiments, and conclusions.

Background

Before presenting research findings, it may be useful to describe the research which led investigators at the University of Southern California (USC) to conduct posture experiments. Original studies at USC (Woods, 1981) were initiated to resolve a debate among gerontologists on whether older humans have slower responses than the young because their central nervous system (CNS) is less aroused. As early as the late 19th century, such behavioural scientists as Galton (Koga and Morant, 1923) investigated the importance of arousal levels to human performance. In the late 1940s, arousal theory received major substantiation with the exploration of the ascending reticular activation system (RAS; e.g., Lindsley *et al.*, 1950; Moruzzi and Magoun, 1949).

The RAS of the brain is located in the brainstem, where it serves as a direct sensory pathway to the cerebral cortex. It has been claimed to control arousal levels of sensory information. Researchers showed that direct stimulation of the RAS increased the level of arousal in lightly anaesthetized or sleeping cats. Furthermore, Lindsley *et al.* (1950) found that a lesion through the midline of the RAS put test animals in a permanent sleeping state and also, that stimulation elicits arousal states. These findings led many researchers to hypothesize that the integrity and proper functioning of the RAS is vital to maintain optimal arousal states. Any age-related deterioration of the RAS might account for tonic (underlying) states of under-arousal, over-arousal (e.g., Eisdorfer and Wilkie, 1977), or reduced range of optimal arousal (e.g., Bondareff, 1980), and might help explain the wealth of evidence which shows that reaction time increases by approximately 20 per cent (or more) between the ages of 20 and 60 years (e.g., Birren *et al.*, 1990; Birren *et al.*, 1980; Vercruyssen, in press).

Since standing stimulates the RAS more than lying or sitting, posture becomes a variable which could be used to resolve the CNS under-arousal *vs.* over-arousal (or reduced range of optimal arousal) debate. For instance, based on the inverted-U hypothesis of arousal and performance (Hebb, 1955; Yerkes and Dodson, 1908), if the elderly possess an under-aroused central nervous system, their reaction times should improve when they stand. Conversely, if performance is impaired, it suggests that older adults have an over-aroused CNS or one with a reduced range of optimal arousal.

Experiment 1: Age, fitness, and posture-induced arousal

In 1981, Anita Woods examined the effects of posture on reaction time in young (18–28 yrs) and old (60–70 yrs) subjects of high and low levels of physical fitness (*see also* Woods *et al.*, 1993). It was hypothesized that changing posture from lying, to sitting, to standing (*see* Figure 8.1) stimulates the RAS, and would increase brain activity (electro-encephalograms) and cardiovascular functions. This hypothesis was based on previous research which showed that systematically increasing heart rate in patients wearing pacemakers significantly decreased reaction time and thereby improved mental performance (e.g., Lagergren, 1974; Lagergren and Levander, 1975).

Testing the hypothesis that an under-aroused CNS in the elderly was responsible for a large portion of age-related slowing of behaviour and reaction time, Woods speculated that postural stimulation (i.e., changing from lying to sitting to standing) would improve their reactive performance. Woods tested 18 young males (18–28 yrs) and 18 older males (60–70 yrs) on a simple and two-choice visual reaction time (RT) task, in three different postures

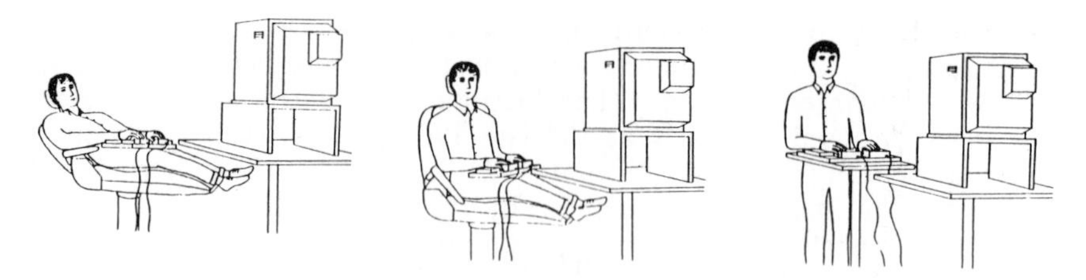

Figure 8.1 Lying, sitting, and standing postures

(lying, sitting, standing). Lying was considered the lowest arousal state, and standing the highest. A significant main effect was found for age; the young were faster than the old across all posture conditions for both tasks.

Posture did not significantly influence reaction time in the young between any of the three conditions. However, the older group showed significant differences in RT between each posture; the fastest RT's were recorded when subjects were standing, for both tasks. These results confirmed Woods' hypothesis that the old have a relatively under-aroused CNS, and that by increasing the level of arousal through posture changes, performance on simple and choice RT tasks improves. Because the young group's reaction times were not affected by posture, underlying differences must exist in the baseline arousal states of young and old.

Woods (1981) was also interested in the overall health status of an organism and its relation to possible changes in behaviour or effects on the RAS. If the elderly CNS is under-aroused due to reduced RAS functioning, health status of the individual must be examined. This hypothesis was also supported by a substantial body of research literature reporting that subjects with cardio-vascular disease showed signs of slowing in information processing relative to healthy subjects of the same age (Eisdorfer and Wilkie, 1977; Reitan, 1954; Speith, 1965). Therefore, Woods decided to include physical fitness as an independent variable and tested the original 18 young males and 18 old males on a bicycle ergometer. Based on a certain workload cut-off point, the groups were further subdivided into young fit ($n = 9$), young unfit ($n = 9$), old fit ($n = 9$), and old unfit ($n = 9$). Experimental treatments were counterbalanced across subjects.

Woods hypothesized that the fit groups would perform significantly faster than the unfit. Her results confirmed that the old fit were significantly faster across all conditions for both simple and choice RT tasks than the old unfit. The young fit were also significantly faster than the young unfit in simple RT conditions, but only when lying. Thus, there were no significant differences in RT between young fit and young unfit in either the standing or sitting choice RT conditions. Posture effects were present in the elderly but not the young; the greatest lying-sitting-standing differences were among the old unfit group. Woods concluded that physical fitness exerts a strong effect on performance, and that the old unfit are most affected by postural stimulation.

Age and fitness level appear to be additive when measuring reactive performance. In other words, old (and sedentary) people will probably be at a greater disadvantage when performing reaction time tasks than their younger (and more physically fit) counterparts.

Experiment 2: Age, fitness, and exercise-induced arousal

Woods (1981) also proposed that exercise could boost performance in individuals with an under-aroused CNS, from ageing or lack of fitness. Phase II of her study included the same 18 young and 18 older males exercising on a bicycle ergometer at 20 per cent and 40 per cent of their VO_2 max, compared to rest and free pedalling (*see also* Mihaly, 1988). This procedure was carried out on Day 2, after the postural conditions were completed on Day 1. Each subject was presented with simple and choice RT trials, with heart rate being measured throughout the session. Woods hypothesized that the old unfit would improve more, relative to the old fit at the optimal physical activity level.

The results did not fully support this hypothesis. The old fit had their fastest reaction times in the free pedalling condition during simple RT, and in the 40 per cent activation state during the choice RT condition. The old unfit results were more consistent, with faster times recorded for both simple and choice RT in the 20 per cent activation condition. Young fit and young unfit were not significantly influenced across activation conditions during simple RT. However, for choice RT, both young fit and unfit were faster in the 40 per cent activation and slowest in the 20 per cent activation condition. Moreover, the 40 per cent activation condition was not significantly faster than free pedalling.

From these results, Woods concluded that the benefits of increased arousal through physical activity depend greatly on age, difficulty of the task, and physical fitness. The old fit may be capable of sustaining both higher arousal levels at higher exertions (relative work intensities).

Experiment 3: Inversion and information processing

In 1982, gravity guidance systems and inverted suspension devices became popular. Manufacturers claimed that use of these products produced such benefits as improved blood circulation, increased oxygen supply to the brain, and slowing of the ageing process (Broatch, 1982). Another benefit was supposed to be an improvement in cognitive functions, particularly the speed and accuracy of information processing (Broatch, 1982; Martin, 1982). A possible explanation for these claims revolved around the RAS, as mentioned in Woods (1981), and its responsibility for maintaining arousal and activation (e.g., Hebb, 1955; Lindsley, 1970).

Diggles *et al.* (1984) decided to further explore the RAS and arousal theory, while at the same time testing advertising claims that touted the benefits of regularly hanging upside-down. After all, if going from lying, to sitting, to standing causes progressive improvement in RT, maybe inversion would improve performance even further. Thus, Diggles *et al.* (1984) used a tilt-table to rotate subjects from right-side up to up-side down. Fourteen healthy college-age students participated (10 males, 4 females) in this experiment. Using an inversion platform with the ankles securely clamped, the subjects inverted themselves by raising their arms to tip the platform 180°. The RT stimulus presented was either a high tone or low tone, to which the subjects reacted by flexing either their right or left wrist. Reaction time was measured with electromyography (EMG) surface electrodes placed on the wrist flexors to receive and record the onset and termination of muscle activity. Both simple and choice RT's were performed prior to, during, and after inversion.

The results showed no significant main effect from postural orientation, despite a slight decrease in SRT during inversion (about 4 per cent). Failing to find evidence of an increase in the rate of information processing appears to refute claims made by manufacturers of inversion devices, as well as the robustness of the posture effect. Either: inversion does not improve processing speed, inversion effects will not reveal themselves using the procedures described, the effect is very small and unreliable, the statistical power (ability of the test to detect a true effect) was low, making it unlikely that significance would be demonstrated, or some combination of these possibilities. The authors speculate that the last three are correct.

Experiment 4: Age, gender, and posture effects

Vercruyssen *et al.* (1988) used variables from Woods' experiments to account for age and gender differences in postural arousal and activation. The psychomotor literature at that time contained several discrepancies as to which sex was faster at different ages. Some reported that older females may have greater attention and response capacity than their male counterparts (e.g., Botwinick and Thompson, 1966). Others, however, have found that males were consistently faster than females across all age groups (e.g., Fozard *et al.*, 1990; Lahtela *et al.*, 1985). Other studies have shown no differences between older males and females (Hodgkins, 1963). Taken as a whole, the effects of gender and their interaction with age in determining speed of response, have not presented a clear, concise picture of what is really happening with age, speed, and gender.

Experiment 4 was intended to address this problem. Forty-four healthy volunteers served as subjects: 14 young women, 14 young men, eight elderly women, and eight elderly men. The tasks included serial four-choice visual reaction time (SCRT; considered a focused task) and variable four-choice visual reaction time (VCRT) with random response-stimulus intervals of 0, 1, 2, 3, and 6 seconds (considered a prolonged or sustained attention task). Stimuli were intact and degraded, as in the earlier study by Vercruyssen *et al.* (1989). Again, postural stimulation, sit to stand, was used to induce changes in level of arousal.

The most pronounced posture effect was such that for all subjects, VCRT was significantly faster standing than sitting. During the degraded task, females were significantly faster than males, while on the intact task, the males performed slightly, but not significantly, faster than the females. Also during the degraded task, males had significantly faster RT's when standing than when sitting, but this posture effect was not present in females. These results led the authors to conclude that postural effects become more important in circumstances involving a low rate of information flow and/or in circumstances which demand more sustained attention spans.

Experiment 5: Gender differences in posture effects

Based on the sex/gender differences obtained in the Vercruyssen *et al.* (1988) study, an additional analysis was conducted by Vercruyssen, Cann, and Hancock (1989) to further explore gender differences in reactive performance. The existence of gender differences has been a hotly debated topic within the psychomotor literature for the past several decades. As noted above, some studies have shown males to be faster than females (Bell *et al.*, 1982; Coles *et al.*, 1975; Ferguson, 1973; Fozard, Vercruyssen, Reynolds and Hancock, 1990), whereas others have concluded the opposite (Fulton and Hubbard, 1975; Landauer *et al.*, 1980; Landauer, 1981; Thomas and French, 1985).

The evidence to date suggests that males have faster reaction times than females on simple and choice RT tasks. A closer look at the gender-related speed of response literature, however, reveals that many of these differences in performance may be due to the type of task involved in the testing procedure. For instance, although males have faster movement times, females actually have faster decision times (Landauer *et al.*, 1980). And in the study described above, females responded faster to degraded stimuli and males were faster responding to intact stimuli (Vercruyssen *et al.*, 1988). The net effect is that no differences are observed between genders because the superior motor skills of the males cancel out the faster processing speed of the females. Consequently, the degree to which the task involves movement time or decision time greatly influences results.

Our 1989 study was unique in that it investigated gender differences in young subjects in a way that sought to determine the locus of the posture effect, i.e., whether the posture effect occurs in an early encoding stage, or in a later response selection stage of information processing, or both (e.g., Sternberg, 1969; Vercruyssen, 1984). Twenty-eight healthy university students (14 males and 14 females) performed four-choice visual reaction tasks while sitting and standing. The stimulus was an arrow that either pointed left or right and was presented in blocks of trials with stimulus image quality either intact or degraded. The response selection stage was loaded by manipulating stimulus-response (S-R) compatibility—low compatibility was produced by having the subjects cross their arms and by mixing the order of finger responses to the stimulus arrows.

Since the literature (discussed earlier) indicates that sex/gender differences in RT depend on the type of task involved, postural stimulation may also differ by gender. The results indicate that females had significantly faster RT's than the males when the stimulus was degraded, but not when intact. Little difference was recorded between genders during the intact task between sitting and standing. However, males (but not females) had significantly faster RT's when standing than when sitting during the degraded stimulus condition. Woods (1981) found young subjects did not benefit from postural arousal. In this experiment, however, the posture effect occurred only for males in the degraded stimulus condition. Females did not appear to benefit from postural stimulation regardless of the task employed.

These results suggest posture affects RT during an early encoding stage of information processing, where there are gender differences in reactions to manipulations of stimulus quality. Thus, as the number of experimental variables increases, so does the available information on the conditions in which posture effects appear and disappear.

Experiment 6: Multifactor posture thesis

In 1990, Cann analysed the effects of age, gender, posture-induced arousal, task loading, task difficulty, and length of response-stimulus interval. Not all of these factors will be discussed here; however, please note that Cann (1990*a*, *b*) included all of the previously investigated factors that were found to be significantly affected by postural stimulation, except fitness level. By design, his experiment was meant to replicate the conditions in previous studies and, at the same time, be more thorough by including a range of variables which could provide a bigger picture of information processing. Cann's hypotheses were similar to those of Woods (1981) and Vercruyssen *et al.* (1988, 1989), with subtle differences. He felt that a good deal of evidence supported the under-arousal theory in the old, but he also felt that the elderly may, at times, suffer from an over-aroused CNS as well, which would become apparent when task difficulty was manipulated in addition to posture. Cann further hypothesized that the tasks which were more difficult would be more sensitive to posture-induced arousal. These were his two most significant hypotheses related to posture.

Cann tested 32 subjects: eight young males, eight young females, eight old males, eight old females. Subjects performed a variety of tasks, including a finger tapping task, a simple visual reaction time task, and several four-choice visual RT tasks with varying response stimulus intervals and intratask factors such as stimulus degradation and stimulus-response compatibility. Each task was performed in each of three postural conditions: lying, sitting, and standing. The results supported the hypothesis that tonic arousal states vary as a function of age. Older subjects were significantly slower than younger subjects such that they took disproportionately longer to perform the more difficult tasks (e.g., the degraded stimulus, low S-R compatibility condition). Posture affected performance for all groups, but only during a moderately difficult task.

Cann (1990*a*, *b*) thus concluded that the benefits of posture are related to task difficulty in that postural stimulation benefits performance most during tasks of relatively moderate difficulty. Cann's over-arousal hypothesis for the old was supported somewhat by the fact that during the most difficult task, standing actually impaired performance in the older males but had virtually no effect on the older females. Therefore, Cann further concluded that difficulty level of the task, plus the postural stimulation, combined to over-arouse the organism, thereby causing performance decrement (e.g., arousal theory by Duffy, 1962, 1972) and, in addition, older subjects may become over-aroused more easily than younger subjects. To summarize, Cann's experiment provides strong evidence that young and old benefit differently from postural arousal, with the old benefiting less or not at all during the performance of more difficult tasks.

In closing, this chapter has presented six posture experiments, five of which showed instances where RT was faster when subjects were standing or cycling than when they were sitting. One study even flipped subjects upside-down, but found no effect on RT. Collectively, this research supports the conclusion that the optimal posture for reactive capacity is standing (or pedalling at 20–40 per cent of each subject's aerobic capacity).

Conclusions

The following qualifications are needed to make accurate conclusions about our research:

1. Only reaction time was tested, not memory, creativity, reasoning, or higher-order complex thinking. Therefore, any generalizations to mental performance should be made cautiously.
2. Postures were maintained for 8–15 minutes each with sessions lasting 20–120 minutes. Statements made about posture effects for intervals shorter or longer are mere speculations.
3. None of the experiments presented herein were conducted for the purpose of explaining the posture effect. They were designed to test other hypotheses and phenomena, mostly those related to individual differences in information processing. Thus, little attempt has been made to explain the physiological basis or underlying mechanisms responsible for this postural phenomenon.

From the experiments presented in this chapter we advance the following tentative conclusions:

- Depending on the individual and specific circumstances, it appears that rapid decision-making is optimal in an erect posture. Performance suffers as one becomes more reclined and comfortable.
- Posture effects appear to be significant for the elderly (especially the sedentary elderly), and in the young when the task causes sufficient increases in arousal such as, degraded stimuli, low S-R compatibility, and complex sustained tasks.
- Those most likely to benefit from a posture effect are physically unfit elderly individuals performing tasks that are somewhat monotonous.
- Those individuals least likely to experience a posture effect are young physically fit females.

This chapter provides evidence that posture can influence reaction time. However, such evidence can be easily misinterpreted or exaggerated. Because the investigators conducting this research consider the posture effect to be small at best and highly dependent on subject demographics, test conditions, and the dependent measures employed, we are reluctant to make sweeping generalizatons. Furthermore, we hope our findings will not be exaggerated or taken out of context as they have been in the past. For example, one article, misquoting an earlier publication, concluded that *standing yields a 2000 per cent improvement over sitting!* – though this sounds very impressive, and even scientific, it is utter nonsense. 'We think better on our feet than on our seat' is an easy-to-remember catch-phrase and is also intriguing trivia for small talk at cocktail parties, but be careful how you tell the story. In some circumstances, like those described in this chapter, posture effects are real, but in many cases, statements such as that catch-phrase, are completely erroneous.

It is amazing how some information is taken completely in the wrong way, without substantiation, and implemented into practice, with great confidence that this new knowledge gives its possessor an advantage over his or her adversaries. For instance, at this early stage in posture research, it would be a waste of money and time to implement static hanging into training programmes in order to enhance mental performance, or provide an ergonomic aid, or in some way create a competitive edge . . . but, pseudo ergonomists somewhere are likely to do something like this. (e.g., Figure 8.2).

Yes, humans function differently when lying *vs.* sitting *vs.* standing, and even hanging upside down, but we do not yet know how these differences manifest

Figure 8.2

themselves in mental performance, particularly work productivity during the course of a work-day, -week, -month, or -year. Nonetheless, it will be interesting to see how future work stations are ergonomically designed as well as how subsequent investigations further our understanding of the effects of posture on mental performance. "Many argue that humans were not designed to spend most of their time sitting (e.g., Grimsrud, 1990) and that we are likely to be less effective in this posture than others".

Acknowledgements

The authors wish to thank Prof James E. Birren for initiating the posture studies at USC (beginning with Woods, 1981) and for sharing insights since this period. Others making major contributions to the evolution of our posture/arousal/activation research (in order of participation) include Dr Anita M. Woods, Dr Donna J. Mah, Dr Nina Turner, Dr Virginia Diggles-Buckles, Dr Mark D. Grabiner, Ms Tina Mihaly, Mr Michael T. Cann, Prof Kazuo Hashizume, Prof Ewan Russell, Cpt Gretchen Greatorex, Mr Kevin Simonton (master's student in human factors thesis, posture experiment number 8), Ms Vanessa Murray, and Ms Emelyn Kim. Ms Rani Lueder provided helpful editing of earlier versions of this chapter. Ms Lara Strom also provided editorial assistance and sketched Figure 8.1. Mr Takaya Torii sketched the Top Gun cartoon. Comments on earlier versions of this manuscript were provided by Mr George Brogmus, Lt Sara Reynolds, and Dr Jen-Yi Chang.

References

Bell, P. A., Loomis, R. J. and Cervonne, J. C., (1982), Effects of heat, social facilitation, sex differences, and task difficulty on reaction time. *Human Factors*, **24**(1), 19–24.

Birren, J. E., Vercruyssen, M. and Fisher, L., (1990), Aging and speed of behavior: Its scientific and practical significance, in M. Bergener, M. Ermini, and H. B. Stahelin (eds), *Challenges in Aging: The 1990 Sandoz Lectures in Gerontology* (pp. 3–23). London: Academic Press.

Birren, J. E., Woods, A. M. and Williams, M. V., (1980), Behavioral slowing with age: Causes, organization and consequences, in L. W. Poon (ed), *Aging in the 1980s: Psychological Issues* (pp. 293–308). Washington, D.C.: American Psychological Association.

Bondareff, W., (1980), Compensatory loss of axosomatic synapses in the dentate gyrus of the senescent rat, *Mechanisms of Aging and Development*, **12**, 221–9.

Botwinick, J. and Thompson, L. W., (1966), Components of reaction time in relation to age and sex, *Journal of General Psychology*, **108**, 175–83.

Bozzi, V., (1989, September), Sitting *vs.* standing: Brain assister. 'An upright posture can tone up your gray matter. *Longevity: Guide to the Art and Science of Staying Young, 86.*

Branton, P., (1987), In praise of ergonomics – A personal perspective, in D. J. Oborne, *International Reviews of Ergonomics*, **1**, 1–20.

Broatch, J., (1982), *Better back, better body* (Good 80s Enterprises, Inc., Vancouver, B.C.).

Cann, M. T., (1990*a*), Age and speed of behavior: Effects of gender, posture-induced arousal, and task loading. Master's thesis in Human Factors. Los Angeles, CA: University of Southern California.

Cann, M. T., (1990*b*), Causes and correlates of age-related cognitive slowing: Effects of task loading and CNS arousal. *Proceedings of the Human Factors Society 34th Annual Meeting* (pp. 149–53). Santa Monica, CA: Human Factors Society.

Coles, M. G. H., Porges, S. W. and Duncan-Johnson, C. C., (1975), Sex differences in performance and related cardiac activity during reaction time task, *Physiological Psychology*, **3**, 141–43.

Davidson, M., (1988, Oct 26), *We think better standing up*. University of Southern California News and Features, No. 0788025. USC News Service, PVW205, Los Angeles, CA 90089-1227.

Diggles, V. A., Grabiner, M. D. and Harding, F. V., (1984), Information processing during inversion. Paper presented at the 1984 American Alliance of Health, Physical Education, Recreation, and Dance National Convention, Anaheim, CA.

Duffy, E., (1962), *Activation and Behaviour*, New York: Wiley.

Duffy, E., (1972), Activation, in N. S. Greenfield and R. A. Stanbach (eds), *Handbook of Psychophysiology*, New York: Holt, Rineholt/Winston.

Eisdorfer, C. and Wilkie, F., (1977), Stress, disease, aging and behavior, in J. E. Birren and K. W. Schaie (eds), *Handbook of the Psychology of Aging* (1st edn), New York: Van Nostrand Reinhold.

Elias, M., (1988), *We're tops on our toes*, USA Today, 1 November.

Entrepeneur, (1989, November), *Stand up and think*, 164.

Ferguson, D. P., (1973), Reaction time comparisons by race, sex, and body type, (Doctoral Dissertation, Oklahoma State University), in R. N. Singer and R. A. Weiss (eds) *Completed Research in Health, Physical Education, and Recreation*, **16**, 1974.

Fozard, J. L., Vercruyssen, M., Reynolds, S. L. and Hancock, P. A., (1990), Longitudinal analysis of age-related slowing: BLSA reaction time data, *Proceedings of the Human Factors Society 34th Annual Meeting* (pp. 163–7). Santa Monica, CA: Human Factors Society.

Fulton, C. D. and Hubbard, A. W., (1975), Effect of puberty on reaction and movement times, *The Research Quarterly*, **46**(3), 335–44.

Grimsrud, T. M., (1990), Humans were not created to sit – and why you have to refurnish your life, *Ergonomics*, **33**(3), 291–5.

Health & You, (1989, Spring), *Thinking on your feet*, 51.

Hebb, D. O., (1955), Drives and the CNS (Conceptual nervous system), *Psychological Review*, **62**, 243–54.

Hodgkins, J., (1963), Reaction time and speed of movement in males and females of various ages, *Research Quarterly*, **34**, 335–43.

Kazanjian, K., (1989), Standing hastens reaction, thought, concludes professor, *Daily Trojan*, Los Angeles, CA: University of Southern California.

Koga, Y. and Morant, G. M., (1923), On the degree of association in reaction times in case senses, *Biometrika*, **15**, 346–72.

Lagergren, K., (1974), Effect of exogenous changes in heart rate upon mental performance in patients treated with artificial pacemakers for complete heart block, *British Heart Journal*, **36**, 1126.

Lagergren, K. and Leavander, S., (1975), Effects of changes in heart rate in different body positions upon critical flicker fusion threshold and reaction time performance in patients with artificial pacemakers, *Journal of Psychiatric Research*, **12**, 257.

Lahtela, K., Niemi, P. and Kuusela, V., (1985), Adult visual choice-reaction time, age, sex, and preparedness, *Scandinavian Journal of Psychology*, **26**, 357–62.

Landauer, A. A., (1981), Sex differences in decision and movement time, *Perceptual and Motor Skills*, **52**, 90.

Landauer, A. A., Armstrong, S. and Digwood, J., (1980), Sex difference in choice reaction time, *British Journal of Psychology*,**71**, 551–5.

Lindsley, D. B., (1970), The role of nonspecific reticulothalmocortical systems in emotion, in P. Black (ed), *Physiological Correlates of Emotion*, New York: Academic Press.

Lindsley, D. B., Schreiner, L. H., Knowles, W. B. and Magoun, H. W., (1950), Behavioural and EEG changes following chronic brainstem lesions in the cat, *Electroencephalography and Clinical Neurophysiology*, **2**, 483–98. Cited in Woodruff, D. S., (1985), Arousal, sleep and aging, in J. E. Birren and W. Schaie (eds), *Handbook of the Psychology of Aging* (2nd edn) (pp. 261–95). New York: Van Nostrand Reinhold.

Los Angeles Daily News, (1988), Researchers take a stand on mental quickness. Cited in *The Times-Picayune*, 19 November, E-5.

Martin, R. M., (1982), *The gravity guidance system.* (Gravity Guidance, Inc., Pasadena, CA).

Mihaly, T., (1988), Exercise and cognition: Effects of exercise-induced arousal (elevated heart rate), limb movement, and practice on four speed of response measures, Master's thesis, Los Angeles, CA: University of Southern California.

Moruzzi, G. and Magoun, H. W., (1949), Brain stem reticular formation and activation of the EEG, *Electroencephalography and Clinical Neurophysiology*, **1**, 455–73.

Readers Digest, (1989, October), *Stay on your toes*, Reuters/adapted from Science News.

Reitan, R. M., (1954), Intellectual and affective changes in essential hypertension, *American Journal of Psychiatry*, **110**, 817–24.

Schroepfer, L., (1989, April), Rise to the occasion: Executive couch potatoes should stand and deliver, *American Health: Fitness of Body and Mind*, 52.

Science Digest, (1989*a*, July), *Keep on your toes*, 60.

Science Digest: Special Supplement, (1989*b*, November), *Keep on your toes*, 16.

Science Impact, (1988, December), *How to think faster*, **2**(7).

Speith, W., (1965), Slowness of task performance and cardiovascular diseases, in A. T. Welford and J. E. Birren (eds) *Behavior, Aging and the Nervous System*, Springfield, IL: C. C. Thomas.

Sternberg, S., (1969), The discovery of processing stages: Extensions of Donder's method, *Acta Psychologica*, **30**, 276–315.

Thomas, J. R. and French, K. E., (1985), Gender differences across age in motor performance: A meta analysis, *Psychological Bulletin*, **98**(2), 260–82.

USA Today, (1989), *Stand up for clearer thoughts*, 12.

Vercruyssen, M., (1984), Carbon dioxide inhalation and information processing: Effects of an environmental stressor on cognition, Doctoral Dissertation, State College, PA: Pennsylvania State University.

Vercruyssen, M., (In Press), Slowing of behavior with age, in R. Kastenbaum (ed), *Encyclopedia of Adult Development*, Phoenix, AZ: Onyx Press.

Vercruyssen, M., Cann, M. T., McDowd, J. M., Birren, J. E., Carlton, B. L. and Burton, J., (1988), Effects of age, gender, activation, and practice on attention (preparatory states) and stages of information processing, *Proceedings of the Human Factors Society 32nd Annual Meeting* (pp. 203–207). Santa Monica, CA: Human Factors Society.

Vercruyssen, M., Cann, M. T. and Hancock, P. A., (1989), Gender differences in postural activation effects on cognition, *Proceedings of the Human Factors Society 33rd Annual Meeting* (pp. 896–900). Santa Monica, CA: Human Factors Society.

Woods, A. M., (1981), Age differences in the effect of physical activity and postural changes on information processing speed, Doctoral Dissertation, The University of Southern California, Los Angeles, California, 90089-0191.

Woods, A. M., Vercruyssen, M. and Birren, J. E., (1993), Age differences in the effect of postural changes and physical activity on information processing speed, *Proceedings of the Human Factors and Ergonomics Society 37th Annual Meeting* (pp. 177–181) Santa Monica, CA, The Human Factors and Ergonomics Society.

Yerkes, R. M. and Dodson, J. D., (1908), The relation of strength of stimulus to rapidity of habit-formation, *Journal of Comparative and Neurological Psychology*, **18**, 459–82.

9

Our posture dictates perception

Yutaka Haruki and Masao Suzuki

Introduction

Our posture adapts to and reflects our environment. When meeting a ferocious dog, for instance, we adopt motions and postures to face it or escape. When afraid, we crouch to protect ourselves. When angry, we try to look larger to intimidate. When discouraged, our head droops, and our body becomes listless.

Consequently, our feelings may both cause and reflect specific postural patterns. For example, strength is diminished if we continue to droop. A postural pattern may also determine emotions. If so, the chair may affect our feeling and awareness through our posture. For example, the chair may affect our awareness through our vision. The size and shape of our chair reflects our social position; our mood is influenced by the social status connoted by the chair. Seated comfort is also influenced by the chair's structure and material. Because our seated posture influences our awareness, the chair affects our physical and mental states.

This study was conducted to analyze how posture affects emotions. To this end, subjects were instructed to sit on a chair and adopt specific postures. The subjects' mood while sitting was measured with a rating scale. Two experiments were conducted. The first experiment investigated the relationship between postures and emotions. Postures were considered independent variables, and mood states dependent variables. Measures of posture were derived from back and head angles. These six postures which were investigated included straight or hunched back, and downward, straight or upward head orientations. To determine whether specific postures influenced mood state, subjects were asked to imagine a particular posture and report the mood created by this image.

The second experiment examined the effect of particular postures on the mood of subjects exposed to a mood-governing stimulus. For example, music is known to affect awareness; it is also evident that the postures we assume when we listen to music influence our awareness. To test this hypothesis, subjects were instructed to listen to two types of music (cheerful and gloomy) in two extreme postures. These postures were either with upright backs and heads, or with backs moderately hunched and faces looking down.

The purpose of the above two experiments was to examine the relationship between postures and mood state, and the corresponding implications on seating and posture.

Experiment 1: Effects of trunk and head angle on mood state

Methods

Subjects:
32 male and 32 female students participated as subjects, ranging in age from 19–22 years.

Apparatus used in Experiment 1 included audio tapes describing posture, tape recorder, a reclining chair, a subjective comfort check list for the experimenter, and white paper for line drawing.

Postures:
Subjects adopted six different postures with upward, straight or downward oriented head positions and straight or hunched backs, (Figure 9.1).

Posture 1: The subject sat with back upright and with face oriented about 45° higher than the horizontal line of sight.

Posture 2: The subject sat upright and faced straight ahead.

Posture 3: The subject sat upright, with face oriented down so that they could view their thighs.

Posture 4: The subject sat with drooped shoulders, hunched back, and with face about 45° higher than the horizontal line of sight.

Posture 5: The subject sat with drooped shoulders, hunched back, and face straight ahead.

Posture 6: The subject sat with drooped shoulders, hunched back, and face downward so that they could view their thighs.

In each posture, male subjects placed their hands lightly on their knees and sat with slightly opened knees and legs. Female subjects placed their hands lightly on the knees, with closed knees and feet.

Procedure:
This experiment was conducted in two parts. During the image rating stage, the subjects listened to the audio tape describing a posture, imagined this posture, and rated their mood in accordance with their postural image. During the behaviour rating stage, subjects listened to the tape describing a posture, adopted this posture, and rated their mood. Condition 1 comprised image rating and behaviour rating (in that order), whereas condition 2 comprised behaviour rating and image rating (i.e., with the order reversed). Each condition was randomly assigned to 16 subjects at a time. Prior to the experiment, the subjects practised in a standing posture (with the back straight, head upright, legs slightly open, and arms dangling). The subjects adopted the six postures in a randomized sequence during both image and behaviour rating sessions.

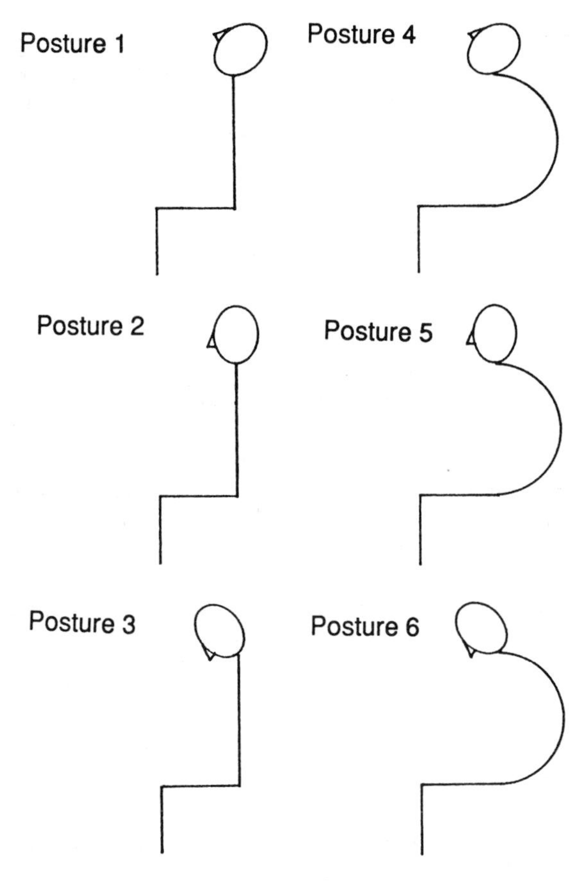

Figure 9.1

Image rating:
The subject reclined on the chair with the eyes closed, and imagined a posture that they heard described on tape. No time limit was imposed. Subjects signalled the experimenter when they achieved this postural image, then rated their awareness of ambience that was evoked by the posture with a three-point scale using adjective pairs. The experimenter verbally asked the subject the rating questions. Seventeen adjective pairs were used to rate awareness, and six adjective pairs were used to rate ambience. After the rating session, the subjects were instructed to draw the posture they imagined. This procedure was repeated for each posture.

Behaviour rating:
The subject sat on the chair, listened to adopted tape descriptions of postures. After emulating this posture, subjects signalled the experimenter. The experimenter verbally asked the subject questions regarding their perceptions associated with this specific posture, using adjective pairs. The subject responded to each question based on a three-point adjective scale.

Results and discussion

The experimental results were factor analyzed as follows. An inter-scale correlation matrix (17 × 17) was calculated from the subjects' image and behaviour ratings of the six postures by the adjective scales (2 × 6 × 64 × 17). Factors were extracted by the Principal Factor method, and the factor axes were rotated by the Varimax method. Twelve scales belonged to Factor I, three scales to Factor II, and two scales to Factor III; these factors accounted for 76.9 per cent, 18.7 per cent, and 4.3 per cent of the total variance, respectively. The scales that loaded high on Factor I reflected liveliness, confidence, lightness and strength or vitality and lightness. The scales that loaded high on Factor II represented tension-relaxation; that is tension. Factor III is related to Factor I, and primarily connotes activity.

To evaluate the postural conditions, a two-by-three analysis of variance (ANOVA) was applied to the behaviour and image ratings (straight or hunched back; upward, straight or downward head position). For both the behaviour and image conditions, the playful-serious adjective pairs alone were not significantly different under the back condition (straight/hunched). However, the back and head conditions were consistently significant ($p < .05$) for all other adjective pairs. The behaviour ratings did not evidence a significant interaction for playful-serious alone, and significant differences of at least

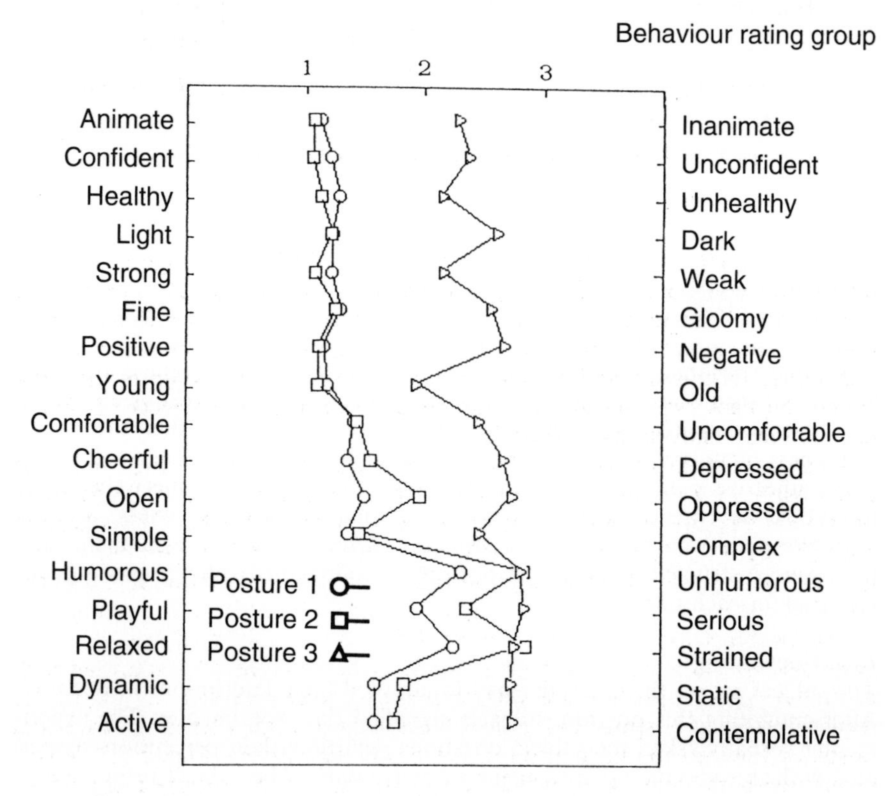

Figure 9.2 Awareness rating profiles (behaviour ratings) of postures 1–3

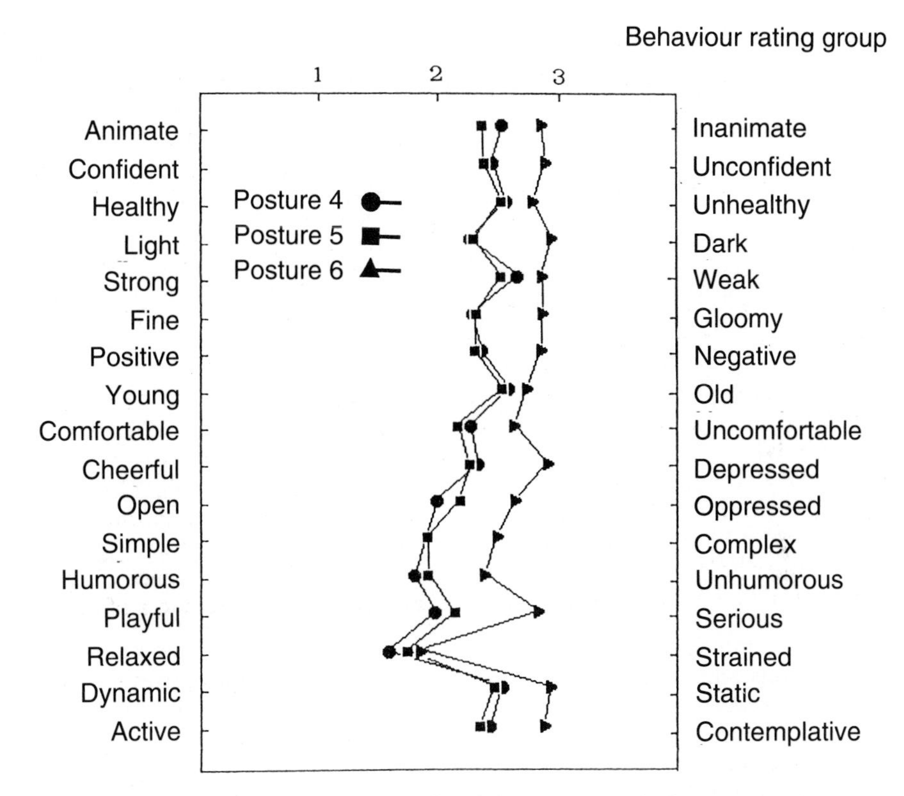

Figure 9.3 Awareness rating profiles (behaviour ratings) of postures 4–6

$p < .05$ were observed with all other adjective pairs. With image ratings, no interaction was observed for simple-complex and dynamic-static; these were significantly different ($p < .05$) than all other adjective pairs, as with the behaviour ratings. When the back was hunched, subjects felt inanimate, insecure, unhealthy, dark, weak, and gloomy, irrespective of the orientation of the head. When the back was upright and the head downcast, the subjects felt inanimate, insecure, unhealthy, and dark.

Analysis of variance was used for the behaviour and image rating conditions and posture conditions (2 by 6) for each adjective pair scale. Differences between the behaviour and image rating conditions for the healthy-unhealthy scale ($F(1.756) = 5.095$) and young-old scale ($F(1.756) = 6.212$) were consistently significant ($p < .05$). The other adjective pairs did not differ significantly, and the behaviour ratings were not clearly different from the image ratings.

The postures for all adjective pairs did differ significantly ($p < .05$). The mean ratings of the adjective pairs of subjects' perceptions after adopting each of the six types of posture are summarized in Figures 9.2 and 9.3. Figure 9.2 depicts ratings of postures 1–3, in which the subjects straightened the back and raised or lowered the inclination of the head, respectively. Figure 9.3 depicts ratings with postures 4–6, which are characterized by drooped shoulders, hunched back, and raised or lowered head positions (respectively). As is evident from the two figures, the 12 scales belonging to factor I exhibit similarities, although some differences exist as well.

Analysis of variance was used to analyse postures associated with each adjective pair. Significant differences were observed among the postures for 'animate-inanimate' [$p < .01$ $F(5,378) = 98.74$]. When the mean pairs of differences between the six postures were individually analyzed by Tukey comparisons, posture 6 was found to differ significantly from all other postures ($p < .05$), and rated inanimate. Postures 3–5 differed significantly from postures 1 and 2. Differences in subjective ratings were evident with drooped shoulder postures when the back is hunched as straightbacked. For 'light–dark', [$F(5,378) = 84.48$], postures were significantly different ($p < .01$). Analysis of variance for the other adjective pairs resulted in significant differences for each scale ($p < .01$). When the differences among the postures were multiply compared by the Tukey method, a significant difference of at least 5 per cent was observed for each scale, except that posture 6 was no different from posture 4 for strong-weak, from postures 4 and 5 for healthy-unhealthy and comfortable-uncomfortable, and from posture 3 for young-old, cheerful-gloomy, open-oppressed and simple-complex.

When analysis of variance was applied to the three scales belonging to factor II, $F(5,378) = 34.21$ for humorous unhumorous, and the other two scales similarly exhibited a significant difference of 5 per cent. When the postures were multiply compared by the Tukey method for the humorous-unhumorous and relaxed-strained scales, significant differences ($p < .05$) were observed between postures 2 and 3 and postures 1, 4, 5, and 6. This means that awareness is affected by the orientation of the head and back.

When analysis of variance was performed for the two scales belonging to factor III, significant differences of 1 per cent were observed among the postures. For dynamic-static ($F(5,378) = 33.87$), differences were observed between posture 6 and postures 1, 2, 4 and 5, between posture 3 and postures 1 and 2, and between postures 4 and 5 and postures 1 and 2. These results also show that awareness is affected by face and back orientations.

Awareness characteristics of each posture are described below. When subjects straightened the back and looked upward (posture 1), they felt positive, young, healthy, and fine. When subjects sat upright with head oriented straight ahead (posture 2), they felt confident, animate, strained, unhumorous, and young. When subjects sat upright with head oriented downward (posture 3) they felt serious, strained, unhumorous, oppressed, and static. When subjects had drooped shoulders, hunched back, and looked upward (posture 4), they felt weak, inanimate, unhealthy, old, and negative. When subjects sat with drooped shoulders, hunched back, and faced straight ahead (posture 5), they felt unhealthy, static, weak, inanimate, and old. When subjects sat with drooped shoulders, hunched back, and looked downward (posture 6), they felt dark, depressed, inanimate, insecure, and static. These findings indicate that mood state is influenced by the head and trunk posture. When the back was hunched, the subjects felt inanimate, unhealthy, or weak, irrespective of the orientation of the head. When the back was straight and the face downcast, the subjects felt serious and oppressed.

The mean values of the image ratings of the six postures (see Figure 9.4) resemble those of the behaviour ratings. Postures for each scale were significantly different ($p < .01$). The results of multiple comparisons among the pos-

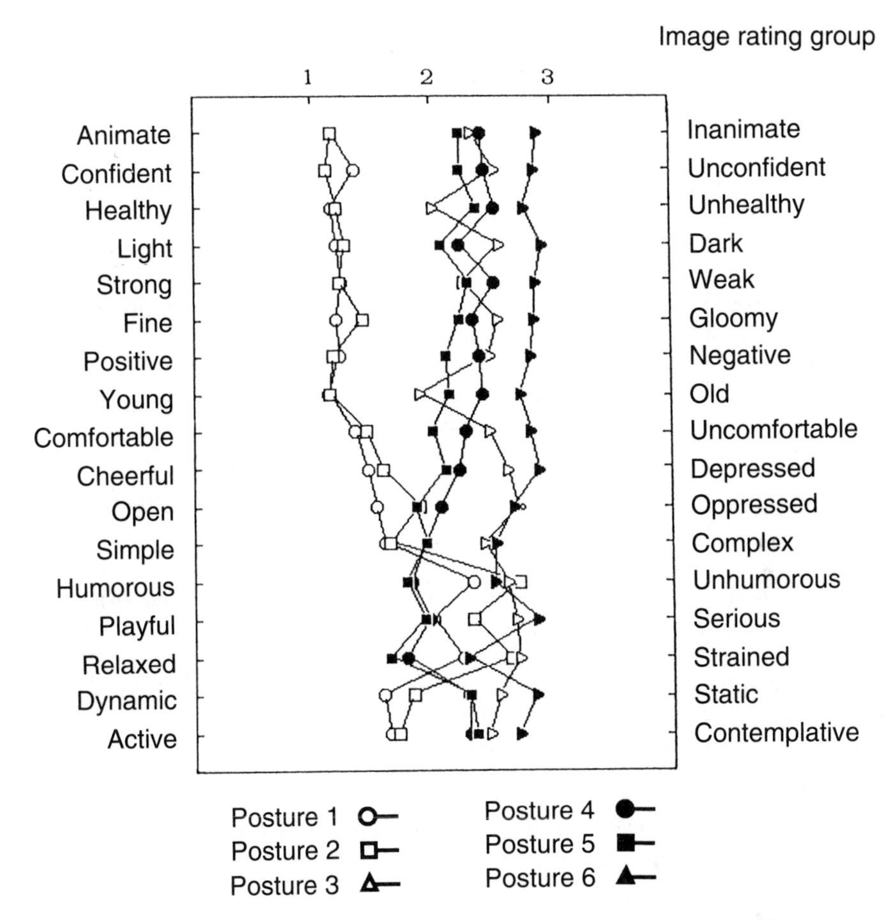

Figure 9.4 Awareness rating profiles (image ratings) of postures 1–6

tures for each scale are similar to those obtained when the subjects actually adopted the specific postures.

Ambience denotes the vividness of subjects' mood state rating. ANOVA revealed that behaviour and image ratings were significantly different from each perspective (see Figure 9.5). [$p < .01$ $F = 276.78$]. Namely, subjects felt that a given posture was more real, powerful or clear when they adopted, rather than imagined, the posture.

No consistent perceptual differences were found between the image and behaviour ratings. Ratings of mood states when postures were adopted differed from those when they were imagined.

These results agreed with those of previous studies conducted with seven-point scales (Suzuki, 1984; Suzuki, 1986; Suzuki and Haruki, 1988). Subjects' ratings did not differ when postures were assumed and imagined. This probably resulted because the postures studied in this experiment are commonly assumed daily. Consequently, by adopting these postures, subjects evoked associations that frequently occur in conjunction with these postures. It may

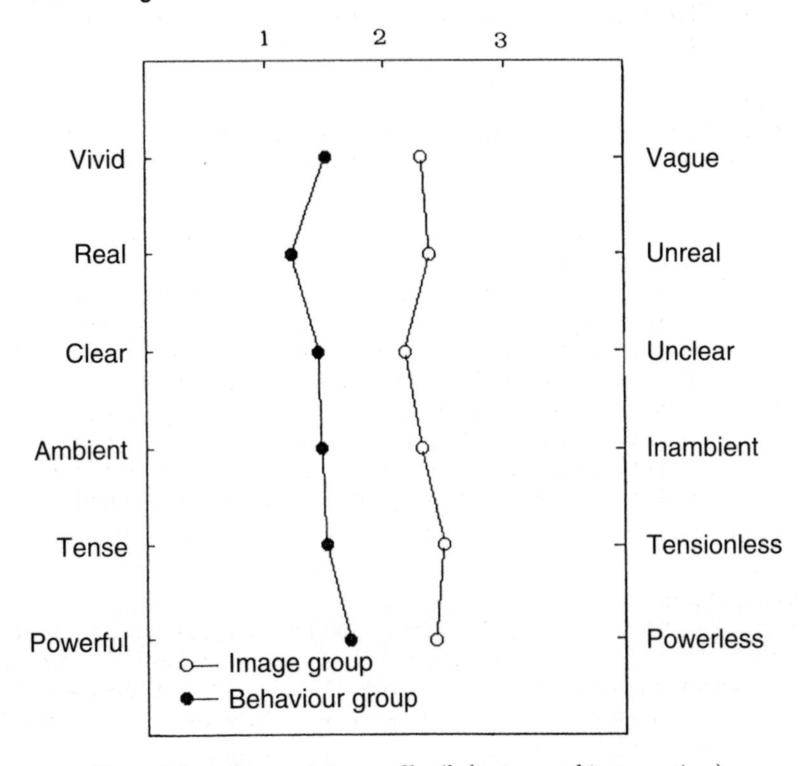

Figure 9.5 Ambience rating profiles (behaviour and image ratings)

be that certain moods are conceptually connected to postures in accordance with past experiences, and that people assume postures that befit their mood. It is difficult to determine whether the subject ratings were based on the ambience of the situation or previous experiences. Cognitive processes may also influence ratings. An understanding of these phenomena requires that current methods for measuring perception be improved.

Experiment 2: Effect of posture on awareness evoked by music*

Methods

Subjects:
30 female students with a mean age of 19.1 years participated in this experiment.

---

* Experiment 2 was conducted with co-operation of Miss Rika Kawano.

Equipment:
A stereo set for stimulus presentation, chair with back rest, a reclining folding chair, music tape, and awareness rating forms were used.

Procedure:
Subjects were instructed to listen to two types of music in three postures and to rate their mood states with 17 adjective pairs. The postures are described below:

Posture 0: Subjects lay on the reclining chair with the hands alongside the body and raised head.

Posture 2: Subjects sat straight and shallow on the chair with back rest. The subject slightly opened the chest to relax the shoulders, lightly placed the hands on the knees, and looked straight ahead. Legs were bent to about 90°, feet supported on the floor.

Posture 6: Subjects sat with shoulders drooping, and back hunched. Subjects relaxed the body with hands supported lightly on the knees. They relaxed the neck and oriented their head downward to such an extent as to view the thighs, with legs and feet as in posture 2.

With all of the above postures, subjects kept their eyes open.

Music condition I:
Carmen Suite, No. 1 Les Toréadors, composed by Georges Bizet. This music is composed of the lilting two-four time allegro vivace portion and the three-four time andante moderato portion. It is a lively A major march played with wind and string instruments, which suggest brilliant and exciting moods when the toreadors enter the bull ring.

Music condition II:
Adagio from the movie Platoon and composed by Samual Barber. This screen music with the theme of the Vietnam war in the 1960s, played with string instruments, is slow and melancholy.

Adjective pairs for rating postures
The subjects rated the postures and music pieces on seven-point scales, using 17 adjective pairs employed by Suzuki and Haruki (1987). There were six combinations of posture and music conditions, which were randomly assigned to the subjects.

Results and discussion

The experimental data were examined using factor analysis. An inter-scale correlation matrix (17 × 17) was calculated from the subjects' adjective pair ratings of the three postures and two music pieces (3 × 2 × 60 × 17). Factors were extracted by the principal factor method, and the factor axes were rotated by the Varimax method. Eleven scales belonged to factor I, four to factor II, and two to factor III. These factors accounted for 62.58 per cent, 19.64 per cent, and 17.78 per cent of the total variance, respectively.

The factors may be interpreted by using these clues. The scales loaded high on factor I (activity factor) were considered to represent vitality and motion, as illustrated by the adjectives powerful, animate, positive, confident, and cheerful. The scales loaded high on factor II (openness factor) were open, humorous, and comfortable. The scales loaded high on factor III (humour factor) were playful and humourous.

The adjective pairs were rearranged in the order of factor loadings obtained, and the mean values of the rating scales for the three postures and music condition I are depicted in Figure 9.6. Under music condition I, posture 6 represented weak, inanimate, negative, lacking in confidence, depressed, static, old, unhealthy, oppressed, and dark moods as compared with posture 2 or 0. Under music condition II, posture 6 is shifted to the opposite side at the boundary of score 4, but the relationship of posture 6 with posture 2 or 0 is the same as observed under music condition I.

Analysis of variance was used to analyse the three postures and two music pieces for each adjective pair. Significant differences ($p < .01$) were observed between music condition I and music condition II, except for strained-relaxed. Music condition I was perceived as powerful, animate, positive, confident, cheerful, dynamic, young, healthy, fine, active, and light. In contrast, music condition II was perceived as weak, inanimate, negative, lacking in confidence, depressed, static, gloomy, dark, oppressed, and serious. Different pieces of music evoked different moods.

No significant differences between the postures were observed for dynamic-static and playful-serious, but significant differences were observed ($p < .05$

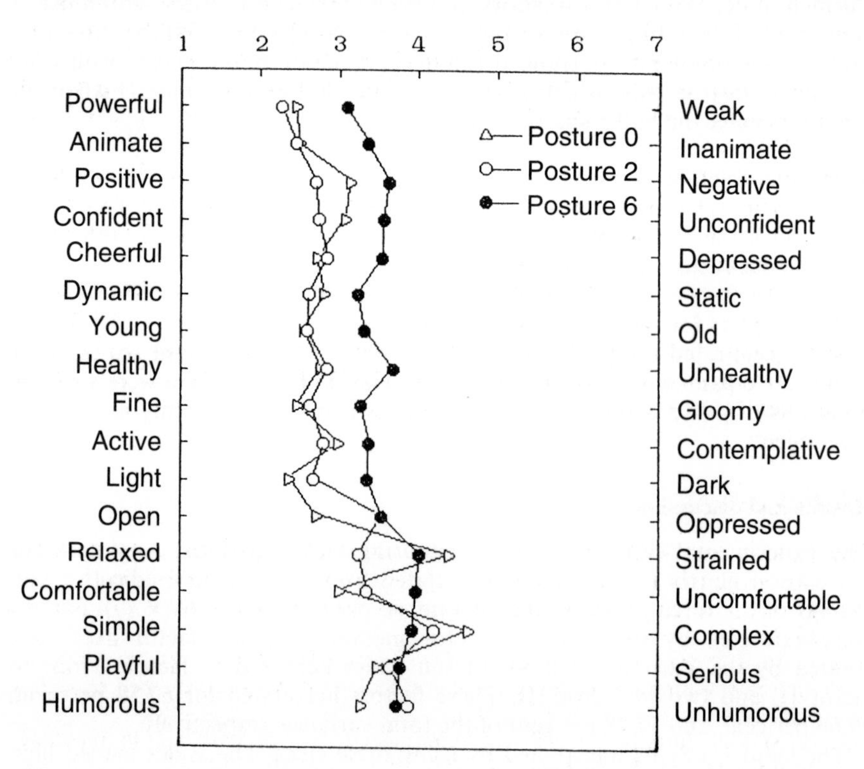

Figure 9.6 Awareness rating profiles of postures 0, 2, and 6 under music condition I

and $p < .01$, respectively) for active-contemplative and simple-complex and for other adjective pairs. When the mean pairs were multiply compared by the Ryan method, significant differences under music condition I were observed ($p < .05$) between postures 6 and 0 and between postures 6 and 2 for many of the adjective pairs. Posture 6 was perceived as weak, inanimate, depressed, old, unhealthy, gloomy, dark, and uncomfortable as compared with postures 0 and 2. Differences between the postures under music condition II were not as large as observed under music condition I.

As discussed above, posture 6 evoked more depressed, weaker, and more inanimate moods than the other two postures, even under lilting music. This underscores the impact of downcast postures as reported by Suzuki (1984, 1986) and Suzuki and Haruki (1988).

General discussion

The results of the above two experiments confirmed the hypothesis that postures affect moods. The results of Experiment 1 indicates that specific postures are associated with specific moods. The results of Experiment 2 indicate that mood state is influenced both by exposure to certain stimuli, in this case, music, and posture.

It is not clear why different moods are produced by specific postures. Because our postures frequently reflect our mood state, our posture may be conceptually related to our awareness. This hypothesis is supported by the results of Experiment 1; when subjects drew their imagined postures, these drawings differed appreciably from how they actually sat. This suggests our awareness and mood are affected more by how we imagine we sit than by the associated physical action.

When we imagine that we assume a posture, we produce the mood we associate with this posture. On the other hand, when we adopt a posture, the resultant mood is based on actual physical sensations, although the postural image may enter in our awareness.

Our thoughts are intimately related to our posture. Correspondingly, particular states of mind can engender specific postures, and particular postures can evoke specific states of mind. The chair directly affects awareness by its shape, colour or feel. At the same time, the chair affects our posture, and this posture influences our awareness.

References

Suzuki, M., (1984), Studies of the relation between postures and awareness, *Bulletin of Graduate Division of Literature, Waseda University*, **11**, 9–21, (in Japanese).

Suzuki, M., (1986), An experimental study of postures, *Waseda Psychological Reports*, **18**, 27–36, (in Japanese).

Suzuki, M. and Haruki, Y., (1987), A study of posture (V), paper presented at the 51st Annual Convention of the Japanese Psychological Association. Tokyo, 409, (in Japanese).

Suzuki, M. and Haruki, Y., (1988), An experimental study of awareness in four bodily postures, *Waseda Psychological Reports*, **20**, 1–7, (in Japanese).

PART V

Back pain

10

Low back pain and seating

Tom Bendix

Epidemiology

The incidence of low back pain (LBP) increases with seating duration. The distribution of LBP in relation to work load is depicted in Figure 10.1. Magora (1972) divided people into three groups in accordance to the amount of time they sat at their workplace. One group sat there for more than four hours a day (Figure 10.1, right). Another group sat infrequently (left) and many of them adopted physically heavy work. The third group sat 2–4 hours; in effect, they alternated between sitting, standing, walking, and even lifting.

Although the group to the left suffered the most from LBP, apparently, the incidence of the predominantly seated group had almost as much LBP, and LBP reduces as postural variation is increased. As a result, the primary principle of seating is: avoid sitting for extended durations. The incidence of LBP in seated tasks can best be addressed by promoting movement, rather than optimizing the design of chair and table.

It has also been demonstrated by Kelsey (1975) that people who drove for more than four hours a day have a three times greater risk of incurring herniated disc. This observation involves – just as for Magora, described above – some elements of a healthy-worker effect: people with infirm backs tend to choose sedentary work. However, various other studies also suggest that the seating task is in itself a risk indicator, as will be discussed below.

The incidence of LBP has increased considerably during the last decades, although back pathology may not have worsened (Waddell, 1987). Several reasons for this increase exist; these will be discussed.

Behavioural aspects, such as expectations of welfare and health, may contribute to this incidence: We have, in these decades, acquired better houses and excellent cars. Health programmes have also improved: hip joints can be replaced; cataract operations allow people to see again and we can even get new hearts. Correspondingly, people expect to be successfully treated for back

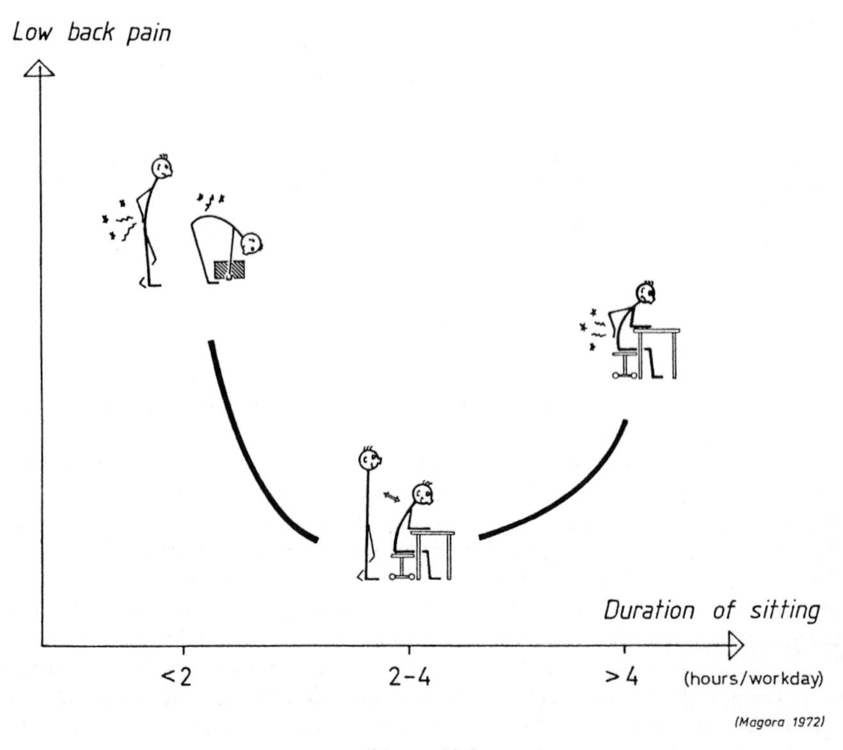

Figure 10.1

pain and are more likely to visit physicians, chiropractors, etc. for back pain. These health specialists are, however, generally not much better able to treat it than they were some decades ago.

Pathogenesis

The pathological aspects of LBP given below have particular relevance for seating.

Disc and facet joints

In most cases, the causes of LBP are not known. However, there are hints. The most widely accepted explanation for most cases of LBP may reside in the interplay between the disc and the posteriorly-positioned facet joints.

The disc consists of an outer layer of lamellas surrounding a soft nucleus. During late childhood and young adulthood, ruptures appear in the lower lumbar discs in most people (Hirsch and Schajowitz, 1952), as illustrated in Figure 10.2. The *smallest* ruptures may not have any significance, because nerve endings are absent in the central lamellas. The *larger* ones (right) correspond to a herniated disc, and represent a definite cause of pain – if nerve roots are compressed. *Middle-sized* ruptures may (Vanharanta *et al.*, 1990) or may not cause any harm in themselves. However, the backs of young people –

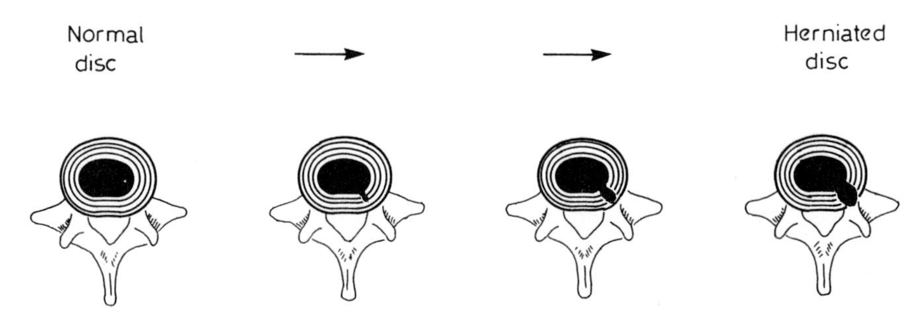

Figure 10.2

who have died from other causes – do not show evidence of any other pathologic features (Nachemson, 1976).

It is therefore likely that the disc ruptures are relevant. LBP may also have something to do with the interplay between this early disc degeneration and the facet joints (Miller *et al.*, 1983; Yang and King, 1984). A normal disc has very firm and stabilizing properties (Figure 10.3, left): even heavy lifting has little influence over disc height. However, once the disc has incurred middle-sized ruptures, the situation changes. In that case, the normal disc height and stability may change during hours loading in the upright position, standing or sitting. The significant issue is that such a compression involves the facet joints as well (Figure 10.3, upper right). In forward bending, shear forces also appear, as presented in exaggerated form on Figure 10.3 (lower right). Thus, in these cases the disc is the origin of the pain, but the facet joint is the agent.

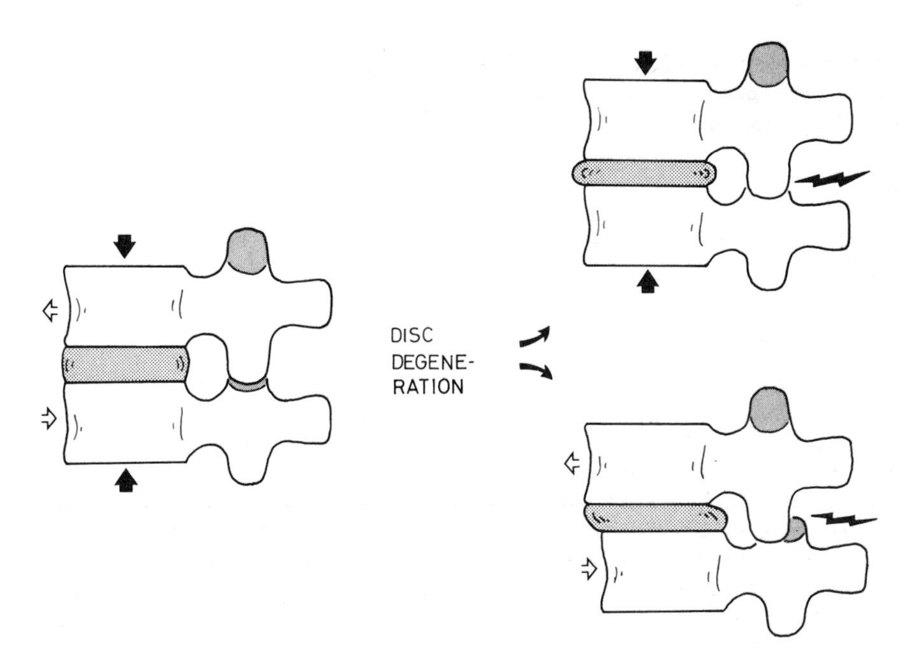

Figure 10.3

Back pain is most likely to originate at the joints from ligament/capsule strain or bone compression, rather than from the muscles. However, muscle pain may appear later because joint-pain may induce muscle spasms in the surrounding area.

Why do disc ruptures take place? Most research has been conducted on people exposed to *heavy loads*, such as with intra-discal pressure measurements, Nachemson (1982) and Andersson *et al.* (1977). Disc degeneration in *sedentary* living (particularly while driving) may be less influenced by the load, (Andersson, 1974; Andersson *et al.*, 1974) as by disc nutrition, and the stress-relaxation phenomenon in collagen structures.

(*A*) *Disc nutrition*

The disc is not provided with blood vessels: consequently disc nutrition occurs by osmosis and may be compared to a sponge with water: When compressed, waste materials in the disc are squeezed out, as pressure is released all nutritious matter is sucked in.

This hypothesis was substantiated by Holm and Nachemson (1983). Four groups of dogs were living in different ways in the last eight months of their lives (Figure 10.4). One group lay in their cages all the time and moved only while eating and cleaning. Three other groups performed various levels of motion. One group ran slowly for two hours a day; another for only $\frac{1}{2}$ an hour a day, but fast; a third group exercised for $\frac{1}{2}$ an hour a day. Within this range of motion, these three groups obtained the same relative increase in disc nutrition as sedentary dogs depicted in Figure 10.4.

It is not surprising that movements improve disc nutrition. All other anatomic structures evidence this phenomenon as well. However, the disc is the largest structure in the body without blood vessels, and has a particular tendency to rupture. Although other factors contribute to LBP as well, movement exerts a substantial influence on disc nutrition.

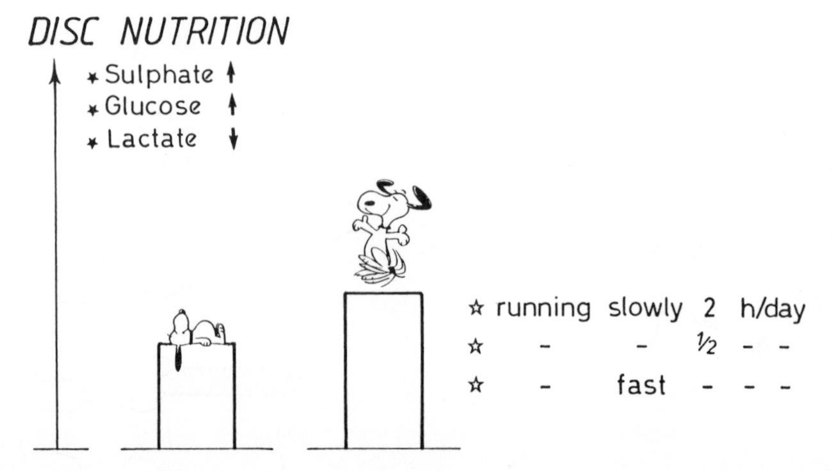

Figure 10.4

(*B*) *Collagen-fibre elasticity*

Ligaments and joint capsules are formed of elastic collagen fibres. When these fibres are stretched, they elongate (see Figure 10.5); this elongation continues to increase during subsequent hours. When this pull force is removed, an additional time span must pass until the original length is obtained (Sanjevi, 1982). This time span varies from 20 min (Bojsen-Møller, personal communication) to hours (Weismann *et al.*, 1980), in accordance with the initial stretching force.

It is relevant that when one sits for hours, *some* ligaments will usually be tensed, depending on the posture. Slumped postures will tense the posteriorly positioned fibres. After hours of sitting in a car or office, stability is slightly reduced, and the joints become stressed performing such movements that normally would not cause harm.

Psychologic aspects

Another reason for the increase in LBP during the last decades may be found in behavioral issues, and in the way the health-care system treats patients:

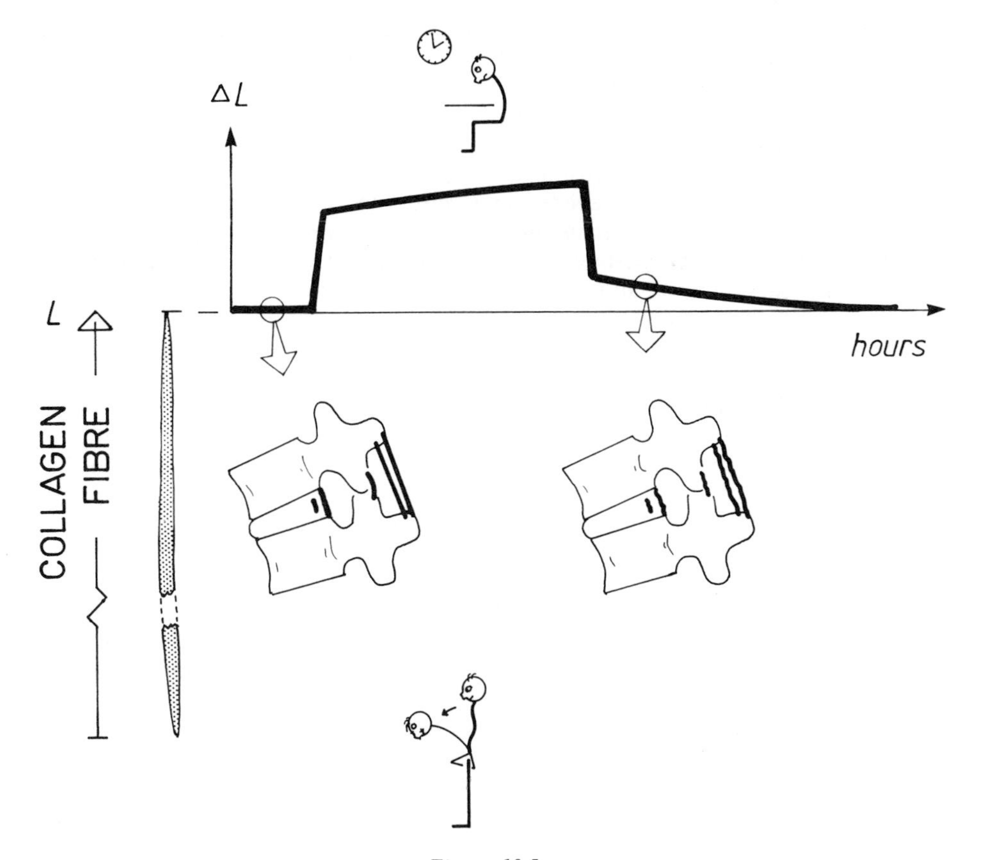

Figure 10.5

Many cases of LBP benefit from physical treatment. However, in other cases no treatment is needed and spontaneous cure may arise. Moreover, although the patient may have some genuine pathology, the main problem is psychological behavioral. In these cases, the patient may presume that if the physician prescribes a treatment, he must be sick. The next time pains occur he again feels that treatment is necessary. As a result, he increasingly focuses on the pain problem.

Instead of making LBP patients passive, we should rather become proactive, train them intensively, and teach them how to cope with pain (Mayer *et al.*, 1987; Hazard *et al.*, 1989). Even training alone seems effective (Mannicke, 1988), at least in untrained people (Hansen *et al.*, 1990).

Seating

Traditionally, attention has focused on what is the best posture and the best chair? It is easy to conduct measurements on these issues, but are we really measuring the right thing? Of course, a well-designed work station is preferable to a poor and uncomfortable product. Unfortunately, however, it is not as easy to solve LBP for long-term sitting – whether in office, industry or in car.

In this chapter, only the time factor, the backrest, and dynamic seating will be discussed.

The time factor

The seating problem is not just a question of either force or time, but a combination of those two, as depicted in Figure 10.6. Forceful tasks (e.g., heavy lifting) may induce injury quickly (left part of the curve). On the other hand, when sitting in an extremely comfortable chair, very small forces on the back (right part), are induced, and much time must pass before injury may result.

This relationship between force and time is extremely important to LBP – and complex. Sitting in an uncomfortable chair, which introduces great loads to various parts of the body, may not be a health issue if one only sits for a short period of time. Tasks and habits, however, force us normally to sit there for a sufficiently long period of time to cause injury. Consider the middle of the ordinate axis on Figure 10.6; the time representing a point 1 cm into the injury area. If the chair is improved – i.e., the force is reduced – the point moves down the curve, and with the same duration of sitting, the hatched area is only barely in contact. If the chair is improved even more, it could reduce the likelihood of back and other locomotor-system injuries (the point moves further down below the hatched area). However, the user instead tends to adjust alternatively, these users may spend more time seated; again the hatched area is touched. Thus, it is naive to believe that optimizing the work station can allow people to spend 8 hours a day in a chair.

Maybe computers should dictate every hour that the computer program is switched off for five minutes and the user taking a break. Such approaches have been attempted with little success; users were stressed from trying to find a suitable time in the program to make a break. By the way, computer workers really tend to have natural breaks, but they are always five minutes ahead!

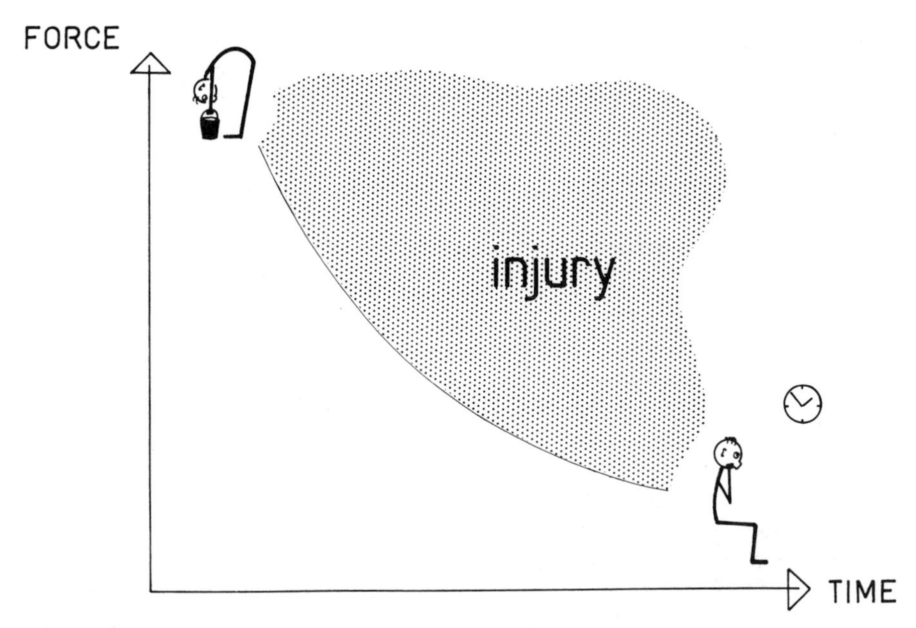

Figure 10.6

Backrest

The design of the lumbar support also affects seating ergonomics. In office seating, lumbar supports do not always increase lordosis. If, whilst sitting without a back support, someone pushes a backrest against the back, either lordosis is initially increased – or kyphosis is reduced. However, most users push the buttocks forward, presumably to press the lumbar spine against the backrest and stabilize posture (Bendix *et al.*, 1990).

This phenomenon is evident with office seating with a traditional vertically-flat backrest. Further, an anteriorly curved backrest (which at least for some activities reduces kyphosis/increases lordosis) apparently increases the stress on the spine because spinal shrinkage is greater (Bendix *et al.*, 1990). Under similar circumstances, however, Andersson (1974) observed a reduced intra-discal pressure.

Obviously, backrests should be used. However, the biomechanical effects of these backrests need to be reconsidered. Further investigations should elucidate whether an anteriorly curved backrest actually induces lordosis in a slant-back car or easy chair, and also whether stress on the spine is thereby being relieved.

Dynamic seating

LBP while seated should be reduced by promoting work-related movements.

A sit-stand work place represents one such significant posture-changing feature for office and industrial seating (Figure 10.7). This encourages changes between standing and sitting during a specific period with a certain task (Aarås, 1987; Winkel and Oxenburgh, 1990).

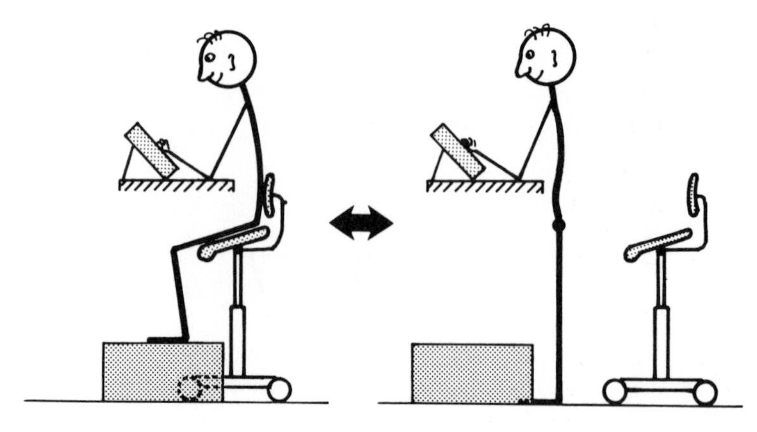

Figure 10.7

It is not realistic to encourage regular (e.g., every 45 min) exercise breaks during long-distance vehicle driving. Alternatively, the backrest can be designed to move passively during driving - Manually INflatable lumbar pads are available. The author believes these are used no more than three times during the first drive, a few times subsequently, and thereafter only infrequently.

Suggested future research

Knowledge of how to prevent LBP and other locomotor-system diseases induced by seating, requires more than biomechanical measurements. Although medical practitioners often emphasize new drugs or other treatments, controlled clinical trials sometimes demonstrate that these are ineffective. In the area of seating, much excellent basic research exists. However, research is needed that compares various seats in blinded trials. These seats should be equivalent in colour and other parameters except those tested, and should be placed in similar surroundings (e.g., the same car) during the various testing periods. Such research would *relevant* further alternative basic research.

References

Aarås, A., (1987), Postural load and the development of musculo-skeletal illness. STK/Institute of Work Physiology, Oslo, Norway, (thesis).

Andersson, G., (1974), On myo-electric back muscle activity and lumbar disc pressure in sitting postures. Göteborg, Gotab (doctoral dissertation).

Andersson, G., Örtengren, R., Nachemson, A. and Elfström, G., (1974), Lumbar disc pressure and myo-electric activity during sitting. IV. Studies on a car driver's seat, *Scand. J Rehabil. Med.*, **6**, 128–33.

Andersson, G., Örtengren, R. and Nachemson, A., (1977), Intradiscal pressure and myo-electric back muscle activity related to posture and loading, *Clin. Orthop. Rel. Res.*, **129**, 156–64.

Bendix, T., (1987), Adjustment of the seated workplace – with special reference to heights and inclination of seat and table, *Danish Medical Bulletin*, **34**, 125–39. (doctoral dissertation).

Bendix, T., Jessen, B. and Winkel, J., (1986), An evaluation of a tiltable office chair with respect to seat height, backrest position, and task, *Eur. J Appl. Physiol.*, **55**, 30–6.

Bendix, T. and Bloch, I., (1986), How should a seated workplace with a tiltable chair be adjusted? *Appl. Ergonom.*, **17**, 127–35.

Bendix, T., Poulsen, V., Klausen, K. and Jensen, C. V., et al. (1993), What does a backrest actually do to the spine? in (eds Andersson, G. *et al.*) *The 1990 annual meeting of the International Society for the Study of the Lumbar Spine*, Boston, Abstract. p. 3.

Hansen, F. R., Bendix, T., Skov, P. and Jensen, C. V., (1990), A controlled trial comparing intensive dynamic back exercises, physiotherapy and placebo. in (eds Andersson, G. *et al.*) *The 1990 annual meeting of the International Society for the Study of the Lumbar Spine*, Boston, Abstract. p. 2.

Hazard, R., Fenwick, J., Kalish, S., Redmond, J., Reeves, V., Reid, S. and Frymóyer, J., (1989), Functional restoration with behavioural support. A one-year prospective study of patients with chronic low-back pain, *Spine*, **14**, 157–61.

Hirsch, C. and Schajowitz, F., (1952), Studies on structural changes in the lumbar annulus fibrosus, *Acta Orthop. Scand.*, **22**, 184–231.

Holm, S. and Nachemson, A., (1983), Variation in the nutrition of the canine intervertebral disc induced by motion, *Spine*, **8**, 866–74.

Kelsey, J., (1975), An epidemiological study of the relationship between occupations and acute herniated lumbar intervertebral discs, *Int. J. Epidemiol.*, **4**, 197–205.

Magora, A., (1972), Investigation between low back pain and occupation. 3, Physical requirements: sitting, standing and weight lifting, *Ind. Med.*, **41**, 5–9.

Mannicke, C., Hesselsøe, G., Bentzen, L., Christensen, I. and Lundberg, E., (1988), Clinical trial of intensive muscle training for chronic low back pain, *Lancet II*, 1473–6.

Mayer, T., Gatchel, R. and Mayer, H. *et al.*, (1987), A prospective two-year study of functional restoration in industrial low back injury, *JAMA*, **258**, 1763–7.

Miller, J., Haderspeck, K. and Schultz, A., (1983), Posterior elements loads in motion segments, *Spine*, **8**, 331–7.

Nachemson, A., (1976), The lumbar spine. An orthopedic challenge, *Spine*, **1**, 59–71.

Nachemson, A., (1982), Disc pressure measurements, *Spine*, **6**, 93–7.

Sanjevi, R., (1982), A visco-elastic model for the mechanical properties of biological materials, *J. Biomechanics*, **15**, 107–9.

Vanharanta, H., Guyer, R., Ohnmeiss, D., Hochschuler, S., Regan, J. and Rashbaum, R. *et al.*, (1990), The number of painful and degenerated lumbar disc levels in low back pain patients, in *International Society for the Study of the Lumbar Spine*, Boston, Abstracts.

Waddell, G., (1987), A new clinical model for the treatment of low back pain, *Spine*, **12**, 632–44.

Weisman, G., Pope, M. and Johnson, R., (1980), Cyclic loading in knee ligament injuries, *J. Sports Med.*, **8**, 24–30.

Winkel, J. and Oxenburgh, M., (1990), Toward optimizing physical activity in VDT/office work, in (eds Sauter, S., Dainoff, M. and Smith, M.) *Promoting health and productivity in the computerized office*, London: Taylor & Francis.

Yang, K. and King, A., (1984), Mechanism of facet load transmission as a hypothesis of low-back pain, *Spine*, **6**, 557–65.

11

Continuous passive lumbar motion in seating

S. M. Reinecke and R. G. Hazard

Introduction

Sitting is the most common posture in today's work-place, particularly in industry and business. Three-quarters of all workers in industrial countries have sedentary jobs. Office workers make up about 25 per cent of the work-force in Japan and 10–15 per cent of the workforce in Singapore. Today, an estimated 45 per cent of all American workers are employed in offices, and this number is projected to grow significantly through the year 2000 (Cornell, 1988). As organizations become increasingly dependent on information processing and industrial automation, sitting in the work-place will become even more prevalent.

As sitting has increased in the work-place, so has the worldwide incidence of occupational low back pain and related disability. This chapter examines the relationship between these trends and describes a promising intervention: continuous passive lumbar motion.

Low back pain: A modern epidemic

Low back pain (LBP), long recognized as the most common musculoskeletal disorder in industrial countries, is a major cause of disability and absenteeism from work. In the adult population, 60–80 per cent of the American public will at some point suffer from low back pain. Moreover, it is a source of staggering social and economic costs stemming from medical treatments, lost earnings and compensation payments. In the U.S., 11.7 million people, or 5.2 per cent of the population, are impaired and 2.3 per cent are disabled by low back pain. Between 1971 and 1981, the number of people in the U.S. with back or spine disabilities increased by 168 per cent, while the country's population increased by 12.5 per cent. Seventy-nine per cent of those with disabilities are in the age group 17–64; at this age, people are in their most productive, wage-earning years (National Centre for Health Statistics, 1981). Frymoyer *et al.*

(1983) estimated the costs in lost earnings alone to be 11 billion dollars per year for males aged 18–55.

In addition to the financial burden to individuals and society, low back impairment and disability cause severe restrictions to personal productivity and quality of life. In the work-place, LBP has been closely associated with heavy manual lifting, repetitive lifting, lifting with simultaneous twisting and bending, and exposure to vehicular vibration. However, it has also been reported as an occupational hazard for sedentary workers.

Sitting and low back pain

Traditionally, work-places have been arranged so that workers sit through most of the work day. Sitting is less fatiguing than standing and reduces physiological loading (Grandjean, 1980*a*; Konz, 1983). Sitting requires less muscular activity, heart rate, oxygen consumption and hydrostatic blood pressure in the lower extremities than does standing (Grandjean, 1973; Grandjean *et al*., 1982). Sitting also lowers the trunk's centre of gravity and increases the base of support and stability of the upper body (Asatekin, 1975; Carlsoo, 1972). Such stability enhances one's capacity for precision tasks or fine movements (Ayoub, 1972, 1973; Asatekin, 1975; Grandjean, 1980).

However, seated work stations have some disadvantages, among which low back pain is the most commonly reported (Eklundh, 1967; Hult, 1954; Kelsey, 1975; Lawrence, 1977; Magora, 1972). Wood and McLeich (1974) found unexpected intervertebral disc morbidity in insurance and bank workers; their work is characterized by long periods of continuous sitting. Helbig (1978) observed that people frequently maintain a slouched kyphotic lumbar posture when seated, and that this posture, perhaps the most common seated posture, is also probably the least healthy. It is commonly believed that prolonged sitting, in a slouched, kyphotic posture, is closely associated with incidence of LBP (Keegan, 1953; Kottke, 1961; Cyriax, 1975; McKenzie, 1981). A slouched posture stresses the posterior fibrous wall of the disc and posterior ligaments of the spine, and increases intravertebral disc pressure.

Disc pressure

Andersson *et al*. (1974*a*) found that intervertebral disc pressure is greater when seated than when standing. Among seated subjects, disc pressures were highest during forward trunk flexion and lowest when subjects reclined against a backrest.

Disc pressure is also affected by different degrees of lumbar curvature. Disc pressure increases as the lumbar spine moves from lordosis to kyphosis (Andersson *et al*., 1974*b*, 1975). In unsupported seating the pelvis tends to rock posteriorly (Keegan, 1953, 1964; Carlsoo, 1972); this causes the lumbar spine to move into kyphosis, and the centre of gravity of the upper trunk to shift anterior to the lumbar spine. This shift creates a greater moment arm on the spine, which is counteracted by tension in either the erector spinae muscles or posterior ligaments. Such tension creates a greater force on the spine which acts as a fulcrum point between the two forces, thus creating an increase in disc pressure (Lindh, 1980).

During normal lumbar lordosis, the facet joints are in contact between adjoining vertebral bodies, and support a portion of the upper body's weight. Sitting in forward flexion and/or kyphosis disengages the adjacent facet joints and transfers the load anteriorly to the disc. Adams and Hutton (1980) found a 16 per cent increase in compressive forces on the disc under these conditions. Conversely, hyper-lordotic postures may transfer the force from the disc to the facet joints; when prolonged, such extreme compressive forces may initiate low back pain (Adams and Hutton, 1983).

Static postures

Sedentary occupational tasks may require varying degrees of forward flexion, lateral bending and axial rotation. Vehicular and computer operators, assembly workers and others must often assume static postures for long periods of time. Static work postures involving contorted or constricted postures, static muscle loading and long-term joint loading, especially in the extreme range of motion, can cause discomfort. Those who maintain static postures over long periods are more prone to low back pain than those whose jobs allow them to change positions (Hult, 1954; Kelsey, 1975; Kelsey, 1988; Lawrence, 1977; Magora, 1972). Reinecke *et al.* (1985) demonstrated that posture is correlated with back discomfort and that subjects are unable to tolerate unsupported, static, seated postures for long. The authors concluded that individuals should be allowed to change their posture throughout the day.

Motion during sitting

Holm and Nachemson (1983) investigated the effects of various types of spinal motion on metabolic parameters of canine invertebrate discs. Spinal movements over a long period of time change the loads on the spine, thereby providing nourishment. Nutrients transports in and out of centrally located areas of the disc improved with lumbar movement. It was concluded that moderate motion should be sufficient to cause solute and metabolic transport and to stimulate the aerobic metabolic pathway, at least in the most mobile intervertebral disc of the canine spine. The reported changes might also apply to the human lumbar disc, since previous studies have demonstrated nutritional similarities between canine and human discs.

Two mechanisms exist for nourishing the disc: diffusion and pumping. Grandjean (1980*a*) has argued that alternately loading and unloading the spine (through movements) is ergonomically beneficial, because fluid is pumped in and out of the disc by osmosis, thereby improving nutritional supply. Bendix (1987) has investigated the effect of motion on seated comfort and has postulated that discomfort can be measured in terms of the frequency of movement or change in posture. A seated individual might move to a new posture or position for a number of reasons. Blood flow might be cut off, producing a feeling of numbness. Muscle fatigue might also cause discomfort. Bendix has shown that people move when static postures become uncomfortable, providing the chair permits them to move.

The use of continuous passive motion (CPM) and early mobilization for treatment of ligament injuries to the knee has been found to provide such benefits as improved ligament strength and joint stability. Early mobilization produces stronger, stiffer and better vascularized tendons in total or partially immobilized tendons. We hypothesize that CPM to parts of the spine could be used to shorten the period of rehabilitation, speed recovery time and enhance mechanical properties and the healing of soft tissue.

Back support

Many seating studies have focused on physiologic issues, spine shape, muscle fatigue, intra-discal pressure, and determination of optimal seated postures. However, there is limited knowledge of how a chair supports the body in different postures. Seating comfort and body posture are greatly affected by chair contouring, padding, and the mechanical motions of the backrest and seat pan. Chaffin and Andersson (1984) commented that the two most important considerations in seating are adequate back support and allowance for movement or postural change. The means by which a chair supports the body can be determined by the pressure distribution between the body and chair. This information illustrates how the body is supported and what forces the chair imposes on the body. Jurgens (1969) noted that, while pressure distribution is not the sole determinant of seating quality, it should be an integral part of any investigation of human seated support. The optimal chair should allow a worker to maintain a relaxed, but supported, posture and should allow for freedom of motion over the course of the day.

Excessive contouring and padding can be detrimental if motion is restricted. There is a common misconception that sitting is a static and inactive, as opposed to a dynamic, activity (Zacharkow, 1988). An individual needs to move in order to transfer pressures throughout the buttocks over time. Jones (1969) determined that if seating comfort is to be maintained, localized pressure discomfort to the lower back and buttocks should be avoided.

Zacharkow (1988) recommended proper sacral and pelvic support to prevent the posterior rotation of the pelvis and lumbar kyphosis. Keegan (1953, 1962) pointed out that the ideal shape of the spine should involve lumbar lordosis. The most common postural support used to maintain lumbar lordosis in a backrest is a lumbar pad.

Specific recommendations for lumbar supports have focused on the placement of a lumbar pad and the radius of horizontal and vertical supports. Kroemer (1971) recommended pad placement 18–20 cm above the seat, while Schobert (1962) suggested placement at 16–20 cm and Grandjean (1980*a*) at 10–20 cm. Recommendations for the horizontal radius of the support include 31–46 cm (Diffrient *et al.*, 1974) and 40–50 cm (Grandjean, 1980*a*). Diffrient *et al.* (1974) added that the vertical radius of such a support should be approximately 25 cm, and Grandjean (1980*a*) has recommended the use of a slightly convex lumbar pad.

The kind of support provided by the chair can have a direct effect on how individuals change their postures. To study support points and the amount of support afforded by different chairs, Reinecke *et al.* (1987) measured pressure distribution on the seat pan. Six different commercially available office chairs were evaluated in terms of the amount of force between the body and chair surface. Chairs tested were found to provide inadequate support to the back when the subject sat in a forward flexed position, as is typical of a person working at a desk or bench. The chairs appeared to support the lordotic curve of the spine best when individuals maintained an upright erect posture. As subjects reclined, the pressure distribution shifted, and support displaced upward, from the lordotic, lumbar region, toward the scapular region of the back.

In conclusion, seated individuals need adequate back support and the support must accommodate fluctuations in body position. Movement or postural change are inevitable, and in fact desirable, throughout the day. Seating should support a relaxed position yet allow freedom of motion at all times.

Spinal motion during sitting might have some advantages over static sitting. Schobert suggested the possibility of changing postures relative to a relaxed, upright, seated posture in order to minimize muscular activity and the static muscular load needed for sitting. It is generally agreed that seating should promote movement while the body is being supported in different postures.

Continuous passive motion

A device that provides CPM to the lumbar spine during sitting has been developed. This pneumatic system causes continuous motion at the lumbar region by increasing and decreasing pressure against the low back. Several design criteria were established prior to constructing the device. First, it had to promote postural change in the lumbar region. Second, the CPM unit had to induce motion that would follow that of the normal spine. Third, the system had to accommodate any movements or postural shifting, to continue to facilitate motion when the subject changed postures.

The CPM unit includes an inflatable bladder that increases and decreases the amount of lumbar support. Air or fluid is pumped into, and bled out of, the bladder in a cyclic fashion; cycle rates are determined by the users. During a simple cycle, the bladder inflates, causing lumbar lordosis, then deflates to provide a more kyphotic posture.

The pressure within the bladder is controlled by the individual through a variable pressure regulator. If the pressure exceeds the individual's set limit, air is bled out of the support so its volume does not increase further. This allows users to control the amount of lumbar motion received by matching the inflation pressure to the compliance of the spine. At the end of the cycle, a valve turns off the pressure supply from the pump and allows the bladder to bleed off air; this, in turn, deflates the bladder and either reduces the extent of lumbar lordosis or generates kyphosis.

The system is dynamic because, as an individual changes posture (for instance, by leaning forward to answer a phone), the bladder increases in size until it reaches the preset maximum pressure, and thus changes the curvature of the back. When one leans back in the chair, pressure in the bladder increases; this causes the air in the bladder to be bled, deflating it to accommodate the new posture. When the cycle resumes, it will then facilitate motion from that body posture.

Several prototypes were designed and tested to ensure that: (a) maximum pressure in the bladder was sufficient to induce motion (not just to provide soft tissue massage); (b) air flow rates to the bladder were appropriate for optimal inflation and deflation cycle times; (c) noise was adequately controlled; and (d), bladder size would accommodate variations in subjects' height, size and weight.

Evaluations

Testing
The first test involved 20 subjects who used the chair at their work-places for at least one week. The majority of people desired a flow rate to the bladder of 6–8 litres per minute of air and a flow rate out of the bladder of 6–7 litres per minute. The maximum bladder pressure for any individual in this study was 1.5 lbs per square inch. Subjects complained about the noise made by the clicking of the solenoid valves; these were modified in subsequent prototypes. The first office prototype device used a flexible 1/4 in × 36 in rubber tube to connect the pneumatic system to the chair. Individuals experienced difficulties by becoming entangled with the tubing. To alleviate this problem, the tubing has been replaced with a precoiled extension tube.

Auto Testing
In the automobile environment a single bladder design was used. Subjects reported that cornering caused the bladder to shift or swim; that is, one side tended to deflate or squeeze down while the other side increased in size. This gave the driver a feeling of instability. A three-bladder system, in which each bladder is separated from the other two, is currently being developed. This allows for control of the flow rate and pressure into all the bladders, while bladder systems remain independent.

Office Testing
A third study was conducted with 10 subjects who used the device for four consecutive days. During each of the first three days, the subjects used the chair under the following pre-set conditions:

1. a fast cycle of 4–30 seconds;
2. a slow cycle of 60–120 seconds;
3. no cycle (pump shut off).

On the fourth day, the subjects were allowed to select one of the above conditions: fast, slow or no CPM. The order of these sitting conditions was randomized. Overall, people preferred the flow cycle rate, choosing a 67-second average inflation period and a 45-second deflation period. All subjects opted to use CPM on the fourth day. The clear preference for slow CPM suggests that its benefits are derived from spinal postural changes and not from whatever massage effect might result from fast CPM.

Measurement of Lordosis and Pelvic Motion
To determine whether the CPM device was actually changing posture during use, tests were performed to determine lordotic change. Two devices were developed to measure changes in lumbar spinal posture during CPM: a Lordosimeter and a pelvic motion device (PMD). The Lordosimeter measures amount of change in lumbar curvature. The pelvic motion device (PMD) measures pelvic position in a seated posture (the device is described in more detail in Chapter 14). Pelvic tilt is directly correlated to lumbar curvature. To determine whether lumbar curvature changed during CPM use, 10 subjects were asked to sit in a chair fitted with the CPM device for 30 minutes while the Lordosimeter measured the changes in lordotic curve. During this test period, lordotic curve changed by as much as 19° through one cycle of CPM.

Conclusion

A continuous passive motion device has been designed and fabricated for use while seated. The device promotes continuous motion of the lumbar spine in the sagittal plane. Judging from clinical evaluations to date, the device promises to be effective in relieving the effects of static posture during prolonged sitting.

References

Adams, M. A. and Hutton, W. C., (1980), The effect of posture on the role of the apophysial joints in resisting intervertebral compressive forces, *Journal of Bone and Joint Surgery*, **62**-B, 358–62.

Adams, M. A. and Hutton, W. C., (1983), The effect of posture on the fluid content of lumbar intervertebral discs, *Spine*, **8**(6), 665–71.

Andersson, G. B. J., Jonsson, B. and Ortengren, R., (1974*a*), Myo-electric activity in individual lumbar erector spinae muscles in sitting: I. A study with surface and wire electrodes, *Scandinavian Journal of Rehabilitation Medicine, Supplementum*, **3**, 91–108.

Andersson, G. B. J. and Ortengren, R., (1974*b*), Lumbar disc pressure and myo-electric back muscle activity during sitting. II. Studies on an office chair, *Scandinavian Journal of Rehabilitation Medicine*, **6**, 115–21.

Andersson, G. B. J., Ortengren, R., Nachemson, A. L., Elfstrom, G. and Broman, H., (1975), The sitting posture: An electro-myographic and discometric study, *Orthopaedics in North America*, **6**(1), 105–20.

Asatekin, M., (1975), Postural and physiological criteria for seating. A review. *M.E.T.U., Journal of the Faculty of Architecture*, **1**, 55–83.

Ayoub, M. M., (1972), Sitting down on the job (properly), *Industrial Design*, **19**, 42–5.

Ayoub, M. M., (1973), Work place design and posture, *Human Factors*, **15**, 165–8.

Bendix, T., (1987), Adjustment of the seated workplace, Ph.D. Dissertation, Laegeforeningens Forlag.

Carlsoo, S., (1972), *How Man Moves* (London: Heinemann).

Chaffin, D. B. and Andersson, G. B. J., (1984), *Occupational Biomechanics* , (New York: John Wiley & Sons).

Cornell, P., (1988), The Biomechanics of Sitting. Form Number S-065, Steelcase.

Cyriax, J., (1975), *The Slipped Disc*, 2nd edn (Epping: Gower).

Diffrient, N., Tilley, A. R. and Bardagjy, J. C., (1974), *Humanscale 1/2/3* (Cambridge: MIT Press).

Eklundh, M., (1967), Prevalence of musculo-skeletal disorders in office work, *Socialmedicinsk*, **6**, 328–36.

Frymoyer, J. W., Pope, M. H. Clements, J. H., Wilder, D. G., MacPherson, B. and Ashikaga, T., (1983), Risk factors in low back pain, *Journal of Bone and Joint Surgery*, **65**-A, 213–8.

Grandjean, E., (1973), *Ergonomics of the Home*, (London: Taylor & Francis).

Grandjean, E., (1980), Sitting posture of car drivers from the point of view of ergonomics, in *Human Factors in Transport Research*, Vol. 2, (eds D. J. Oborne and J. A. Levis) (London: Academic Press).

Grandjean, E., (1980*a*), *Fitting the Task to the Man*, 3rd edn (London: Taylor & Francis).

Grandjean, E., Hunting, W. and Nishiyama, K., (1982), Preferred VDT work station settings, body posture and physical impairments, *Journal of Human Ergology*, **11**, 45–53.

Helbig, K., (1978), *Sitzdruckverteilung beim ungepolsterten sitz*, Anthropologischer Anzieger, 36.

Holm, S. and Nachemson, A., (1983), Variations in nutrition of the canine intervertebral disc induced by motion, *Spine*, **8**(8), 866–74.

Hult, L., (1954), Cervical, dorsal and lumbar spine syndromes, *Acta Orthopaedica Scandinavica* (Supplement 17).

Jones, J. C., (1969), Methods and results of seating research. in *Proceedings of the Symposium on Sitting Posture*, (ed E. Grandjean) (London: Taylor & Francis, 57–67).

Jurgens, H. W., (1969), Die verteilung des korperdrucks auf sitzflache und reckenlehne als problem der industrieanthropologie. in *Proceedings of the Symposium on Sitting Posture*, (ed E. Grandjean) (London: Taylor & Francis, 84–91).

Keegan, J. J., (1953), Alterations of the lumbar curve related to posture and seating, *Journal of Bone and Joint Surgery*, **35**-A, 589–603.

Keegan, J. J., (1962), Evaluation and improvement of seats, *Industrial Medicine and Surgery*, **31**, 137–48.

Keegan, J., (1964), The Medical Problem of Lumbar Spine Flattening in Automobile Seats. Society of Automotive Engineers, Publication 838A.

Kelsey, J., (1975), An epidemiological study of the relationship between occupations and acute herniated lumbar intervertebral discs, *International Journal of Epidemiology*, **4**, 197–205.

Kelsey, J. L. and Golden, A. L., (1988), Occupational and workplace factors associated with low back pain, *Occupational Medicine*, **3**, 7–16.

Konz, S., (1983), *Work Design. Industrial Ergonomics*, 2nd edn (Columbus: Grid).

Kottke, F. J., (1961), Evaluation and treatment of low back pain due to mechanical causes, *Archives of Physical Medicine and Rehabilitation*, **42**, 426–40.

Kroemer, K. H. E., (1971), Seating in plant and office, *American Industrial Hygiene Association Journal*, **32**, 633–52.

Lawrence, J., (1977), *Rheumatism in Populations*, (London: William Heinemann Medical Books Ltd).

Lindh, M., (1980), Biomechanics of the lumbar spine, in *Basic Biomechanics of the Skeletal System* (eds V. H. Frankel and M. Nordin) (Philadelphia: Lea and Febiger, 255–90).

Magora, A., (1972), Investigation of the relation between low back pain and occupation. 3. Physical requirements: Sitting, standing and weight lifting, *Industrial Medicine*, **41**, 5–9.

McKenzie, R. A., (1981), *The Lumbar Spine: Mechanical Diagnosis and Therapy* (Walkanae: Spinal Publications).

National Center for Health Statistics, (1981), Prevalence of selected impairments. United States, 1977. DHHS Publication (PHS) 81-1562, Series 10, #134 (Hyattsville, MD: DHHS).

Reinecke, S., Bevins, T., Weisman, J., Krag, M. H. and Pope, M. H., (1985), The relationship between seating postures and low back pain. Rehabilitation Engineering Society of North America, 8th Annual Conference, Memphis, Tenn.

Reinecke, S., Weisman, G. and Pope, M., (1987), Effect of seating posture on pressure distribution. Rehabilitation Engineering Society of North America, 10th Annual Conference, San Jose, CA.

Schobert, H., (1962), *Sitzhaltung, Sitzchaden, Sitzmobel* (Berlin: Springer-Verlag).

Wood, P. H. N. and McLeish, C. L., (1974), Statistical appendix: Digest of data on the rheumatic diseases: 5. Morbidity in industry and rheumatism in general practice, *Annals of the Rheumatic Diseases*, **33**, 93–105.

Zacharkow, D., (1988), *Sitting, Standing, Chair Design and Exercise* (Springfield, Illinois: Charles C. Thomas).

12

Effects of body position and centre of gravity on tolerance of seated postures

S. Reinecke, G. Weisman and M. H. Pope

Introduction

Schoberth (1962) describes three unsupported seated postures (middle, anterior, and posterior) in which the body's weight is solely supported by the buttocks and feet. These postures are characterized by the location of the upper trunk's centre of gravity relative to the ischial tuberosities.

In the centred posture, the centre of gravity is located over the ischial tuberosities. If the erector spinae muscles are relaxed in this position, the lumbar spine is straight of slightly kyphotic. If these muscles are active, the pelvis rocks forward and the lumbar spine is lordotic.

In an anterior posture, the upper trunk flexes forward. Weight shifts to the feet and the centre of gravity shifts forward or anterior to the ischial tuberosities. When the erector spinae muscles are active, the spine remains flat with a forward rotation of the pelvis. When the muscles are inactive, the pelvis rocks back and maximizes kyphotic flexion of the spine.

In posterior sitting, the centre of gravity shifts rearward to the ischial tuberosities and the pelvis rotates backward with kyphosis of the lumbar spine. Alternatively, when the muscles are relaxed and the lumbar spine is kyphotic, the spine is supported by posterior ligaments (Akerblom, 1948). This posture is described as slouched. Helbig (1978) observed that this slouched kyphotic sitting posture predominates; further, it is probably the most unhealthy of prolonged postures because it increases spinal flexion and kyphosis. Such lumbar kyphosis has been implicated as a major cause of low back pain (Keegan, 1953; Kottke, 1961; Cyriax, 1975; McKenzie, 1981) because slouching stresses the posterior fibrous wall of the disc and posterior ligaments of the back and increases pressure within the disc.

When sitting forward, most people slouch. Many tasks, such as writing, eating, typing and assembly work, require a forward-flexed posture. This posture is maintained by gravitational forces of the upper trunk. Kellogg (1927) noted that the moment relaxation occurs, gravitation takes control and the body is pulled into whatever position the supporting structure may cause it to assume.

The spine works as the fulcrum point for the upper body, and is supported by ligaments, muscles, and the skeletal system. As the upper trunk moves into forward flexion, back muscles initially counteract the forward shift of mass anteriorly to the spine. Subsequently, the posterior ligaments begin to counteract the force and the muscles turn off. At this point, the spine is hanging on the ligaments.

In a previous study (Reinecke *et al.*, 1985), tolerances were determined for various postures representing different combinations of forward flexion (FF), lateral bending (LB) and axial rotation (AR). The outcome measures were tolerance time, centre of gravity (CG) displacement, pain location and intensity. Subjects used a visual analogue scale to indicate pain intensity and a pain drawing task to indicate pain location. Results showed that the length of time that a posture can be sustained decreases as centre of gravity shifts away from that of the neutral upright posture. It was not evident, however, whether onset of pain resulted from the alteration of posture, which might reduce muscular efficiency, or from the change in CG, which might require the muscles to work harder.

The purpose of the present study is to determine the extent to which posture and CG displacement affect seating tolerance in both the middle and anterior position.

Methods

Subjects:

Ten subjects (5 male, 5 female) with no previous history of low back pain participated in the study. The subjects' ages ranged from 19–34 years (mean 24). Among the men, heights ranged from 170.1–183 cm (mean 180.8 cm) and weights ranged from 74 to 81 kg (mean 79.6 kg). Among the women, heights ranged from 157.5–172.7 cm (mean 166.1) and weights ranged from 52.2–68 kg (mean 58.6).

Procedure:

Subjects were tested in two postures, 0° forward flexion (middle position, 0° FF, 0° CG) and 30° forward flexion (anterior position, 30° FF, 30° CG). Subjects were also tested in two postures with artificially altered centre of gravity (Figure 12.1). At 0° forward flexion, the subjects' CG was altered to simulate that of a posture with 30° forward flexion (0° FF, 30° CG). At 30° forward flexion, the subjects' CG was altered to simulate an upright posture (i.e., 0° FF, 30° CG). All subjects were tested twice in each of the four conditions, representing a total of eight test sessions each. Subjects were tested in only one posture per day and were asked not to engage in any strenuous activity for two hours before each test session.

Testing was conducted on an office chair, from which the backrest had been removed. The chair was bolted to a force plate large enough to accommodate the chair and the subjects' feet, to determine centre of force in the vertical

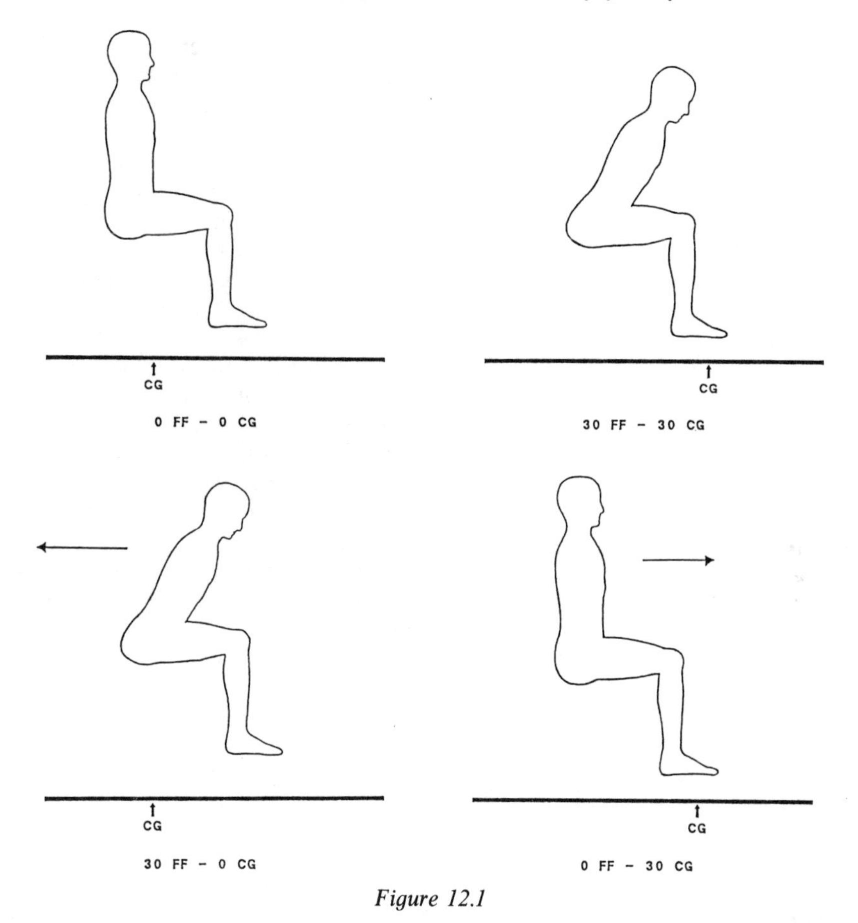

Figure 12.1

plane. The seat pan was adjusted so that each subject's lower leg was oriented at 90° to the thigh.

To produce an artificial shift in CG of the upper trunk, each subject wore a wind-surfers' chest harness during all test conditions. This harness was modified by adding hooks to both the front and back of the harness. The test apparatus was positioned in the middle of a room. On opposite sides of the room, directly in front of and behind the subject, were pulleys adjustable to the same height as the harness.

To simulate a shift in CG, a rope was attached to the harness and then passed though the pulley; this allowed the rope to hang vertically and the weights to be attached and hang freely. Weights were added until the desired shift in CG was achieved (Figure 12.2). CG shift was determined by monitoring the force plate until the proper shift was achieved for the given test position.

The degree to which the trunk deviated from the erect sitting posture was measured by the force plate. Non-restraining tactile feedback clamps helped subjects maintain the proper position. These clamps, which simply encompass

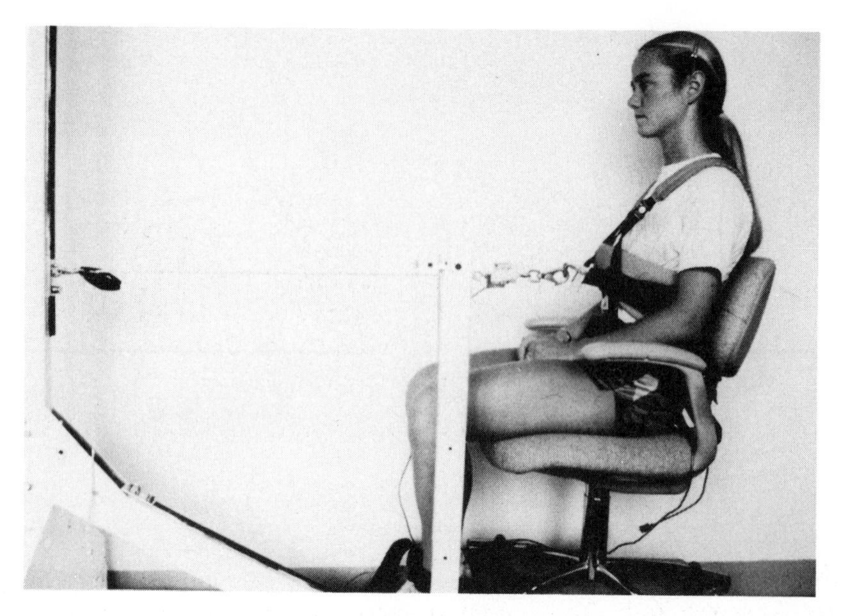

Figure 12.2

the shoulder, were first set to a predetermined position that allowed subjects to be positioned within 30 seconds. A calibration jig that mimicked the anatomical dimensions of the subject made it possible to preset the clamps without using the subject as a model.

Test sessions:
Subjects were instructed to sit in each test position for as long as they could tolerate that given posture. Excruciating pain was described as a point beyond discomfort, at which one can no longer tolerate the static position. Test order was selected randomly. Subjects were tested only in one test position per day. Testing for each subject was performed at the same time of day to minimize time of day effects, such as fatigue.

A pain drawing task, in which subjects marked location of discomfort on an outline of a human form, was administered throughout each test session at five-minute intervals until the termination of the test.

During the test sessions, subjects watched a film to minimize any possible effects of boredom on the perception of pain. It was also recognized that subjects might prolong a test session in order to see more of the film. They were therefore informed at the outset that it would be continued from one session to the next.

Results

The neutral, upright posture (0° FF with 0° CG displacement) was tolerated longer than any of the other three postures ($p < 0.01$), and the flexed position with a naturally occurring CF shift was the least tolerated. Figure 12.3 shows

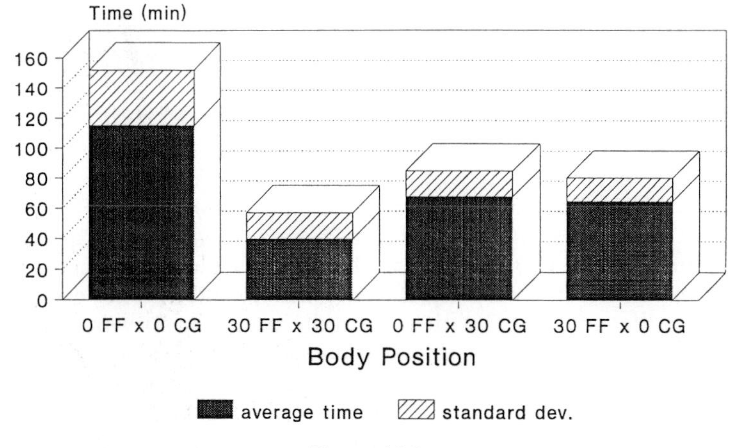

Figure 12.3

the means and standard deviations of tolerance time for each of the four postures. As shown in the figure, mean tolerance times ranged from 39.8 minutes (for 30° FF with 30° CG) to 115 minutes (for 0° FF with 0° CG). A repeated measures ANOVA showed that the postures differed in terms of the length of time they were maintained (F = 16.19, $p < 0.001$). Duncan's multiple range test showed that the two positions involving artificial CG displacement (30° FF with 0° CG and 0° FF with 30° CG) did not differ significantly from one another in tolerance time (Figure 12.4), although both differed from the two natural positions (upright and natural CG displacement).

In the pain drawing task, subjects most often indicated the low back as the location of discomfort experienced during the test sessions. The average rate of low back complaints was 4 per cent; average rates of middle and upper back complaints were 34 per cent and 20 per cent, respectively.

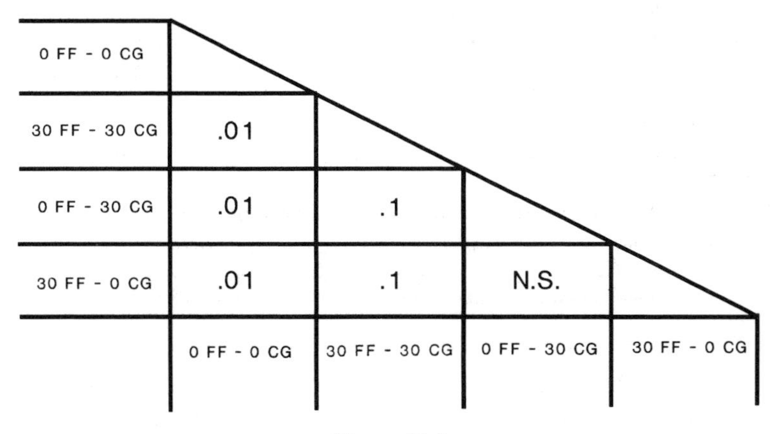

Figure 12.4

Tolerance times for the four postures and number of pain responses related to the low back varied considerably among subjects, as demonstrated by the high standard deviations.

Discussion

The results of this study indicate that a seated posture with 0° FF and 0° CG is the most tolerable sitting position. Positions involving either a shift in CG (while maintaining 0° FF) or a shift in the flexion angle (while maintaining the same centre of gravity) were significantly less tolerable. The least tolerable posture was that of 30° FF with a naturally occurring forward shift of CG, that is, an anterior position or slouched posture. These findings suggest that a forward shift in centre of gravity, which requires support from the back muscles, results in a reduction of tolerance time. Changing one's posture while maintaining a constant CG also results in decreased tolerance time. The greatest effect on tolerance time resulted from the combination of forward shift of CG and forward flexion. Thus, altering CG and/or posture may adversely affect sitting posture tolerance time. However, there was no apparent difference between the ideal posture (0° FF) with the altered CG (30° CG) and the ideal CG (0° GG) with the altered forward flexion (30° FF). This suggests that both posture and CG shift have substantial and equal effects on sitting tolerance.

Subjects reported pain in the low and middle back for much of the testing time, suggesting that posture tolerance time may be a function of the muscles. Biering-Sorensen's (1984) prospective study on low back pain in men revealed that the two most important risk factors for first time occurrence of low back trouble were poor isometric endurance of the back muscles and hypermobility of the lumbar spine in flexion. He suggested that good back muscle endurance should, to some extent, protect one against low back trouble. He further pointed out that back muscles, since they maintain the erect posture of the spine throughout the day, must require a certain isometric endurance – a degree of endurance probably also essential to many manual handling tasks.

Chaffin (1984) and others have shown that the further from the body a weight is lifted, the more force is generated at the discs and the more tension occurs in the lumbar muscles. A similar phenomenon is suggested by the present findings: the more poorly tolerated postures were those characterized by greater shifts in CG from the upright position.

Conclusions

1. Sitting upright (0° forward flexion, i.e., the middle posture) was tolerated better than postures that combined varying degrees of forward flexion and forward shifts of centre of gravity.
2. An anterior sitting posture at 30° forward flexion with a forward shift of CG were tolerated least.
3. Changing either or both the CG and posture affected sitting posture tolerance adversely.
4. Posture and CG shift affected sitting tolerance equally.
5. Complaints of low back pain were more prevalent than either middle or upper back pain.

Acknowledgments

This research was funded by a grant from the National Institute of Disability and Rehabilitation Research. The authors acknowledge the editorial assistance of Antonia Clark, Vermont Rehabilitation Engineering Center.

References

Akerblom, B., (1948), *Standing and Sitting Posture*, Stockholm, Nordiska Bokhandeln.

Biering-Sorensen, F., (1984), Physical measurements as risk indicators for low back trouble over a one-year period, *Spine*, **9**, 106–19.

Chaffin, D. B. and Andersson, G. B. J., (1984), *Occupational Biomechanics*, New York, John Wiley & Sons.

Cyriax, J., (1975), *The Slipped Disc*, 2nd edn, Epping, Gower.

Helbig, K., (1978), Sitzdruckverteilung beim ungepolsterten sitz, *Anthropologischer Anzieger*, **36**, 194–202.

Keegan, J. J., (1953), Alterations of the lumbar curve related to posture and seating, *Journal of Bone and Joint Surgery*, **35**-A, 589–603.

Kellogg, J. H., (1927), Observations on the relations of posture to health and a new method of studying posture and development, *The Bulletin of the Battle Creek Sanitorium and Hospital Clinic*, **22**, 193–216.

Kottke, F. J., (1961), Evaluation and treatment of low back pain due to mechanical causes, *Archives of Physical Medicine and Rehabilitation*, **42**, 426–40.

Mandal, A. C., (1984), The correct height of school furniture, *Human Factors*, **24**, 257–69.

McKenzie, R. A., (1981), *The Lumbar Spine. Mechanical Diagnosis and Therapy*, Waikanae, Spinal Publications.

Reinecke, S., Bevins, T., Weisman, G., Krag, M. and Pope, M., (1985), Relationship between sitting postures and low back pain, *Proceedings of the 8th Annual RESNA Conference*, Memphis, Tennessee.

Schoberth, H., (1962), *Sitzhaltung, Sitzchaden, Sitzmobel*, Berlin, Springer-Verlag.

13

Influence of furniture height on posture and back pain

A. C. Mandal

Introduction

Much of our knowledge about the seated posture is unfortunately based on short-time laboratory experiments under conditions which bear little resemblance to real work. Subjects are often seated with a horizontal axis of vision and saddled with needles, wires etc., and are under more or less constant supervision or manipulation by the investigators. Recommendations based on this type of experiment are of little value. Further, the investigators have seldom bothered to check whether their recommendations can be applied to real life.

The traditional office

The present research was based on an analysis of the flexion of the lumbar region during daily work in a traditional office by means of time lapse photography. This was accomplished by stitching the end of a 13 cm long nylon ruler to a pair of medium tight jeans (at positions corresponding to the hip joint and 4th lumbar disc) and by marking the shoulder joint and knee joint with tape. In the reading/writing posture, the lumbar flexion decreased 17.1° on average, when the table height was increased from 72 cm, as recommended by CEN to 87 cm ($p < 0.005$).

A Sedeomatic tilting chair was used; its height was increased accordingly and the seat and table top were sloping against one another. Ten secretaries were investigated during 3 × 15 minutes. Photographs were taken every two minutes (Figure 13.1A and 13.1B).

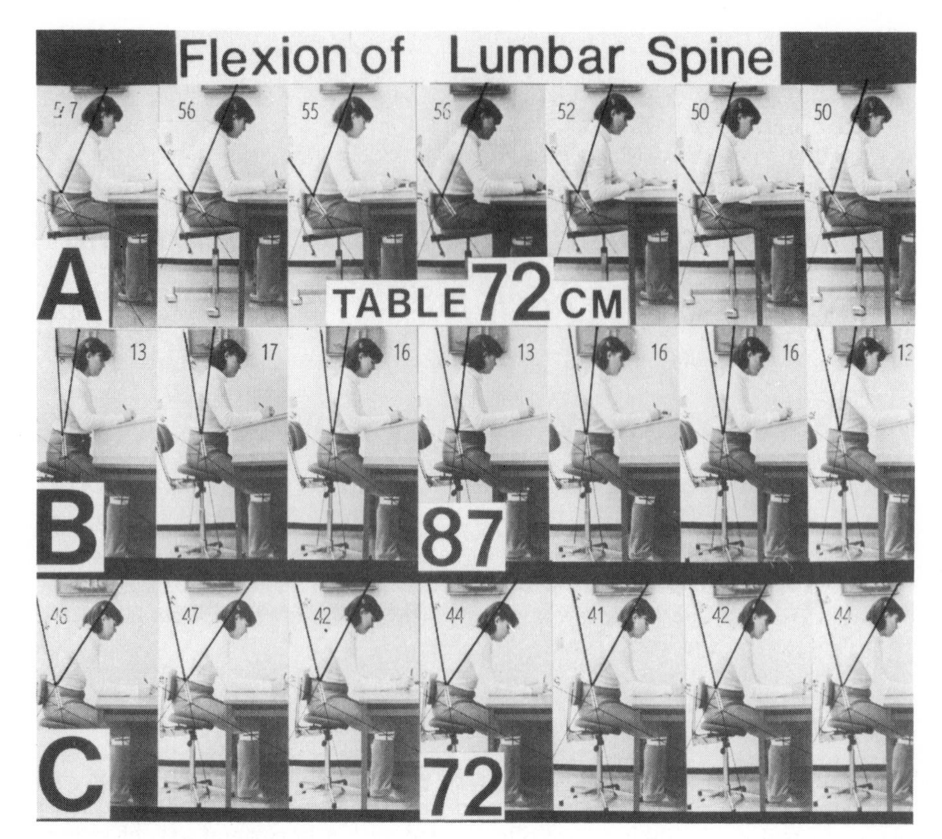

Figure 13.1 Registration of lumbar flexion in office work by means of marks on the clothes

Earlier experiments have shown that the majority of subjects prefer this height of furniture for reading/writing (Mandal, 1982). Nine out of the 10 secretaries investigated found the higher position considerably more comfortable.

Over the last years there have been attempts in Scandinavia to use the tilting chair in a different way. T. Bendix (1983) used a distance of 21 cm between the table top and chair and positioned the person on the back of the seat. The seat consequently tilted to an almost horizontal position similar to that of traditional office seats. Table height 72 cm.

When the tilting chair was used in the TB-way (Figure 13.1C), an average decrease in the lumbar flexion of only 1.4° compared to the CEN chair (-N.S.) and an increase of 15.7° (!) compared to the tilting chair with the higher table ($p < 0.001$).

Bendix found in his experiment a reduction of only 0.8° in the preferred position when changing from the CEN standard to the tilting chair. He also used the Sedeomatic chair, which was constructed in 1975 by the author. The mean comfort rating of the T.B.-position was 1.0 (1 = miserable).

The subjects gave the tilting chair at a high table a comfort rating of 4.6 (5 = excellent). The comfort rating of the CEN height was 2.0.

The electronic office

In the modern electronic office, the importance of height adjustment increases as many workers will have to assume constrained postures for many hours with highly repetitive work. The traditional recommendations include low furniture and "effective" lumbar support, but results have been disappointing.

Grandjean (1984) found that conventional terminal work resulted in about 5 times more complaints (and data entry work in about 13 times as many complaints) from the neck and shoulders as traditional office work. The number of back and neck sufferers is apparently increasing rapidly. These figures are unacceptable and indicate that hours spent with data entry work must be reduced unless working conditions can be improved.

Experiment with adjustment of furniture height

In a large bank in Copenhagen, many data entry personnel complained of severe neck, shoulder and back problems. In a monthly magazine for the employees I offered a correction of postures to those suffering from chronic back pain (daily). The 13 who wanted to participate ranged in height from 158 to 175 cm and in age from 24 to 54.

On the first day these subjects were positioned on a hydraulic chair with a tilting seat. While they were sitting and keying-in, the height of the chair and table were altered to the height at which they felt pain was minimized.

They all wanted to sit higher than before and most of them were sitting with a more upright position and with a straight back. The height of the chair and desk were measured and their daily work-place was adjusted to this height. They were told only to sit in the higher position for 10–20 minutes during the first days, as they would get tired.

Although the furniture in the work-place was easily adjustable, most had previously been used at standard height. Most of the participants had used tilting chairs for several years.

After 4–5 weeks most of them preferred to sit in the higher position for most of the day.

After a 2 month period, during which they could easily change the height of the furniture, I measured the heights the 13 participants had permanently been using. The average preferred desk height was 71.6 cm compared to CEN (European Standardisation Organisation) recommendations of 65 cm. Preferred chair height was 54.3 cm (CEN: 42–50 cm). Only 2 of the shorter subjects, both 160 cm tall, preferred to use the CEN standard desk height of 65 cm. These two are not included in the following investigation, as comparison of posture was not possible.

The postures of the remaining 11 were registered by means of an automatic camera. The participants were asked to wear semi-tight sweaters and jeans and a 13 cm long white nylon ruler was stitched to the jeans with one end over the hip joint (trochanter major) and the other end over the 4th lumbar disc (the point mid-way between the anterior and posterior iliac spine). Besides that, the 7th cervical vertebra (vertebra prominens) and the knee joint were marked on the skin and on the clothes respectively.

First, the subjects were seated at their preferred height while doing normal data entry work for 15 minutes and photographs were taken by automatic camera at a rate of one per minute. Immediately afterwards, the furniture height was reduced to CEN standard i.e., desk height 65 cm and chair height 50 cm. The seat was locked in a 5° backward sloping angle and the lumbar support adjusted.

This position was also maintained for 15 minutes and another 15 photographs were taken.

During the investigation the partipants were left alone working in a shielded corner of a large office. Lines were drawn on the photographs between the marks on the clothes and the angles measured.

Results

In the higher, preferred sitting position the average flexion of the back (lumbar plus hip joint) was 11° less than in the recommended CEN position (Figure 13.2, $p < 0.01$). Pain indication (with a visual analogue scale, Huskisson, 1974) was 35 mm in the higher position compared to 67 mm in the recommended CEN standard position (Figure 13.3, $p < 0.001$).

Seven out of 11 users were able to sit with a permanently straight back or a lumbar lordosis in an upright balanced position without using the lumbar support (Figure 13.4). The position is similar to the 'horse-riding' position. This posture reduces the strain on the back and allows abdominal respiration. When users wanted to rest, they lowered the seat and leaned back against the lumbar support. This posture is normally considered a good working position.

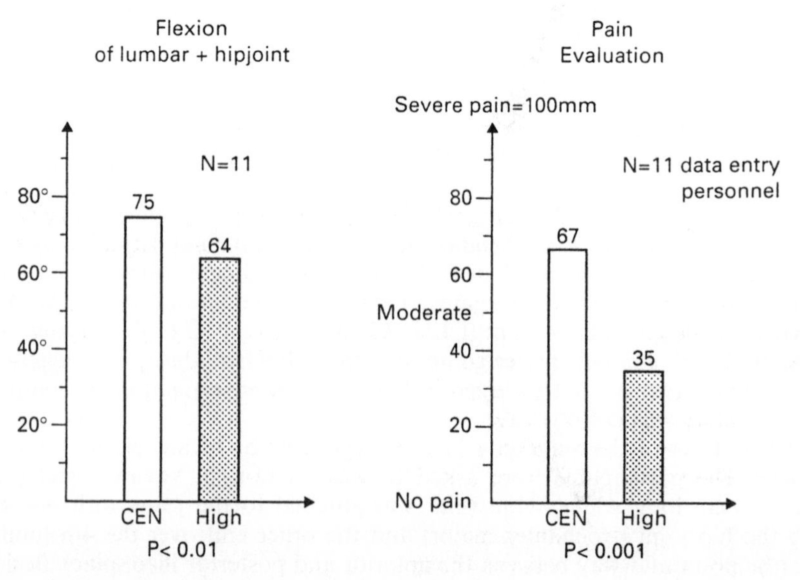

Figure 13.2 and 13.3 The CEN standard furniture resulted in increased flexion of the lumbar region and higher pain ratings

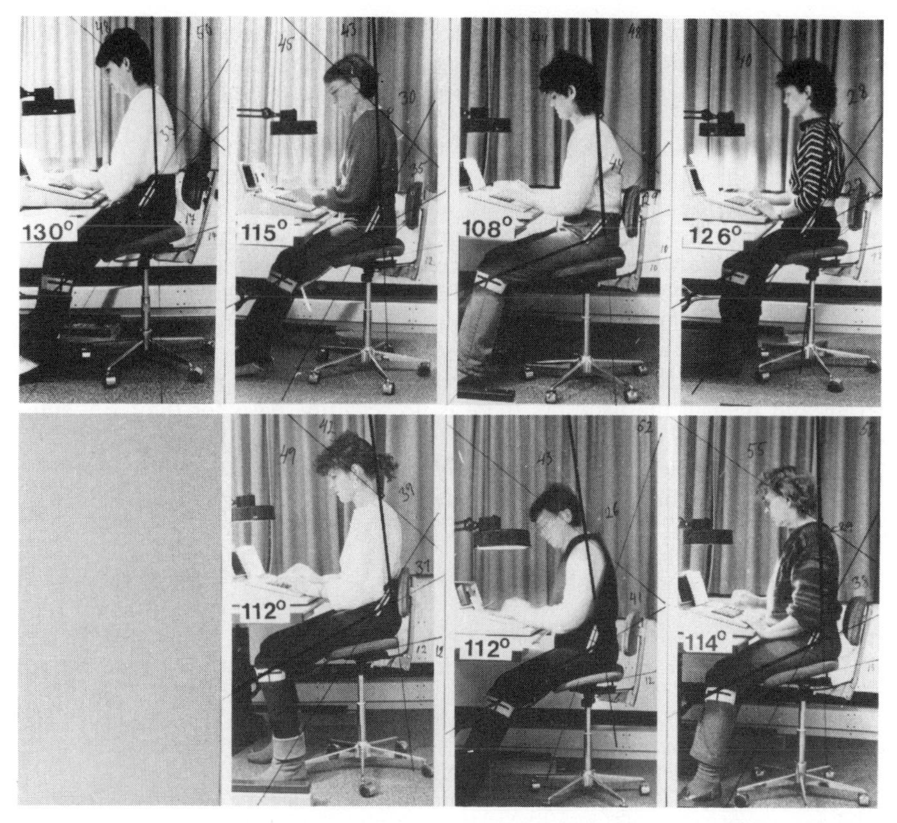

Figure 13.4 Seven out of 11 preferred to sit permanently with a straight back in an upright balanced position without using the lumbar support. Average Trunk/Thigh angle 117°

The four people using lumbar support sat further back on the seat and with more pronounced flexion of the lumbar region than the rest. Such a position naturally rotates the seat pan and the pelvis to a more horizontal position; correspondingly, the lumbar flexion increases almost to the same extent as on traditional chairs. These four had all been instructed at the end of an 8-hour working day; this proved to be a bad time as they were tired and more interested in getting home. For the latter, the seat sloped an average of 5.7° forward, compared to a 12.7° forward slope with the seven more upright persons. A forward slope of at least 10° appears to be necessary to achieve a lumbar lordosis.

Three out of 11 users had previous problems with swelling of the legs – these all reported improvement.

Discussion and conclusions

Earlier research indicates that most people prefer sitting 10–20 cm higher when reading and writing than recommended by the standards (Mandal,

1982). In the higher position, they felt considerably less back pain. The natural consequence of this was to invite the back-sufferers to adjust the furniture to exactly the height where pain was least pronounced. For traditional office workers, a 15 cm higher desk combined with a higher forward sloping chair was found much more comfortable and reduced lumbar flexion by 17.1°.

For data entry personnel a 6.6 cm higher desk than recommended by CEN standards was found to be optimum after a two month trial period. In the higher position, seven out of 11 users were able to sit with a straight back in a balanced position without using the lumbar support. The ratings of pain from 67 mm (visual analogue scale) on CEN furniture was reduced to 35 mm at the preferred height. Three users had previously been suffering from swelling of the lower legs; they all reported improvement.

The present CEN standards will apparently result in unnecessary flexion, strain and pain of the back. As there is no scientific basis for the CEN standards, they adversely affect the health of millions of people.

In the future preferably we should rely more on the wishes of the consumers, as they often have to sit for many hours with back pain every day.

References

Bendix, T., (1983), *Scandinavian Journal of Rehabilitation Medicine*, **15**, 197–203.

CEN (Comitée Européen de Normalisation) 1982: Pr En, (Afnor, Tour Europe CEDEX 7, 92080 Paris, La Defense).

Grandjean, E., (1984), *Ergonomics and Health in Modern Offices.* (London and Philadelphia: Taylor & Francis), p. 447.

Huskisson, E. C., (1974), Measurement of pain, *The Lancet*, 1127.

Mandal, A. C., (1982), Correct height of school furniture, *Human Factors*, **24**, 257–69.

PART VI

Biomechanics

14

Sitting (or standing?) at the computer workplace

K. H. E. Kroemer

Introduction

At the beginning of this century (when clerks were all male) it was common to stand while working in the office. Since then it has been commonly accepted that one should sit while working in the office. The reasons for this change in attitude are not clear since *homo erectus* is biomechanically designed for moving around, but not standing still or sitting still. Grieco (1986) believes that the human spinal column is suited for four-legged motion and has not adapted to upright bipedal locomotion. He also fears that the current rapid transformation to *homo sedens* promotes disorders of the spine. Indeed, musculo-skeletal pain and discomfort (together with eye strain, and fatigue) constitute at least half (in some surveys up to 80 per cent), of all subjective complaints and objective symptoms of computer operators in both Europe and North America (Helander, 1988).

Theories of healthy sitting

In the 19th century, Staffel (1984) and other physiologists/orthopaedists provided recommendations for seat and furniture design that were based on the assumption that sitting with an upright trunk means sitting healthily.* This idea has remained virtually unchallenged until recently, when various (sometimes contradictory) theories led to rather radical proposals for sitting

* For reviews, see for example Akerblom (1948), Bradford and Prete (1978), Grandjean (1969), Grieco (1986), Keegan (1952), Kember (1985), Kroemer and Robinette (1968), Kroemer (1988, 1991), Mandal (1982), Schoberth (1962), and especially Zacharkow (1988).

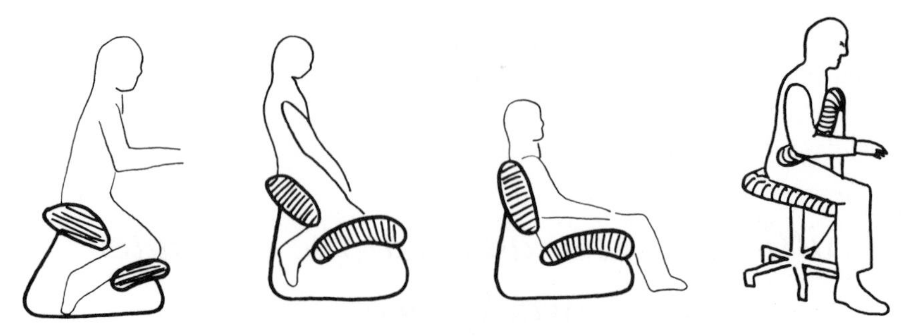

Figure 14.1 Examples of unconventional seats proposed in the 1980s

and seats at work. These include semi-sitting on high forward-tilted supports with or without knee pads; saddle-type supports; and use of a belly rest instead of a backrest – see Figure 14.1.

Sitting upright

During the 19th century, it was commonly presumed that the spinal column of the sitting person should be erect or upright (a contradiction in terms because it is actually curved) and resemble the healthy normal upright standing posture. Special emphasis was placed on maintaining a normal lumbar lordosis; this posture was believed to minimize strain on the spinal column and its supportive structures, including the musculature. This posture was also considered socially proper and used to develop recommendations for the design and use of seats (Grimsrud, 1990). These ideas were generally accepted without question until the middle of the 20th century. In retrospect, this is quite surprising given the lack of experimental support for the appropriateness of the upright posture.

Pressure in the intervertebral discs

Innovative and courageous measurements of pressure in the inter-vertebral discs were performed in Scandinavia at about the middle of the 20th century, as reviewed by Chaffin and Andersson (1984), Grandjean (1969), and Grieco (1986). EMG measurements indicate most muscle activity is exerted by five major muscle pairs of the trunk; these are erector spinae, latissimus dorsi, internal and external obliques, and rectus abdominus. They exert pull forces along the length of the trunk, stabilize the spinal column, and are influenced by variations in postures, by external loads, and by seat features (Nag, Chintharia, Saiyed and Nag, 1986).

Proponents of the upright posture used the experimental findings to emphasize the presumed benefits. However, careful examination of the published data indicates that there are really no pressure benefits associated with sitting erect as opposed to sitting slumped. In fact, leaning on a backrest can reduce disc pressure significantly below the values found while sitting erect without support (see Figures 9.14 and 9.16 in Chaffin and Andersson, 1984). Since it is difficult to measure disc pressures directly, the spinal compression recently has been calculated from biomechanical models (Kroemer, Snook, Meadows and

Deutsch, 1987) which, unfortunately, usually ignore shear, bending and twisting loads.

Trunk-leg interactions and seat pan position

The biomechanical interactions among lower spinal column, pelvic girdle, and thigh and knee angles were also investigated in the middle of the 20th centry. Of particular interest were the angular positions (in the sagittal view) of the pelvic girdle, which is elastically connected to the sacrum and the lumbar spine. Thus, forward rotations of the pelvic girdle about the ischial tuberosities (resting on the seat pan) increases the lumbar lordosis if the trunk is erect. Conversely, rotating the pelvic girdle backward about the ischial tuberosities flattens out the lumbar curvature and may induce a kyphotic condition, see Figure 14.2. This reversal of lumbar lordosis takes place when associated muscles relax; muscle activity or changes in trunk tilt can counter these effects.

The pelvis also rotates forward and lumbar lordosis increases when the knees drop below the height of the tuberosities. However, muscles that span

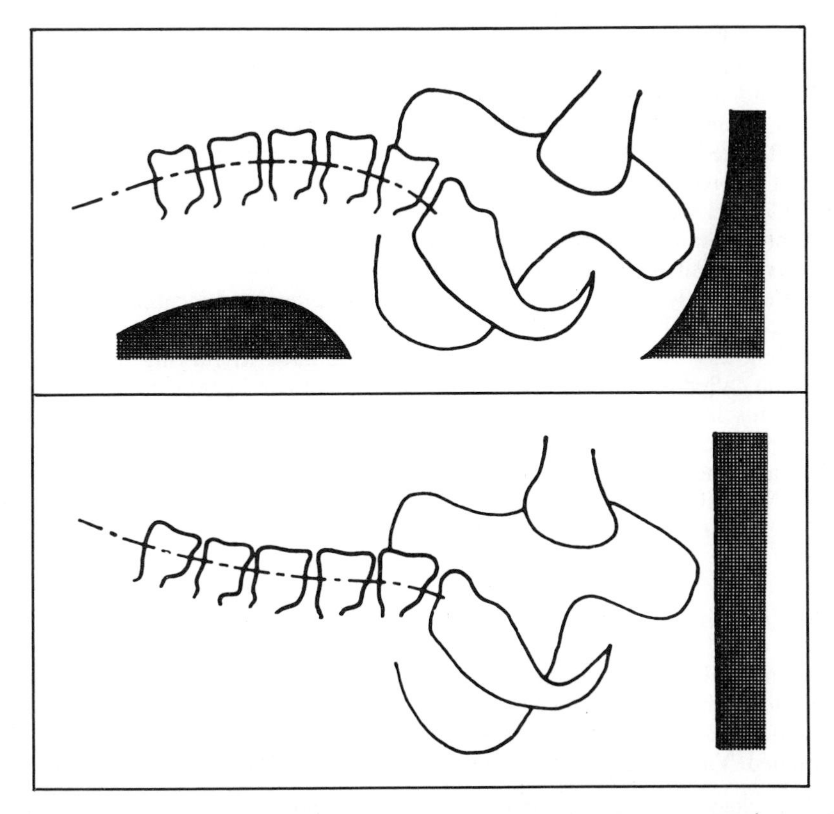

Figure 14.2 Lumbar kyphosis generated by body weight and muscle tension on a flat horizontal seat; lordosis provoked by a pad in the lumbar region or by tilting the pelvic girdle on a forward-downward declined seat which also – if the knees can be lowered below seat height – opens the trunk-hip angle beyond 90°

the hip joint may affect pelvis and spine positions: the effect of opening the hip angle on the lumbar spine depends on whether muscle tension is present (Bridger, Wilkinson and Van Houweninge, 1989; Keegan, 1952).

Such anatomic considerations gave rise to recommendations that, to facilitate lumbar lordosis, the seat pan tilt down at the front, or tilt up at the rear. The first idea was already expressed by Staffel in the 19th century, the second was commercially used in the 1950's known as Schneider Wedge (Kroemer and Robinette, 1968) and recently promoted again by Mandal (1975, 1982), by Congleton *et al.* (1985) and for various semi-seats (Chaffin and Andersson, 1984). The traditional opposing suggestion is that the seat pan decline rearwards to encourage leaning against the backrest. If this backrest is shaped appropriately, then the desired curvature of the spine, especially lumbar lordosis, is promoted.

Back support

Schools of thought regarding whether and how to support the back through a backrest have changed dramatically throughout the last 100 years (Corlett and Eklund, 1984). It is recognized now that, if one sits upright, a backrest provides little or no benefit unless one presses the back against it (which requires muscular effort) to promote the desired spinal configuration.

Around the middle of this century, it was widely presumed that a backrest should conform to and support the lumbar curvature; this furniture feature was often referred to as the Akerblom pad. For several decades it was considered appropriate and sufficient to support this lumbar region in isolation, as evidenced by the low lumbar support that was attached to the so-called secretarial chair. This design did not support the upper back; thus, a taller backrest for full back support is now preferred.

More recently, two contradictory ideas have been advocated. One is that the muscles of the trunk should be continuously stimulated (possibly by not providing any backrest) to move and hold the back in varying postures; this has been called dynamic sitting. While movement can benefit the user by exercising muscles, the associated longitudinal muscle forces in the trunk also generate spinal compression. A tall backrest that reclines allows the users to lean backward and to let the backrest carry some of the weight of the upper body. This diminishes trunk muscle activities and also reduces the compressive loading of the spinal column.

Evaluation of 'suitable' postures at work

Experimental studies treat body postures as independent variables. If all other conditions and variables are controlled (such as work task, environment, etc.), dependent variables associated with postures can be observed, measured and evaluated. Various dependent variables have been used in different disciplines.

- Physiology – oxygen consumption, heart rate, blood pressure, electro-myogram, fluid collection in lower extremities, etc.
- Medicine – acute or chronic disorders, including cumulative trauma injuries
- Anatomy/biomechanics – X-rays, CAT scans, changes in stature, disc and intra-abdominal pressure, model calculations

- Psychophysics – interviews and subjective ratings
- Engineering – observations and recordings of posture; forces/pressures on seat, backrest, or floor; amplitudes of body displacements; productivity

Some of these techniques have become standard procedures (Corlett, 1989; Kroemer, 1991). Yet, their appropriateness, or the interpretation of their results, have recently been questioned. For example, widely used data on disc pressure either calculated or measured, were questioned by Jaeger (1987). Adams and Hutton (1985) noted that the contributions of the facet joints (apophyseal joints) to load bearing of the spinal column have been largely neglected. After Boudrifa and Davies (1984) investigated the relationships between intra-abdominal pressure and back support, McGill and Norman (1987) as well as Marras and Reilly (1988) researched the relationship between intra-abdominal pressure and spinal compression: they found that abdominal pressure may not relieve the spinal column as had been assumed. (Although these studies were done on lifting, they are also relevant to office postures.)

During the last decade, electro-myography has been used extensively to assess trunk muscle activity levels during sitting. Most EMG techniques assume a static (isometric) loading of the observed muscles, such as to establish the reference amplitude at a maximum voluntary contraction. Yet, for dynamic conditions, where muscle lengths change, the recording and interpretation of EMG data is difficult and often tentative (Basmaijan and DeLuca, 1985). Furthermore, the relevance of small changes in electro-myographic events is subject to question. For example, in upright sitting, the observed electro-myographic signals indicate that longitudinal trunk muscles use only 10 per cent or less of their capacity. Yet, even small changes of these low magnitudes of activity, caused by variations in posture or seats, have been considered significant.

Observations of body motions (voluntary and unconscious) while sitting are difficult to interpret: high levels of movement may result from discomfort or from a chair that promotes postural change; sitting still may be enforced by a confining chair design or suggest that the user is comfortable.

Many studies use subjective ratings (by the subject or the experimenter) to evaluate seated posture (Bendix *et al.*, 1985; Bridger, 1988; Drury and Francher, 1985; Gale *et al.*, 1988; Helander *et al.*, 1987; Yu *et al.*, 1988, 1989). While the procedures vary widely among researchers, most rely on the initial work by Shackel *et al.* (1969) and by Corlett and Bishop (1976). Standardized discomfort or pain questionnaires have been developed, in Italy by Occhipinti *et al.* (1985); and in Scandinavia by Kuorinka *et al.* (1987).

Table 14.1 lists observation and recording techniques and contains this author's 'subjective assessment' of their status and usefulness. Frequently, threshold values, separating suitable from unsuitable conditions, are unknown or unclear. Thus, it is difficult or impossible to interpret the results obtained by many of these techniques.

The current status

The review of analytical procedures shown in Table 14.1 suggests that the currently most useful techniques are based on subjective assessments by seated users. These ratings presumably encompass the phenomena addressed in

Table 14.1 Current techniques for posture assessment and their interpretation (adapted from Kroemer, 1991; with permission by Elsevier, 1991)

Observation/techniques	Measurement procedures	Assessment criteria	Threshold values	Relevant
Oxygen consumption	E	E	V	probably not
Heart rate	E	E	V	perhaps
Blood pressure	E	E	V	yes
Blood flow	E-D	E-D	V	yes
Innervation	D-V	D-V	U	yes
Leg/foot volume	E	V	U	perhaps
Temperature, skin or internal	E-D	U	?	yes
Muscle tension	E-D	V	U	yes
Electro-myography	D	D	V	yes
Joint diseases	E-D	D	V	yes
Musculo-skeletal disorders	D	D	V	yes
Cumulative trauma disorder	D	D	V	perhaps
Spinal phenomena:				
disc pressure	E-V	D	V	yes
disc disorders	D-V	D	V	yes
disc shrinkage	D	D	U	yes
facet disorders	D-V	V	U	yes
alignment of vertebrae	D-V	V	V	yes
spine curvature	D-V	V	U	yes
mechanical stresses, including model calculations	D-V	V-U	V	yes
Intra-abdominal pressure	E-D	V	U	perhaps
Surface (skin) pressure at				
buttocks	E-D	V	U	yes
back	E-D	V	U	yes
thighs	E-D	V	U	yes
Upper extremity posture	E-D	V	U	yes
Posture				
of head/neck, trunk, legs	E-D	V	?	yes
changes in posture	E	?	?	yes
change in stature	E	E	V	perhaps
Sensations (ratings) of				
ailments	E-D	V	D	yes
pain	E-D	V	D	yes
discomfort	E-D	V	D	yes
comfort, pleasure	E-D	D	D	yes

Legend: well established: E; being developed: D; variable or unknown: V, U; questionable: ?

physiological, biomechanical and engineering measurements, and they appear most readily subject to scaling and interpretation.

Currently, no single theory about the proper, healthy, comfortable, efficient, etc., sitting posture at work prevails (Lueder, 1983). The traditional assumption is no longer accepted that everybody should sit upright (and that seats should be designed for this posture). Instead, it is now recognized that many postures may be comfortable (healthy, suitable, efficient, etc.) depending on the

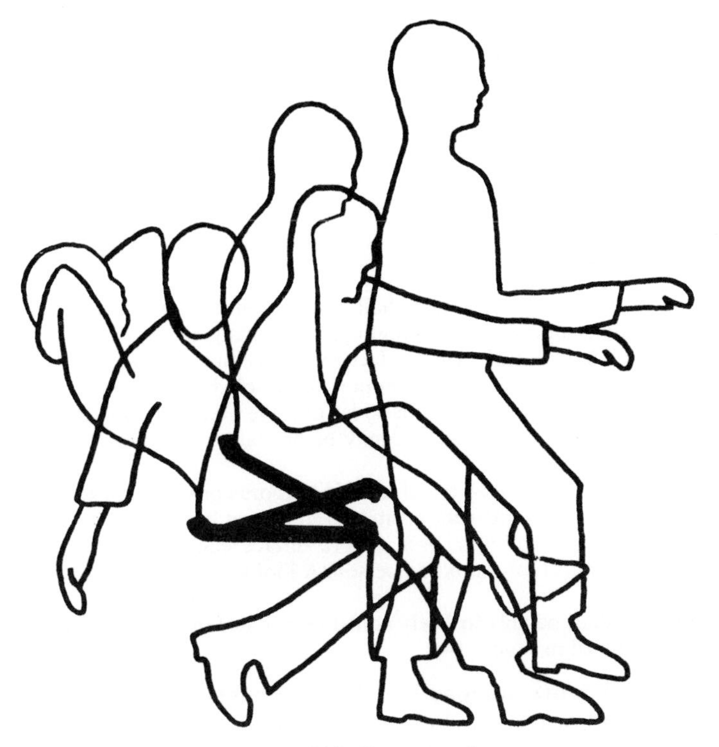

Figure 14.3 Free posturing

individual's body, preferences, and work activities. Consequently, it is now generally presumed that furniture should allow many postural variations and, for this, permit easy adjustment in seat pan height and angle, backrest position, or knee pads and footrests. For computer work stations, it is widely accepted that the user should control the location of the input devices and of the display. In particular, the display should be distinctly below eye height (Kroemer, 1988, 1991). Thus, change, variation and adjustment to fit the individual are considered central to well-being (Kroemer, 1988). If any label can be applied to current theories about proper sitting, it may be that of 'free posturing' as depicted in Figure 14.3.

The principles of free posturing include:

1. allowing the operator to readily assume a variety of sitting (or standing) postures, to adjust the work station, and even to get up and move about as part of the job;
2. designing for a variety of user dimensions and for a variety of user preferences;
3. new technologies develop quickly and should be incorporated at the work station. For example, radically new keyboards and input devices (including voice recognition) may be available soon; display technologies and display placement are undergoing rapid changes.

Many recommendations for the design of computer work stations exist (e.g., by Eklund, 1986; ANSI/Human Factors Society, 1988; Kroemer, 1988, 1991; Tougas and Nordin, 1987). Yet, given the rapid pace of change in computer technology, such recommendations may quickly become obsolete.

Conclusions

Attitudes about proper working postures and office work tasks are changing rapidly. Traditional sciences like orthopaedics, anatomy, physiology and medicine do not provide clear clues for healthy work postures. Re-evaluation of traditional axioms and establishment of new physiological (biomechanical) and psychological (behavioural) assessment methods are needed.

Accordingly, only generic advice is available for the design of proper computer work stations.

- Work procedures and equipment shall allow, even encourage, frequent and freely chosen variations among sitting, standing, and moving about.
- Seat pans shall have a fairly wide surface to rest the buttocks and proximal thighs.
- Backrests shall be high enough to support the whole back and even neck and head. The backrest shall be able to decline so far that the user can truly rest on it.
- Support for the lower legs and feet may be desirable, by a foot rest, foot stool, or ottoman.
- However, furniture also shall be adjustable to more conventional settings and allow an upright or even forward-bent position if so chosen by the user.
- For standing, if selected by the user, a work surface shall be at approximately elbow height of the comfortably standing person. A foot support at about 1/3 knee height is desirable.
- The operator shall be able to freely adjust height, distance, and angle of the display screen, as depicted in Figure 14.4.

Certainly, such work stations bear little resemblance to those traditionally used in the offices of the last 100 years. Time will show whether future work stations will look as depicted in Figure 14.5.

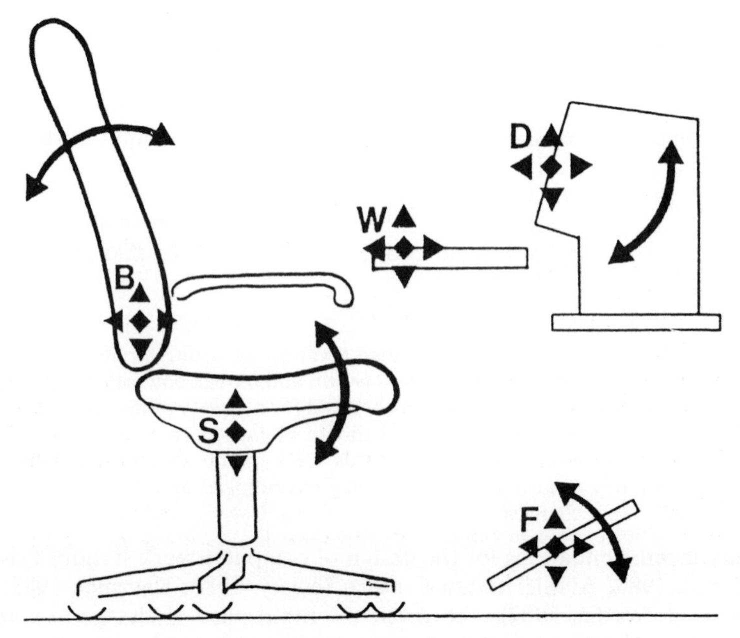

Figure 14.4 Adjustment features of a computer workstation

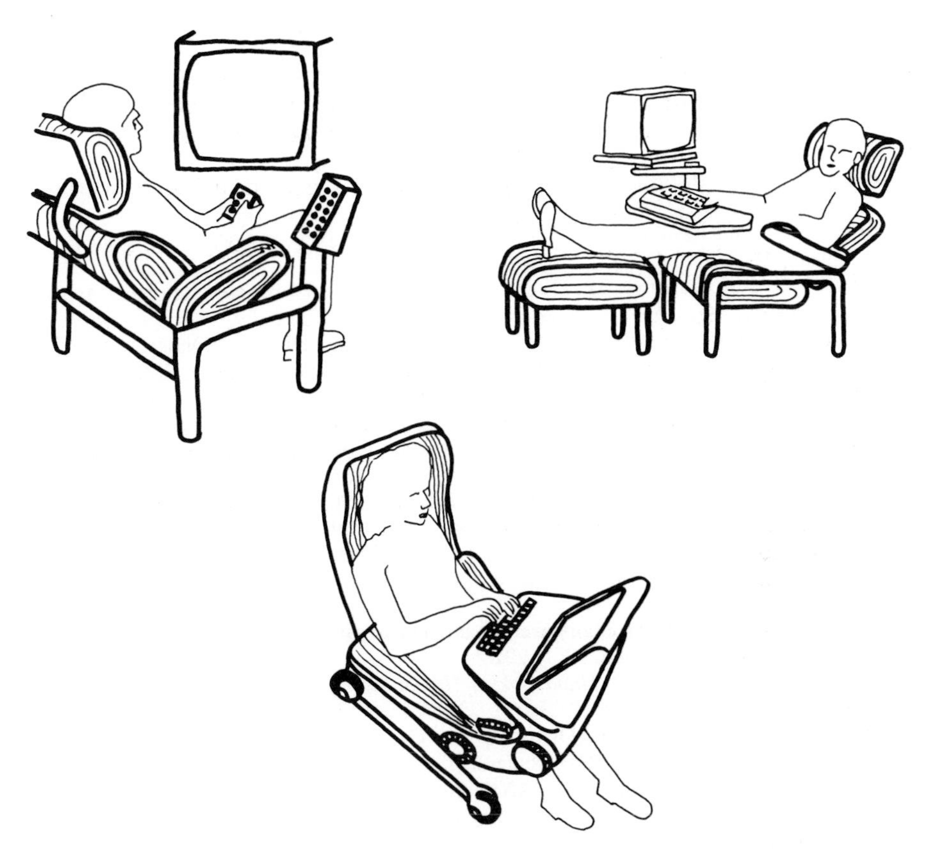

Figure 14.5 Examples of unconventional computer workplaces (sketched from illustrations in popular journals)

References

Akerblom, B., (1948), *Standing and Sitting Posture*, Stockholm: Nordiska Bokhandeln.

Adams, M. A. and Hutton, W. C., (1985), The effect of posture on the lumbar spine, *Journal of Bone and Joint Surgery*, **67**-B(4), 625–9.

Basmajian, J. V. and DeLuca, C. J., (1985), *Muscles Alive* (5th edn), Baltimore: Williams and Wilkins.

Bendix, T., Winkel, J. and Jessen, F., (1985), Comparison of office chairs with forward inclining, or tiltable seats, *European Journal of Applied Physiology*, **54**, 378–85.

Boudrifa, H. and Davies, B. T., (1984), The effect of backrest inclination, lumbar support and thoracic support on the intra-abdominal pressure while lifting, *Ergonomics*, **27**(4), 379–87.

Bradford, P. and Prete, B. (eds) (1978), *Chair*, New York, NY: Crowell.

Bridger, R. S., (1988), Postural adaptations to a sloping chair and work surface, *Human Factors*, **30**(2), 237–47.

Bridger, R. S., Wilkinson, D. and Van Houweninge, T. V., (1989), Hip mobility and spinal angles in standing and in different sitting postures, *Human Factors*, **31**(2), 229–41.

Chaffin, D. B. and Andersson, G. B. J., (1984), *Occupational Biomechanics*, New York, NY: Wiley.

Congleton, J. J., Ayoub, M. M. and Smith, J. L., (1985), The design and evaluation of the neutral posture chair for surgeons, *Human Factors*, **27**(5), 589–600.

Corlett, E. N., (1989), Aspects of the evaluation of industrial seating, *Ergonomics*, **32**(3), 257–69.

Corlett, E. N. and Bishop, R. P., (1976), A technique for assessing postural discomfort, *Ergonomics*, **19**(2), 175–82.

Corlett, E. N. and Eklund, J. A. E., (1984), How does a back-rest work? *Applied Ergonomics*, **15**(2), 111–4.

Drury, C. G. and Francher, M., (1985), Evaluation of a forward-sloping chair, *Applied Ergonomics*, **16**(1), 41–7.

Eklund, J., (1986), Industrial seating and spinal loading, Doctoral Dissertation, University of Nottingham, Linkoeping: University of Technology (ISBN 1-7870-14409).

Gale, M., Feather, S., Jensen, S. and Coster, G., (1988), A multi-disciplinary approach to the design of a work seat to prevent lumbar Lordosis for seated workers, in *Preprints, International Conference on Ergonomics, Occupational Safety and Health and the Environment*, 24–28 October 1988 (pp. 905–915). Beijing: The Chinese Society of Metals/Australian Darling Downs Institute of Advanced Education.

Grandjean, E. (ed), (1969), *Sitting Posture*, Taylor & Francis: London.

Grieco, A., (1986), Sitting posture: An old problem and a new one, *Ergonomics*, **29**(3), 345–62.

Grimsrud, T. M., (1990), Humans are not created to sit – and why you have to refurnish your life, *Ergonomics*, **33**(3), 291–5.

Helander, M. G. (ed), (1988), *Handbook of Human–Computer Interaction*, Amsterdam: North-Holland.

Helander, M. G., Czaja, S. J., Drury, C. G., Cary, J. M. and Burri, G., (1987), An ergonomic evaluation of office chairs, *Office: Technology and People*, **3**, 247–62.

Human Factors Society (ed), (1988), *ANSI Standard HFS 100 American National Standard for Human Factors Engineering of Visual Display Terminal Workstations*, Santa Monica, CA: The Human Factors Society.

Jaeger, M., (1987), Biomechanisches Modell des Menschen zur Analyse und Beurteilung der Belastung der Wirbelsaeule bei der Handhabung von Lasten. Biotechnik Series 17, No. 33. Duesseldorf, Germany: VDI Verlag.

Keegan, J. J., (1952), Alterations to the lumbar curve related to posture and sitting, *Journal of Bone and Joint Surgery*, **35**(A.3), 589–603.

Kember, P., (1985), *Bibliography: Anthropometry Related to Seating, 1975–1984*, Cranfield: Ergonomics Laboratory, Cranfield Institute of Technology.

Kroemer, K. H. E., (1988), VDT workstation design, in M. G. Helander (ed), *Handbook of Human–Computer Interaction* (pp. 521–39). Amsterdam: North-Holland.

Kroemer, K. H. E., (1991), Sitting at work: Recording and assessing body postures, designing furniture for the computer workstation, in A. Mital and W. Karwowski (eds), *Workspace, Equipment and Tool Design* (pp. 93–112). Amsterdam: Elsevier.

Kroemer, K. H. E. and Robinette, J. C., (1968), Ergonomics in the Design of Office Furniture. A Review of European Literature. AMRL-TR 68-90. Wright-Patterson AFB, OH, Also published with shortened list of references (1969), in *International Journal of Industrial Medicine and Surgery*, **38**(4), 115–25.

Kroemer, K. H. E., Snook, S. H., Meadows, S. K. and Deutsch, S. (eds), (1987), *Ergonomic Models of Anthropometry, Human Biomechanics, and Operator-Equipment Interfaces*, Washington, DC: National Academy Press.

Kuorinka, I., Jonsson, B., Kilbom, A., Vinterberg, H., Biering-Sorensen, F., Andersson, G. and Jorgensen, K., (1987), Standardized Nordic questionnaire for the analysis of musculo-skeletal symptoms, *Applied Ergonomics*, **18**(3), 233–7.

Lueder, R. K., (1983), Seat comfort: A review of the construct in the office environment, *Human Factors*, **25**(6), 701–11.

Mandal, A. C., (1975), Work chair with tilting seat, *Lancet*, 642–3.
Mandal, A. C., (1982), The correct height of school furniture, *Human Factors*, **24**(3), 257–69.
Marras, W. S. and Reilly, C. H., (1988), Networks of internal trunk-loading activities under controlled trunk-motion conditions, *Spine*, **13**(6), 661–7.
McGill, S. M. and Norman, R. W., (1987), Reassessment of the role of intra-abdominal pressure in spinal compression, *Ergonomics*, **30**(11), 1565–88.
Nag, P. K., Chintharia, S., Saiyed, S. and Nag, A., (1986), EMG analysis of sitting work postures in women, *Applied Ergonomics*, **17**(3), 195–7.
Occhipinti, E., Columbini, D., Frigo, C., Pedotti, A. and Grieco, A., (1985), Sitting posture: Analysis of lumbar stresses with upper limbs supported, *Ergonomics*, **28**(9), 1333–46.
Schoberth, H., (1962), *Sitzhaltung, Sitzschaden, Sitzmoebel*, Berlin: Springer.
Shackel, B., Chidsey, K. D. and Shipley, P., (1969), The assessment of chair comfort, *Ergonomics*, **12**, 269–306.
Staffel, F., (1984), Zur Hygiene des Sitzens, *Zbl. Allgemeine Gesundheitspflege*, **3**, 403–21.
Tougas, G. and Nordin, M. C., (1987), Seat features recommendations for workstations, *Applied Ergonomics*, **18**(3), 207–210.
Yu, C. Y., Keyserling, W. M. and Chaffin, D. B., (1988), Development of a work seat for industrial sewing operations: Results of a laboratory study, *Ergonomics*, **31**(12), 1765–86.
Yu, C. Y., Keyserling, W. M. and Chaffin, D. B., (1989), Letter, *Ergonomics*, **32**(6), 683–4.
Zacharkow, D., (1988), *Posture: Sitting, Standing, Chair Design and Exercise*, Springfield, IL: Thomas.

15

Measurement of lumbar and pelvic motion during sitting

Steven Reinecke, Kevin Coleman, and Malcolm Pope

Introduction

Sitting is the most common posture in today's workplace; those who spend long periods of time sitting on the job range from office workers to vehicle drivers to assembly line workers. Commuting to and from work, as well as relaxing at the end of the day, most often involves sitting. Many people spend half of their waking hours sitting down. With the increasing prevalence of office and manufacturing automation, sitting for prolonged periods will become even more prevalent.

Because so many people spend so much time sitting down, it is important to understand both the benefits and drawbacks of this posture. Certainly, sitting offers some advantages over standing: muscular activity is reduced, as well as heart rate, cardiovascular demands, oxygen consumption and hydrostatic blood pressure to the feet and lower legs (Grandjean, 1973, 1980; Winkel and Jorgensen, 1986). When one sits, the trunk's centre of gravity moves closer to the supporting surface, and enlarges the base of support; this increases the body's stability (Asatekin, 1975; Carlsoo, 1972), and enhances the capacity for precision tasks of fine movements (Asatekin, 1975; Ayoub, 1972, 1973; Grandjean, 1980).

Disadvantages of sitting exist as well. While seated, it is more difficult to generate high forces, such as while performing a task. Abdominal cavity pressure increases, and prolonged sitting is frequently accompanied by reports of neck, shoulder and back pain.

Low back pain may be the most common problem associated with prolonged static sitting and is the leading cause of activity limitation among those under 45 years of age (National Center for Health Statistics, 1981). This musculoskeletal disorder has become a major occupational health concern in transport, industrial and office settings; it represents the single greatest source of compensation costs and the second most common cause of lost time at work (Kelsey *et al.*, 1984; Kelsey and Golden, 1988).

A better understanding of seated postures would contribute to better seating design. This, in turn, might help reduce low back pain and other disadvantages of prolonged sitting, particularly in the workplace. Two aspects of seated postures important for seating design are pelvic orientation and spinal curvature. The ability to measure these characteristics can provide information useful in seating design.

Proper seating design can help reduce low back pain and other disadvantages of prolonged sitting, particularly in the workplace. Improved seating design, however, requires an understanding of the effects of postural variables, such as pelvic orientation and spinal curvature, and the ability to measure them. Two new devices for measuring postural variables are described below.

Effects of postural variables in sitting

Poor posture has long been suspected as a principal contributor to low back pain among sedentary workers (Eklund, 1967; Hult, 1954; Kelsey, 1975; Lawrence, 1977; Magora, 1972; Majeske and Buchanan, 1984). Poor postures usually involve forward flexion, lateral bending and/or axial rotation, in combination with different pelvic orientations. Helbig (1978) suggested that a slouched, or kyphotic, position, which is the most common seated posture, is the least healthful. Slouching involves prolonged spinal flexion, which is considered a principle cause of low back pain (Keegan, 1953; Kottke, 1961; Cyriax, 1975; McKenzie, 1981). A slouched posture stresses the posterior fibrous wall of the disc and posterior ligaments of the back, increasing intravertebral disc pressure.

Disc pressure is also affected by lumbar curvature, and increases as the lumbar spine moves from lordosis to kyphosis (Andersson and Ortengren, 1974a; Andersson *et al.*, 1975). When one is seated, the pelvis tends to rock posteriorly (Keegan, 1953, 1962; Carlsoo, 1972), causing the lumbar spine to move into kyphosis. The centre of gravity of the upper trunk shifts anteriorly to the lumbar spine. An anterior shift of force creates a greater moment arm on the spine, which is counteracted by either the erector spinae muscles or posterior ligaments. This increased moment arm creates a greater force on the spine which acts as a fulcrum point between the two forces, thus increasing disc pressure (Lindh, 1980; Andersson and Ortengren, 1974*b*).

Upper trunk posture results from a combination of spinal flexion and pelvic tilt. The pelvis is less stable when seated than when standing. When one is seated, the pelvis will tend to rotate or “rock” posteriorly, causing the lumbar spine to become flat or kyphotic (Akerblom, 1948; Coe, 1983); this directly affects the upper trunk’s centre of gravity.

Schoberth (1962) described three seated postures: middle, anterior, and posterior; these were defined by the amount of shift in the upper trunk’s centre of gravity with reference to the ischial tuberosities. In the middle posture, the center of gravity is located directly over the ischial tuberosities. If the erector spinae and iliopsoas muscles are relaxed, the lumbar spine becomes straight or slightly kyphotic. If active, the pelvis will rock anteriorly and the lumbar spine becomes lordotic (Andersson and Ortengren, 1974*b*).

An anterior posture, one in which the upper trunk is flexed forward, is characterized by a shift of force to the feet and a shift of the centre of gravity forward, or anterior to the ischial tuberosities. If the erector spinae muscles are active, the spine is flat with a forward rotation of the pelvis. If the muscles are inactive the pelvis rocks back, maximizing flexion and kyphosis of the spine.

In posterior sitting, the centre of gravity shifts rearward to the ischial tuberosities. The pelvis rocks backward, and stabilizes the body with support from the ischial tuberosities and coccyx. When muscles are relaxed in either the anterior or posterior position and the lumbar spine is in kyphosis, the spine is supported by the posterior ligaments (Akerblom, 1948; Schoberth, 1962; Andersson *et al.* 1974*b*).

Pelvic orientation is integral to seated postures and lumbar spine position. Zacharkow (1988) maintains that the key to proper seat support is proper sacral and pelvic support, which prevents the posterior rotation of the pelvis and lumbar kyphosis. By controlling the position of the pelvis through seat pan tilt (Bendix, 1987) or pelvic supports, an optimal upper trunk position can be maintained.

Seat design should encourage proper posture; but optimizing seating support and seat design requires an understanding of both lumbar and pelvic motion. The ability to measure such motion can improve our understanding of both seating kinematics and good seating.

Measurement of pelvic tilt

Several investigators have studied the role of the pelvis in both standing and seated postures (Akerblom, 1948; Bendix, 1987; Keegan, 1953). Sanders and Stavrakas (1981) and Kippers and Parker (1984) describe a simple technique for measuring pelvic tilt, in which posterior superior iliac spines (PSIS) and the anterior superior iliac spines (ASIS) are located by palpation; these are marked on the skin with a felt pen. The distance of each mark from the floor is measured, and the distance between the left side of the ASIS and the PSIS is measured with bowleg calipers. Pelvic orientation (tilt) is determined by interpolating the resulting trigonometric angles. Majeske and Buchanan (1984) investigated the angular motion of the forearm, upper arm, pelvis, trunk, neck and head during sitting. In this study, pelvic motion was monitored by attaching adhesive data markers to the skin over the greater trochanter and the iliac crest, when subjects were seated. Subjects were photographed in the sagittal plane in two sitting postures, and angles were measured with a standard protractor.

Gajdosik *et al.* (1985) investigated the reliability of measuring pelvic tilt angle with the marking technique described by Sanders and Stavrakas (1981). Gajdosik *et al.* found such markings unreliable since the pelvis moved under the skin during active movement. They found it necessary to determine the location of the PSIS and ASIS by palpation after each change in posture.

Thurston (1985) monitored the angular displacement of the pelvis and the lumbar spine using a motion analysis system comprising three television cameras which were connected to a digital computer. Targets mounted on two lightweight rigs, attached over the upper end of the lumbar spine and over the sacrum, were tracked in various postures. Data were recorded while patients walked at a comfortable speed across the assessment area. However, the effectiveness of the strap arrangement (of the target rigs) is questionable; because of possible skin motion over bony landmarks, the motion of the rig might not correspond precisely with pelvic motion and lumbar curvature. Gajdosik *et al.* (1985), for example, demonstrated that such skin motion can confound readings of pelvic motion.

Day *et al.* (1984) used an objective, non-invasive method to determine the effect of pelvic tilt on lumbar curvature in the sagittal plane. A computerized system, the Iowa Anatomical Position System (IAPS), was used to obtain coordinates of external body surface landmarks from which pelvic tilt measurements were determined. The system consists of a positioning device to find the three-dimensional co-ordinates of a point in space with a pointer. A point is palpated and marked for each reading. For pelvic measurements, the anatomical points of the posterior sacral protrusion, left and right posterior superior iliac, and S2 are marked. One limitation of this system is that each point must be marked without the subject moving between measurements, which precludes continuous readings.

Measuring spinal curvature

Several techniques have been used to monitor lumbar curvature. Schober (1937) measured skin distraction between markings 10 cm above the posterior superior iliac crests and 5 cm below, with subjects standing erect. Skin distraction increased with increased forward flexion. Adrichem and Korst (1973) tested 250 schoolchildren using the skin traction method described by Schober (1937); this approach was found to produce accurate measurement of lumbar spinal flexion independent of age, body length, lumbar length and degree of lumbar lordosis. When subjects are seated, however, this technique is of limited usefulness when backrests are used. The backrest obscures a view of the back and introduces shear forces between back and backrest; that is, the relative motion between the back and backrest changes normal skin distraction.

A two-inclinometer technique, described by Loebl (1967) and modified by Mayer *et al.* (1984), provides a measurement of the difference in inclination between the sacrum and T12-L1; however, use of this technique is limited to unsupported sitting conditions. A photometric technique described by Gill *et al.* (1988) involves the use of fiduciary markers placed on the subject's lateral epicondyle of the knee, greater trochanter, anterior superior iliac spine (ASIS), and over the tenth rib. Sagittal plane photographs are taken to determine the X and Y co-ordinates of each marker, and lines drawn between the markers of the photographs are used to define the angle of the lumbar spine. However this technique is not applicable to measuring lumbar curvature when subjects are seated, since the back cannot be observed.

In summary, a better understanding of the kinematics of the back in seated postures should provide insight into the relationship between posture and low back pain. Additionally, the ability to measure the lumbar spine curvature and pelvic orientation of a seated subject would provide the basis for improving seating, particularly in the work environment.

However, presently-available techniques for measuring pelvic motion and lumbar curvature are of limited utility when subjects are seated. Motion between skin and underlying bone can interfere with the measurement of pelvic orientation. Also, when palpation of the pelvis is required to locate a bony landmark, normal, continuous pelvic motion is interrupted. The measurement of lumbar curvature among seated subjects is particularly difficult when they are leaning against a backrest, since their backs cannot be observed or accessed.

At the Vermont Rehabilitation Engineering Center, two unique tools have been developed to overcome the problems of measuring pelvic tilt and lumbar curvature: The Pelvic Motion Device (PMD) and the Lordosimeter.

Measurement tools

Pelvic motion device (PMD)

The PMD incorporates previously used, clinical methods of locating the PSIS and ASIS. Once they are located, the system allows continuous monitoring of the amount of pelvic tilt, which is not possible with other methods requiring

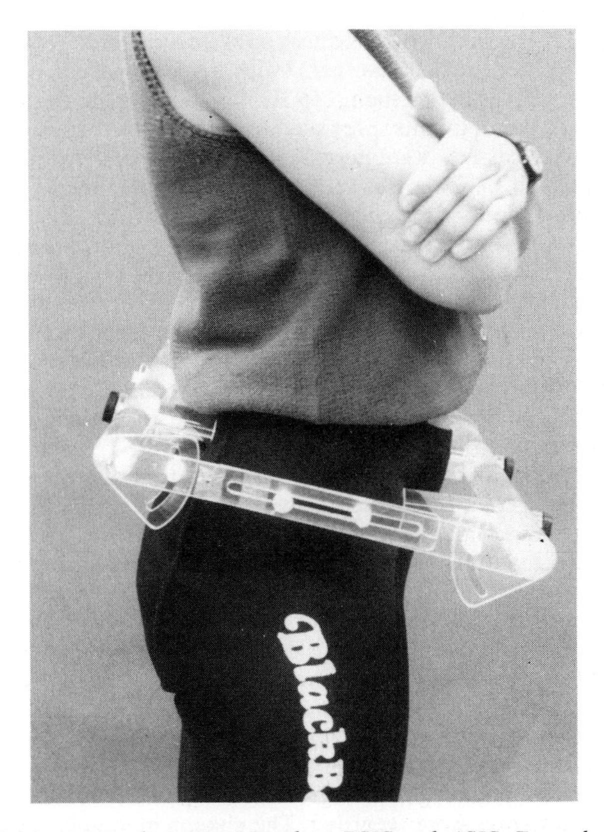

Figure 15.1 Pelvic motion device positioned on PSIS and ASIS. Four plungers, two for the PSIS and two for the ASIS, protrude inward from the hoops to come in contact with the bony landmarks. Clamping the hoops tightly together increases the force on both posterior and anterior superior iliac spines. Once clamped, the only contact between the PMD and the subject is at these points

repeated palpation to locate PSIS and ASIS. The conceptional requirements of the PMD called for a device that could be clamped on the PSIS and ASIS and remain in place throughout a dynamic task. It was important that the device not hinder normal activity when worn and that it be accurate within three degrees of pelvic tilt. The device consists of two half hoops clamped together. One hoop is positioned anterior to the pelvis while the second is positioned posterior to the pelvis. The hoops are attached on both lateral sides of the subject's pelvis with bolts that allow adjustment (Figure 15.1).

Testing

To verify the accuracy of the PMD in tracking pelvic motion, a study was performed to compare pelvic motion (in this case, flexion), derived from x-rays of low back patients with the motion measured by the PMD. Because clinical X-rays would have to be taken, a special PMD device was fabricated that incorporated two principal modifications. First, the device was made translucent to X-rays, i.e., it could not conceal or mask the area from L5 to S2 on the X-ray film. Second, the device allowed for the patient to be positioned adjacent to the X-ray film, which reduces image distortion.

Thirteen subjects (8 men, 5 women), with an average age of 40 years and average weight of 193 lbs for men, 156 lbs for the women, participated in the study. Low back pain patients, previously scheduled for extension and flexion X-rays for other reasons, were recruited to participate. Another reason for recruiting low back patients was to determine whether they could easily use and tolerate the device.

Two small lead BB's were attached to the anterior and posterior ends of the PMD as markers for later readings of the position of the device from the X-rays. Two lateral X-rays were used to measure the change in pelvic rotation, one of the patient standing up in full extension, the second in forward flexion (Figure 15.2). The device was held around the waist while the plungers were adjusted for position and inclination to the PSIS and ASIS. Once the plungers were positioned, the device was clamped tightly to ensure enough pressure on the bony landmarks to hold it in place. Patients were then positioned lateral to the X-ray plate in full extension with arms folded; they were positioned in the centre of the X-ray so that the angle of dispersion of the X-ray particulars were minimal. An X-ray was taken from L1 to S3. Patients were then requested to flex as much as possible, without moving their feet, while supporting themselves with a chair placed in front of them. A second X-ray was then taken.

Once the two X-rays were obtained, the following measurements were made. A single line was drawn between the two BB's which were visible on the X-ray. This line represents the angular position of the PMD to a reference horizontal plane. A second line was drawn between the peaks of the anterior and posterior crest of the sacrum to represent the position of the pelvis relative to the horizontal plane. The changes in degrees between the two lines for the extension and flexion X-rays were recorded. The change in angular tilt of the pelvis was compared to the angular tilt of the PMD. Each X-ray was marked and measured twice. The difference between the recorded angle of the pelvis and the recorded angle of the PMD represents error, introduced by pelvic rotation, in the PMD measurement.

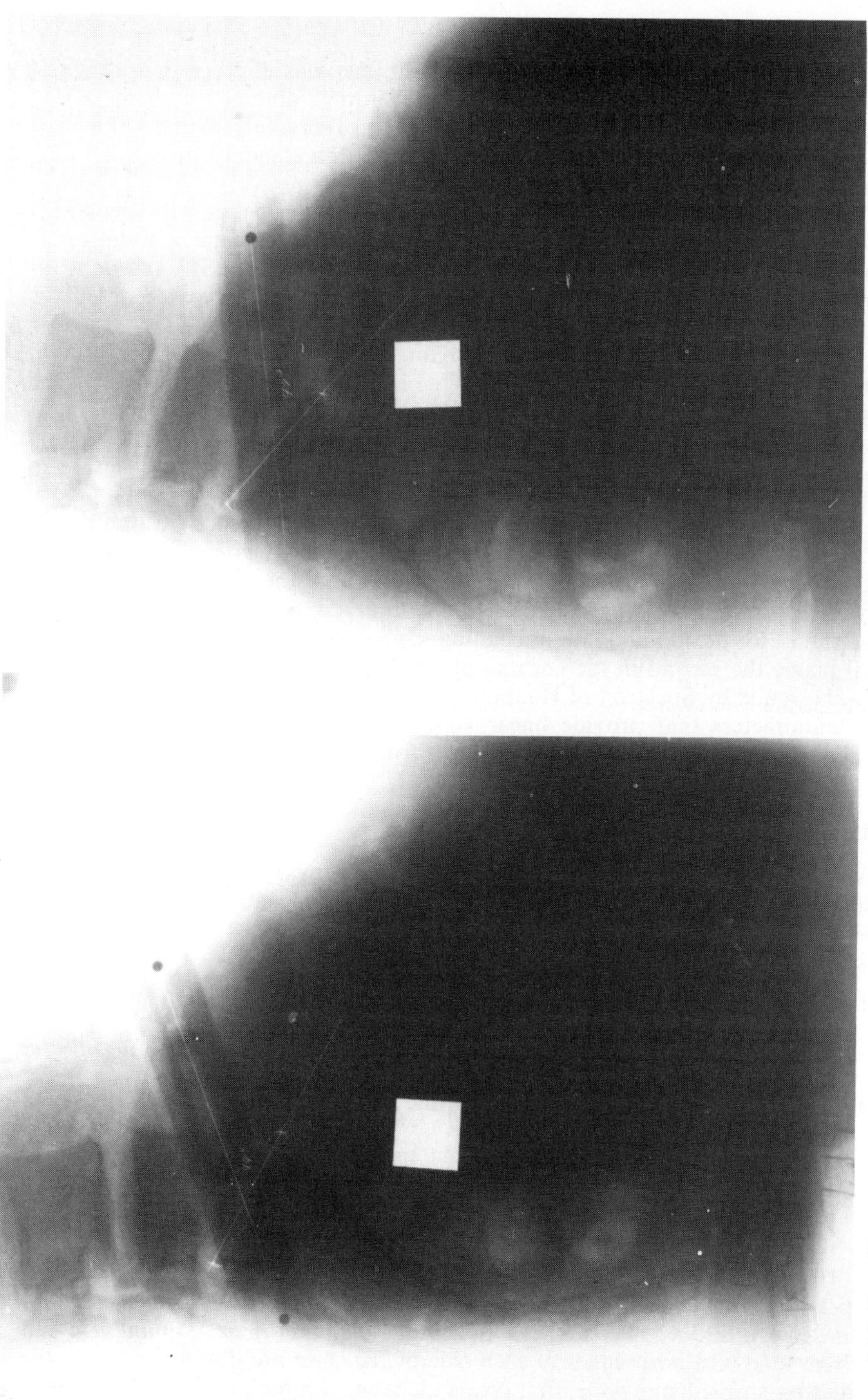

Figure 15.2 Two X-rays, one taken with the subject in extension and the other with the subject in flexion, were compared to determine whether postural changes produced any motion between the PMD and the pelvis

Findings

The average change in pelvic tilt for all subjects was 26.5°, with a range of 15°–40°. The average error in pelvic tilt recordings obtained with the PMD was 2.07°, with a standard deviation of 1.4°. This represents an error of 8.8 per cent in monitoring pelvic tilt with the PMD, with a standard deviation of 7.2 per cent.

The largest error in monitoring the position of the pelvis with the PMD occurred with heavy or obese people. Because of their greater amount of fat under the skin, the PMD tended to move or shift off the PSIS and/or ASIS more readily. Less than a 3 per cent measurement error was recorded for subjects with minimal fat tissue layers.

Several other factors might account for the error rate of 8.8 per cent in monitoring pelvic tilt with the PMD. One is the error inherent in reading the X-rays for pelvic position, since there is some error involved in marking a point on a curve of the sacral crest. Second, a small amount of lateral bending and/or rotation might have occurred when patients moved from extension to flexion; this could have resulted in small inaccuracies.

Once it was clear that pelvic motion could be monitored by clamping the device to the ASIS and PSIS, a second PMD was designed for clinical testing. This device was designed to fit more closely around the patient. Two electrolytic tilt sensors were used to record real-time continuous measurements of pelvic tilt for flexion and lateral bending. One sensor was oriented in the sagittal plane, the second in the coronal plane. The electrolytic tilt sensors, model L-211, made by Spectron of Hauppauge, N.Y., are vertical sensing electrolytic potentiometers that provide linear voltage output proportional to tilt in a single axis. The sensors' tilt angle ranges from $\pm 60°$, repeatable at any angle to within 0.03°.

Lordosimeter

Measurement of lumbar motion requires a non-invasive device that continuously measures lumbar curvature in both seated and standing postures. Practical considerations that affect the design of such a device include the following.

1. The device would have to overcome the problem of measuring lumbar motion when a subject leans against a back rest.
2. It would be necessary to affix a transducer to the back, without its being affected by skin distraction.
3. Shear forces between the subject's back and the backrest (during movement) would have to be minimized.
4. The device would have to be both thin and comfortable, and not affect normal motion.
5. The device could not be affected by pressure deriving from the force of the back against a backrest.

The Lordosimeter consists of several thin layers of material with an overall thickness of 5.1 mm. It is 25 mm wide and 23 cm long. The core of the Lordosimeter is a flexible ruler made of poly-vinyl, 0.67 mm thick, 25 mm wide and 23 cm long. On both edges of each side of the ruler are double-sided flexible adhesive strips, 7 mm wide, that create a channel on both sides of the ruler. In the centre of the channels are two mercury strain gauges, 180 mm long, that

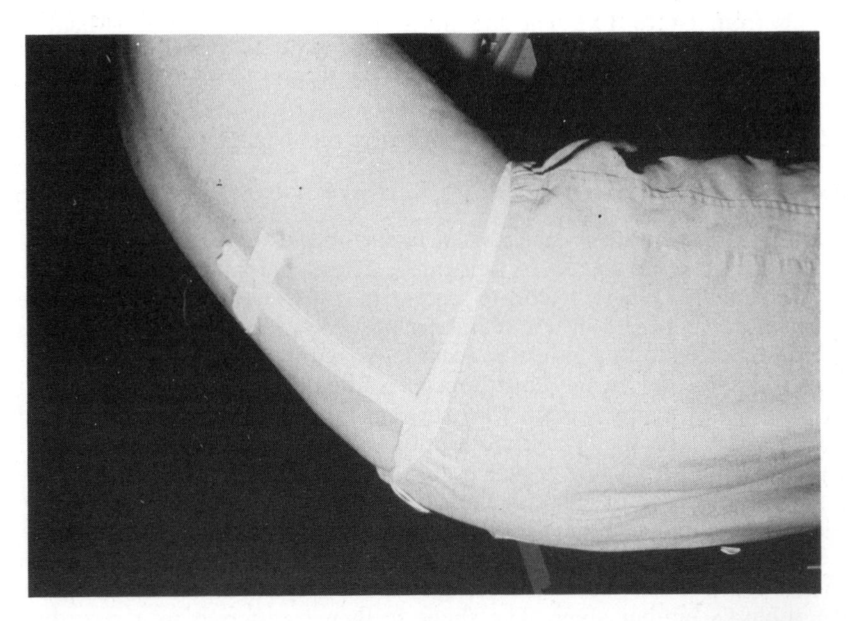

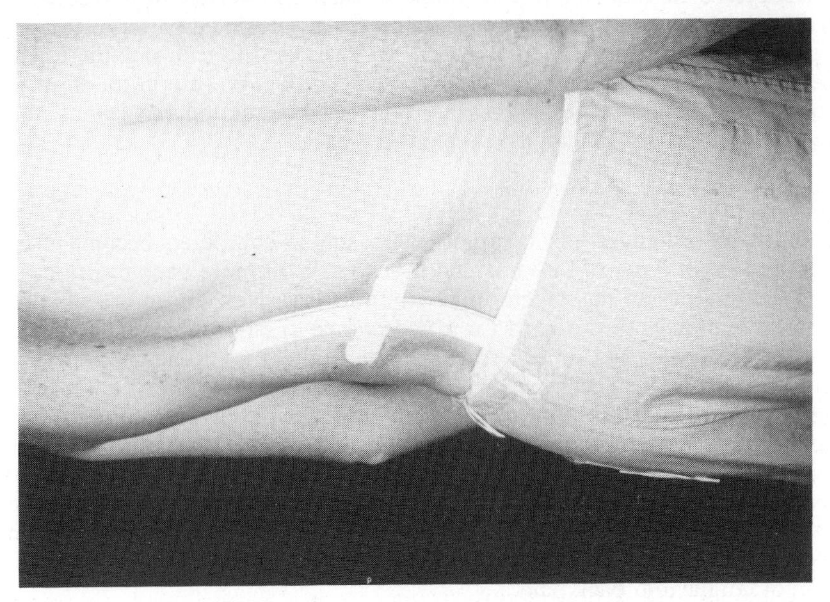

Figure 15.3 The Lordosimeter measures changes in lumbar curve

produce a change in resistance when either elongated or contracted. A lycra covering minimizes adhesion to the back.

When the Lordosimeter is attached to the sacrum of a subject in full flexion, it extends from L1 to T12. The top portion of the Lordosimeter is held in place against the back by a strip of cloth attached to the skin on both sides. This cloth bridge allows the ruler to slide under it and against the skin, therefore minimizing any skin distraction. As the subject's lumbar curve changes, the flexible rule, held close to the back, measures that change (Figure 15.3).

Testing

To verify that the Lordosimeter gives a linear output for variation in curvatures, 10 curved templates were fabricated. These templates ranged from a curvature radius of 150 cm (convex curve) to 150 cm (concave curve). The Lordosimeter was placed on the templates and compressed while output was recorded. Repeated testing indicated accuracy to within 0.05 per cent.

To calibrate the device, the Lordosimeter is fitted on the subject's back, with the base attached at the sacrum with surgical tape and the top portion held against the back with the flexible cloth bridge. Once fitted, the subject is asked to stand and hold a posture in full flexion, full extension, and upright. Measurements are then recorded with the Lordosimeter and a flexible ruler. The flexible ruler technique is the NIOSH standard measurement technique (1973). The examiner locates the spinous process of L5 and T12 and marks both locations. The flexible ruler is placed against the subject's back and conforms to the subject's lumbar spine. The shape is then transferred to a recording form. The trigonometric derivation of the angle is calculated to represent the shape of the lumbar spine. These numbers are then used to calibrate the Lordosimeter which then provides a continuous record of curvature in the form of voltage output. The voltage signals are converted to digital recordings and degree of curvature is calculated on a real-time basis.

Summary

The number of sedentary jobs is rapidly increasing as computers become more widely used in all types of business and industry. With more people sitting on the job, seating design has taken on new importance. New and more suitable chairs and seats will be designed for a wide range of environments, including offices, industrial settings, homes, trucks, boats, and planes, to name but a few. Chairs will likely be designed for specific activities and individual needs. Designing chairs that are both comfortable and functional will require information about seating needs in different environments. Biomechanical analysis is invaluable to further seating development and design. The pelvic motion device and Lordosimeter represent two tools to study the kinematics of the seated body. Measuring changes in posture over the course of a workday can provide information useful in the improvement of seating design and the prevention of fatigue and back pain.

References

Adrichem, J. A. M. and Korst, J. K., (1973), Assessment of the flexibility of the lumbar spine, *Scandinavian Journal of Rheumatology*, **2**, 87–91.

Akerblom, B., (1948), *Standing and Sitting Posture*, (Stockholm: Nordiska Bokhandeln).

Andersson, G. B. J. and Ortengren, R., (1974*a*), Lumbar disc pressure and myo-electric back muscle activity during sitting. II. Studies on an office chair, *Scandinavian Journal of Rehabilitation Medicine*, **6**, 115–21.

Andersson, G. B. J. and Ortengren, R., (1974*b*), Myo-electric back muscle activity during sitting, *Scandinavian Journal of Rehabilitation Medicine*, Supplement No. 3, 73–90.

Andersson, G. B. J., Ortengren, R., Nachemson, A. L., Elfstrom, G. and Broman, H., (1975), The sitting posture: An electro-myographic and discometric study, *Orthopaedics North America*, **6**, 105–20.

Asatekin, M., (1975), Postural and physiological criteria for seating.

A review. (1975), *M.E.T.U. Journal of the Faculty of Architecture*, **1**, 55–83.

Ayoub, M. M., (1972), Sitting down on the job (properly), *Industrial Design*, **19**, 42–5.

Ayoub, M. M., (1973), Work place design and posture, *Human Factors*, **15**, 165–8.

Bendix, T., (1987), Adjustment of the seated workplace, PhD. dissertation, Laegeforeningens Forlag.

Carlsoo, S., (1972), *How Man Moves*, (London: Heinemann).

Coe, J. B., (1983), The influence of community status on work station design, in *Proceedings of the 20th Annual Conference of the Ergonomics Society of Australia and New Zealand*, (ed T. Shinnick and G. Hill), (Sydney), 185–90.

Cyriax, J., (1975), *The Slipped Disc*, 2nd edn, (Epping: Gower).

Day, J., Smidt, G. and Lehmann, T., (1984), Effect of pevic tilt on standing posture, *Physical Therapy*, **64**, 510–6.

Eklund, M., (1967), Prevalence of musculo-skeletal disorders in office work, *Socialmedicinsk*, **6**, 328–36.

Gajdosik, R. L., LeVeau, B. F. and Bohannon, R. W., (1985), Effects of ankle dorsiflexion on active and passive unilateral straight leg raising, *Physical Therapy*, **65**, 1478–82.

Gearhart, J. R., (1978), Response of the skeletal system to helicopter-unique vibration, *Aviation, Space and Environmental Medicine*, **49**, 253–6.

Gill, K., Krag, M. H., Johnson, G. B., Haugh, L. D. and Pope, M. H., (1988), Reproducibility of four clinical methods for assessment of lumbar spinal motion, *Spine*, **13**, 50–3.

Grandjean, E., (1973), *Ergonomics of the Home*, (London: Taylor & Francis).

Grandjean, E., (1980), Sitting posture of car drivers from the point of view of ergonomics, in *Human Factors in Transport Research*, Vol. 2, (eds D. J. Oborne and J. A. Levis) (London: Academic Press).

Helbig, K., (1978), *Sitzdruckverteilung beim ungepolsterten sitz*, Anthropologischer Anzieger, 36.

Hult, L., (1954), Cervical, dorsal and lumbar spine syndromes, *Acta. Orthopaedica Scandinavica* (Supplement 17).

Keegan, J. J., (1962), Evaluation and improvement of seats, *Industrial Medicine and Surgery*, **31**, 137–48.

Keegan, J. J., (1953), Alterations of the lumbar curve related to posture and seating, *Journal of Bone and Joint Surgery*, **35**-A, 589–603.

Kelsey, J. L., (1975), An epidemiological study of the relationship between occupation and acute herniated lumbar intervertebral discs, *International Journal of Epidemiology*, **4**, 197–205.

Kelsey, J. L., Githens, P. B., O'Connor, T., Weil, U., Calogero, J. A., Holford, T. R., White, A. A., Walter, S. D., Ostfeld, A. M. and Southwick, W. O., (1984), Acute prolapsed lumbar intervertebral disc, *Spine*, **9**, 608–13.

Kelsey, J. L. and Golden, A. L., (1988), Occupational and workplace factors associated with low back pain, *Occupational Medicine*, **3**, 7–16.

Kippers, V. and Parker, A. W., (1984), Hand position at possible critical points in the stoop-lift movement, *Ergonomics*, **26**, 895–903.

Kottke, F. J., (1961), Evaluation and treatment of low back pain due to mechanical causes, *Archives of Physical Medicine and Rehabilitation*, **42**, 426–40.

Lawrence, J., (1977), *Rheumatism in Populations*, (London: William Heinemann Medical Books Ltd).

Lindh, M., (1980), Biomechanics of the lumbar spine, in *Basic Biomechanics of the Skeletal System*, (eds V. H. Frankel and M. Nordin) (Philadelphia: Lea and Febiger), 255–90.

Loebl, W. Y., (1967), Measurements of spinal posture and range in spinal movements, *Annals of Physical Medicine*, **9**, 103.

Magora, A., (1972), Investigation of the relation between low back pain and occupation. 3. Physical requirements: Sitting, standing and weight lifting, *Industrial Medicine*, **41**, 5–9.

Majeske, C. and Buchanan, C., (1984), Quantitative description of two sitting postures, *Physical Therapy*, **64**, 1531–3.

Mayer, T. G., Tencer, A. F., Kristoferson, S. and Mooney, V., (1984), Use of non-invasive techniques for quantification of spinal range-of-motion in normal subjects and chronic low-back dysfunction patients, *Spine*, **9**, 6.

McKenzie, R. A., (1981), *The Lumbar Spine: Mechanical Diagnosis and Therapy*, (Walkanae: Spinal Publications).

National Center for Health Statistics, (1981), Prevalence of selected impairments, United States. (1977). DHHS Publication (PHS) 81–1562, Series 10, #134 (Hyattsville, MD: DHHS).

Sanders, G. and Stavrakas, P., (1981), A technique for measuring pelvic tilt, *Physical Therapy*, **61**, 49–50.

Schoberth, H., (1962), *Sitzhaltung, Sitzchaden, Sitzmobel*, (Berlin: Springer-Verlag).

Schober, P., (1937), Lendenwirbelsaule und Kreuzschmerzen, *Munch Med. Wschr.*, **84**, 336.

Thurston, A. J., (1985), Spinal and pelvic kinematics in osteoarthrosis of the hip joint, *Spine*, **10**, 467–71.

Winkel, J. and Jorgensen, K., (1986), Evaluation of foot swelling and lower limb temperatures in relation to leg activity during long-term seated office work, *Ergonomics*, **29**, 313–28.

Zacharkow, D., (1988), *Posture: Sitting, Standing, Chair Design and Exercise*, (Springfield, Illinois: Charles C. Thomas).

16

Does it matter that people are shaped differently, yet backrests are built the same?

R. Lueder, E. N. Corlett, C. Danielson, G. C. Greenstein, J. Hsieh, R. Phillips, and DesignWorks/USA

When sitting, the hamstring and gluteal muscles tighten, causing the pelvis to rotate and flex (flatten) the lumbar curve.* Andersson (1980) has shown that, as a result, disc pressures in that region increase approximately 40 per cent relative to standing.

The implications of these findings are complicated by our limited understanding regarding individual differences in lumbar curvature and pelvic tilt. It has been noted that less than half the population has a so-called normal curve of the lumbar spine (e.g., Branton, 1984; Schoberth, 1962, 1978). Direct observation also underscores the broad spectrum of back shapes that prevail in the population.

Some note, in particular, a tendency for hyper-lordosis among females, and hypo-lordosis among males (Grandjean *et al.*, 1969). If one accepts this tenet, a possible mechanism for such differences may be that (presumably as an evolutionary consequence of childbirth), the female hip socket is forward of the centre of gravity, whereas this tends to be more directly at the hip socket among men (Tichauer, 1978).

Further, studies in space indicate that the torso-thigh angles associated with neutral body postures can vary considerably (128° ± 7° found by Webb Associates, 1978). Such angular characteristics presumably influence the curvatures of the spine as well, by influencing the extent of pelvic rotation.

---

* Specifically, the lumbar lordosis decreases by an average of 38°. Two-thirds of this reduction (28°) is from rotation of the pelvis, and one-third (10°) is from the flattening of the lumbar spine. The angle of the sacrum (sacro-iliac joint) changes by approximately 4° (Andersson, 1986).

Theoretically, if spinal segments are angulated disc pressures of some people with hyper-lordoses may even improve when seated. However, Bendix (1987) suggests that the posterior characteristics of the lumbar spine can be adversely affected from sitting; pain may result from prolonged compression of the facet joints (protruding from the vertebrae).

Such issues have been addressed little in seating research. It is not unusual to find the following limitations in seating studies:

Subjects have not been representative of the general population

Research on actual or inferred disc pressures while seated has been conducted by Andersson (1980), Andersson and Ortengren (1974*a*), Nachemson and Elfstrom (1970), Nachemson (1981), Bendix (1984), Bendix and Biering-Sorensen (1983), Bendix and Hagberg (1984), Boudrifa and Davies (1984), Brunswic (1984) and others. These studies have typically either used young and healthy male college students with no history of back pain, or did not specify the characteristics of the subject population. Individual differences in curvature of the spine have not been specifically addressed, although Grandjean *et al.* (1969) examined subjective comfort ratings of individuals with varying degrees of lumbar lordosis, and Branton (1984) developed back profiles of a group of male and female railway employee volunteers. Andersson and Ortengren (1974) also investigated the disc pressures of wheelchair users in different postures; however, the contours of these individuals' spines were not determined.

As a result, implications for seating of variations in the lumbar profile are not well understood. For example, are the hypo-lordotic individuals better accommodated in a seat with a lumbar curve that fits their contours or in one that provides greater lumbar curvature? What depth best accommodates the hyper-lordotic individual?

Another example arises when sitting with a large trunk-thigh angle. As a person leans back against the chair, the hamstring muscles relax so that the pelvis rotates, and the lumbar curvature increases relative to the upright position. What are the associated implications for the optimum lumbar depth at different backrest angles? Can hyper-lordotic and hypo-lordotic individuals be similarly accommodated?

Research tends not to address the variability of the data

Research has frequently emphasized main effects quoting averages and standard deviations. That is, categories have been compared without examining the implications of individual differences. For example, Andersson's (1980) research indicated that 5 cm lumbar seat depth was significantly better than lesser depths, but did not convey how much of the population was accommodated.

Experimental ranges were limited

Boudrifa and Davies (1984) found that more back rest support at the thorax (inclination 10°) was associated with less apparent stress on the spine. However, this was the highest level of contouring examined. Likewise, Andersson (1980) found decreases of pressures on the spine with increases of seat lumbar depths, up to their maximum of 5 cm. Hosea *et al.* (1986) used EMGs to evaluate depths to 7 cm for specific applications. A general mapping to the area is lacking.

The contribution of the backrest, and of lumbar support, to the effectiveness of a seat for a sitter is well recognized in the literature. Yet there seems little evidence in the literature for a model which would link the influencing factors to enable more informed design to be achieved. It was proposed in a number of studies to address this problem, with the following research questions:

- What is the optimum lumbar depth for the backrest?
- Should the lumbar depth of the backrest change automatically, in conjunction with changes in backrest inclination?
- How do individual differences in the curvature of the spine affect seating needs?
- Can a single lumbar depth accommodate the spectrum of lumbar curvatures of people within the general population?
- Is the seat lumbar depth that induces the least stress (lowest compressive forces) also considered the most comfortable?
- What is the relationship between shrinkage of the spine and discomfort?

It was anticipated that the best lumbar depth would be associated with both fewer and/or lower levels of body part discomforts, and with lower levels of compression of the spine (inferred by changes of stature).

Experimental procedure: Pilot study

The purpose of the pilot study was to establish and validate a procedure for measuring an individual's lumbar curvature and determine inter-examiner and intra-individual reliability. Subjects for the pilot were eight adult male volunteers with no history of lower back pain. These subjects participated in the following procedures:

1. The contours of the spine (T12 to S2) were outlined twice by a flexible (flexicurve) ruler while the person assumed each of eight positions, then traced onto white paper each time (Burton, 1985, 1986). (*see* Figure 16.1) Subjects were asked to relax, and move about after each measurement.

These eight postures were:

1. neutral standing (standing in upright position);
2. neutral sitting (sitting upright, without back support);
3. flexing while standing (bending forward as far as possible while standing);
4. extending while standing (arching the back as far as possible while standing, without extending the neck);
5. flexing while seated (bending forward as far as possible while seated, with arms placed between their knees);
6. extending while seated (arching the back as far as possible by protruding their abdomen while seated, with neck in neutral position and free of extension);

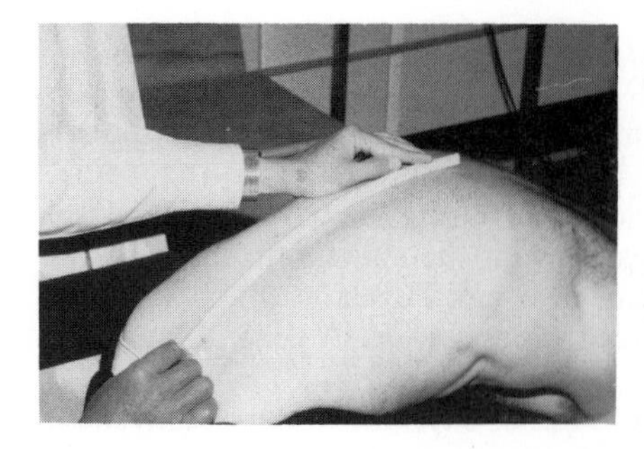
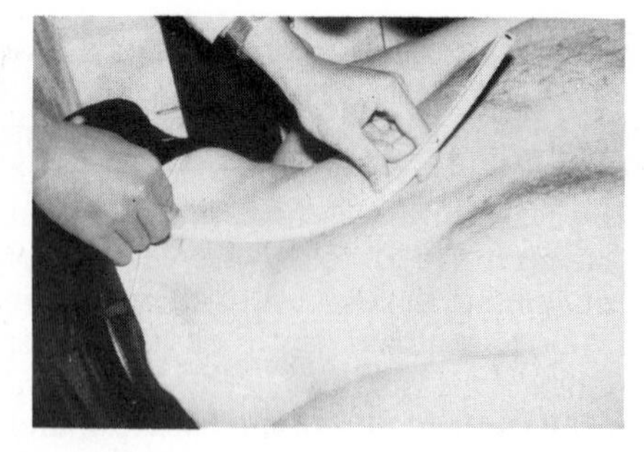

Figure 16.1 The curvature of the spine were traced with flexicurve rulers

7. prone, with flexion (lying with lower back flattened as far as possible); and,
8. prone, with extension (lying with back arched, as far as possible).

2. Tangents as described were drawn for all eight postures, and from each of the two duplicated tracings. The lumbar curve was characterized by measuring the angles between the T12 and L4, and L4 and S2 tangents, giving an upper and lower spinal angle.

The two replications were found to be significantly correlated. The findings were:

Degree of lordosis:

The coefficient of correlation of the measurements of lordosis were $r = 0.9899$ ($p < 0.01$) in the three seated postures, but non-significant at $r = 0.6478$ for the three standing postures.

Measurements of seated lordosis were most highly correlated with lordosis of the various postures ($r = 0.9773$, -0.1899, and 0.9536, respectively, for sitting, lying, and standing postures). However, these values were not significant, perhaps because of the small sample size (eight subjects). When analyzed separately, only the sitting position was significantly correlated with upper and lower lumbar lordosis.

Flexion:

Measures of flexion were also significantly correlated when seated and standing ($r = 0.9901$, $p > 0.001$), and when seated and lying down ($r = 0.9660$, $p < 0.01$). As a result, it was concluded that the most appropriate measure of flexion was in the seated position.

Extension:

None of the extended positions of each of the repeated measurements were significantly correlated, although the sitting position was most highly correlated. Reasons for this lack of significance include the small sample size*, and perhaps because more practice trials were needed.

Range of Motion:

A range-of-motion value was calculated for each individual, representing the sum of their lumbar flexion and extension

Based on these findings, it was decided to use the flexicurve measures of lumbar lordosis in the three seated positions for subsequent evaluations.

Experiment 1

The objective of this study was to evaluate the implications of individual differences in lumbar characteristics, lumbar depth, and backrest angle on subjective discomfort ratings and stature change.

The independent variables are listed below.

- Individual differences in lumbar lordosis.
  The measurement procedure is described above, and based on Burton (1986)
- Lumbar depth of the backrest (3, 5, and 7 cm; see next paragraph)
- Back rest angle (upright/90° and 110°)

Three seat fixtures were developed to test lumbar depth and angle. These three fixtures† were 3, 5, and 7 cm deep, and could each be adjusted to either 90° or 110° (*see* Figure 16.2).

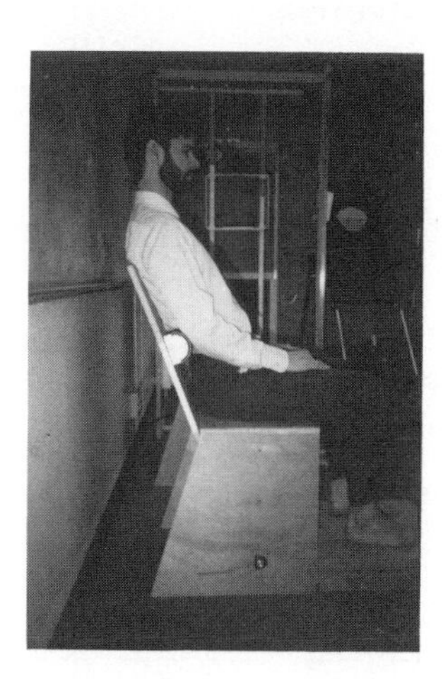

Figure 16.2 Experimental seat fixtures with adjustable back rest angles, and 3 lumbar depths

---

* When one outlier was removed from the scatter diagram due to a calculation error, and four more cases were subsequently added to the data, the correlation coefficient increased to $r = 0.87$ ($p = 0.005$)

† The fixtures were jointly designed and constructed by Humanics and Designworks, with the consultation of E. N. Corlett.

The dependent variables are listed below.

- Body Part Discomfort ratings (*see* Table 16.1);
  These BPD ratings were based on Corlett and Bishop (1976).
- General Comfort Ratings;
- Chair Feature Checklist Ratings (*see* Table 16.2) which gave a more detailed breakdown of the fit between the seat and the subject.
  This form was based on the checklist first developed by Shackel *et al.* (1969), and adapted by Drury & Coury (1982).
- Changes in stature, as measured by a stadiometer (*see* Figure 16.3);
  The stadiometer used was one that measured changes of stature with a reliability of ± 2 mm. Changes in stature provide an index of compressive loads on the spine. It was developed and described by Eklund (1984), Eklund and Corlett (1986a, 1986b), Corlett and Eklund (1986), and reviewed by Corlett (1990). The stadiometer, which was constructed by Designworks/USA and Humanics ErgoSystems, Inc. and based on diagrams provided by Dr. Corlett, is shown on Figure 16.3. Some minor changes were made to the original version, including an electronic "read out" and force plate sensor to increase sensitivity of the measurement process.*

Forty-seven male subjects participated in this first phase of the research. Although 19 of these subjects had a history of low back pain, none were experiencing LBP at the time of the study. Their age ranged from 23 to 51 years

Table 16.1 Body part discomfort ratings. Based on Corlett and Bishop (1976)

Name:__________ **Sex (Please circle): Male Female**

We would like to understand how comfortable you are over the day. Please mark your level of comfort at the beginning and end of three trials.

1 = Very Uncomfortable
2 = Uncomfortable
3 = Some Discomfort
4 = Slight Discomfort
5 = No Discomfort

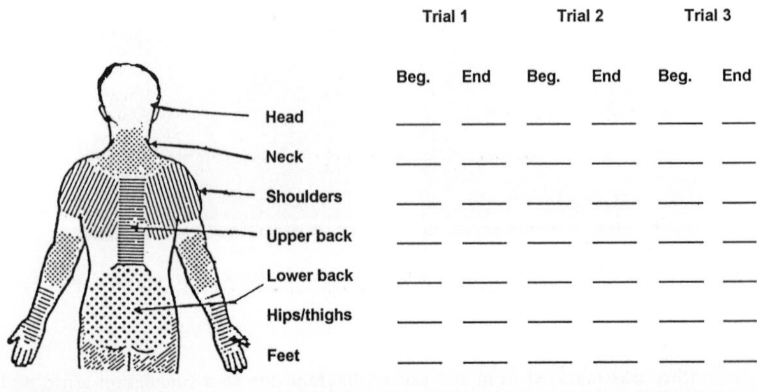

	Trial 1		Trial 2		Trial 3	
	Beg.	End	Beg.	End	Beg.	End
Head	___	___	___	___	___	___
Neck	___	___	___	___	___	___
Shoulders	___	___	___	___	___	___
Upper back	___	___	___	___	___	___
Lower back	___	___	___	___	___	___
Hips/thighs	___	___	___	___	___	___
Feet	___	___	___	___	___	___

* This device is depicted in Figure 16.3. Some minor changes were made to the original version, including an electronic read out and force plate sensor to increase sensitivity of the measurement process.

Table 16.2 Chair feature checklist. Based on the checklist first developed by Shackel et al. (1969), and adapted by Drury and Coury (1982).

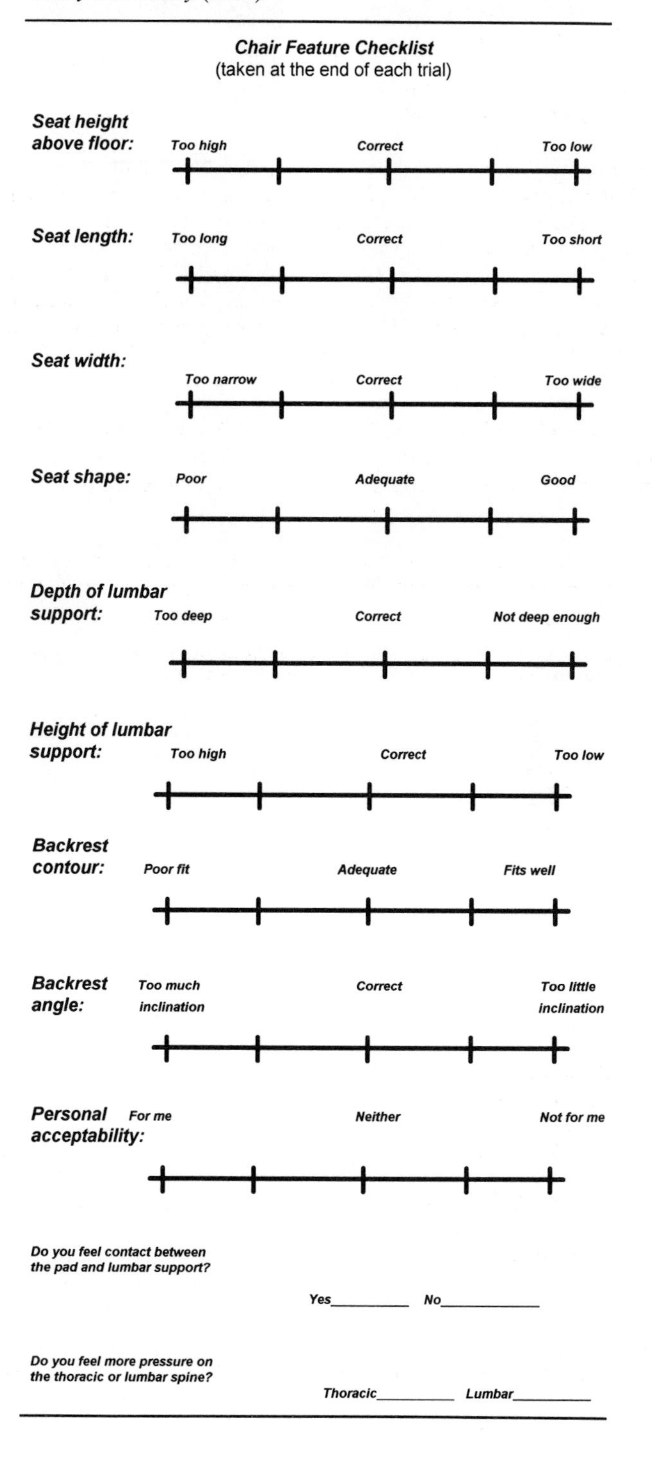

Chair Feature Checklist
(taken at the end of each trial)

Seat height above floor: Too high — Correct — Too low

Seat length: Too long — Correct — Too short

Seat width: Too narrow — Correct — Too wide

Seat shape: Poor — Adequate — Good

Depth of lumbar support: Too deep — Correct — Not deep enough

Height of lumbar support: Too high — Correct — Too low

Backrest contour: Poor fit — Adequate — Fits well

Backrest angle: Too much inclination — Correct — Too little inclination

Personal acceptability: For me — Neither — Not for me

Do you feel contact between the pad and lumbar support?

Yes__________ No__________

Do you feel more pressure on the thoracic or lumbar spine?

Thoracic__________ Lumbar__________

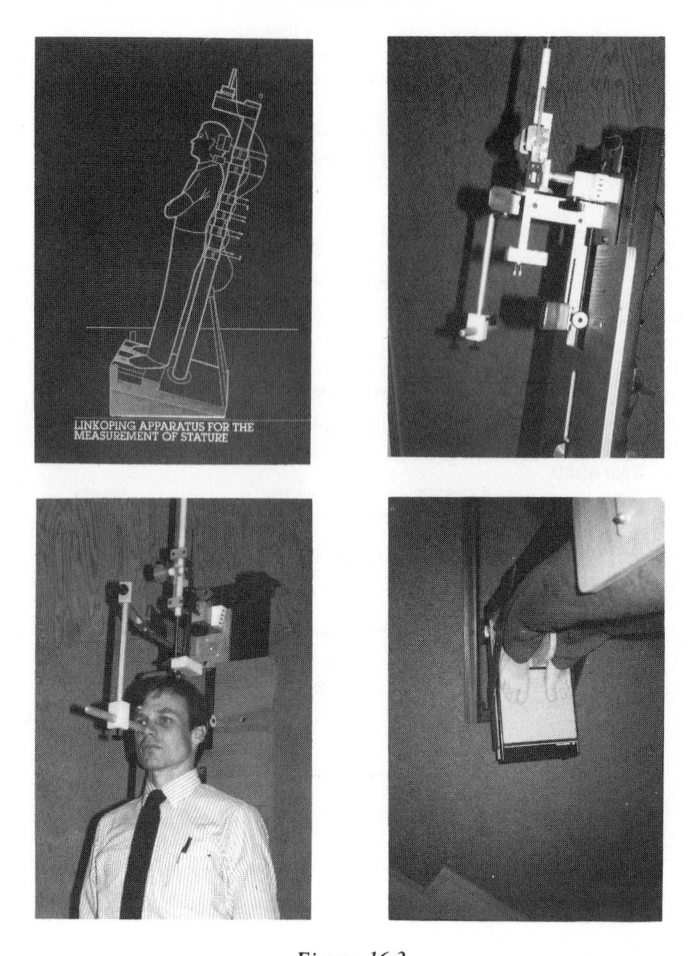

Figure 16.3

(mean = 29.5 ± 6.0 yrs). Body heights ranged from 142.5 to 185.0 cm (mean = 171.3 ± 9.5 cm).

The experimental protocol is as follows:

1. The subjects, wearing gowns, were instructed to lie down on a table for a preset duration. Pillows were placed under the knees, and subjects' hands were folded, and placed on their abdomen. Initially, subjects were asked to lie on a flat table for 30 minutes. Subsequently, the experimental protocol was changed so that subjects lay for 10 minutes, but on a surface tilted at 15°. Initially, subjects lay so feet were higher. Upon discussion, it was subsequently decided to change this configuration to 'heads up', as it was considered more effective at alleviating compressive force. After some subjects complained of the 'hardness' of the table, a thin sheet of cushioning material was used to cover it.
2. After the rest trial, the initial measurements were made. A flexicurve was used to measure subjects' lumbar lordosis upright, flexed, extended. Their sagittal range of motion was measured, in accordance with procedures described above.

3. Prior to the experimental trial, subjects' statures were measured five times on the stadiometer, according to a procedure established by Dr Corlett. An averaged value of stature was ascertained from these measures.
4. Subjects were randomly assigned to one of the three chairs with 3, 5, or 7 cm lumbar depths, and one of the two backrest conditions randomly set at 90° or 110°. Twenty-five subjects were randomized into the 90° backrest angle group, and 22 subjects were randomized into the 110° backrest angle group.
5. After adjusting the lumbar height according to their individual preferences, subjects then sat in the experimental chair for 30 minutes so that hips and knees were at 90° angles, and their buttocks and lower back were in direct contact with the backrest.

   Note: A sample of these subjects sat for 60-minute trials, and their stature was measured three times; pre-trial, after 30 minutes, and after 60 minutes. The data from these subjects indicated that significant differences existed in stature after the first and second half hour, and these differences were both in magnitude as well as in direction of change. Such findings are surprising, and its reason cannot be immediately surmised. If these findings are reproduced in other studies, they warrant attention; however, this analysis was unfortunately beyond the scope of the present study.

   The experimental seats allowed the lumbar pad to be adjusted up to 12 inches from the seat pan. This range was not sufficiently high for some (taller) users, ranging from 6 to 12 in. at the peak of the lumbar curve from the seat reference point.
6. During the seating trial, subjects rated their body part discomfort (Table 16.1) at the beginning, middle, and end of the 30 minute period. At the completion of the sitting trial, subjects' stature was again measured five times, and an averaged value of stature was ascertained.
7. This protocol was repeated for each of the three seat lumbar depth conditions for each subject.

Results

1. Intra-examiner correlation coefficients of lumbar characteristics of the upright, flexed, and extended seated positions were calculated between the flexicurve measures. Correlation coefficients were 0.96 for the flexion measurements, 0.80 in extension, and 0.95 for the upright (neutral) positions. ($p < .05$).
2. Analysis of Variance indicated that neither the lumbar seat depth nor the backrest angle significantly affect stature (measured from the stadiometer), although the effect of seat lumbar depth reached $p = 0.11$. Furthermore measures of lumbar lordosis and range of motion were not able to predict either subjective ratings of discomfort or changes of stature (measured by the stadiometer) (*see* Figure 16.4 for changes in discomfort ratings with time).
3. In general, the 3 cm lumbar support depth was rated as the most comfortable for all subjects, at both backrest angles.

Discussion

The reasons for the lack of significance between groups with different lumbar characteristics, is not clear. Possible reasons for this lack of significance include:

- the variability in stature shrinkage between individuals is too large for the sample size used;
- limitations of the research instruments; for example, since the lumbar supports of the experimental seats were semi-circular, changes in their depth were accompanied by increased thickness;

Discomfort Scale

90 Degree Backrest Angle

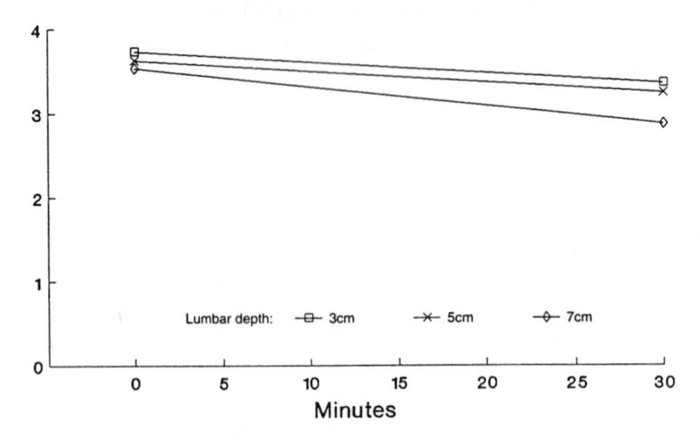

110 Degree Backrest Angle

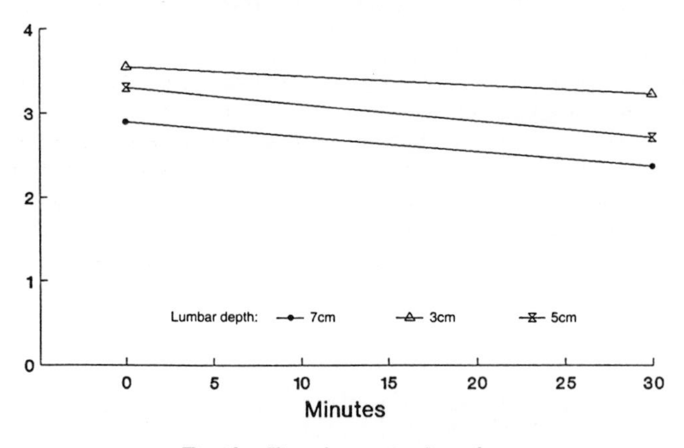

Both Backrest Angles

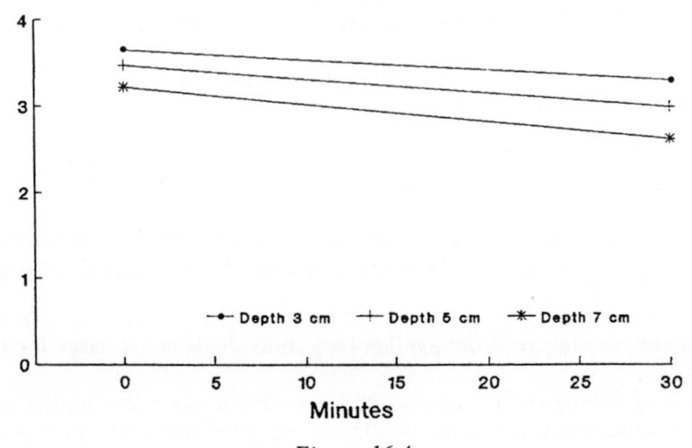

Figure 16.4

Table 16.3 Changes of Stature Associated With Back Rest Angle and Lumbar Pad Depth

	3 cm		5 cm		7 cm	
Lumber Pad depths:	Mean	SD	Mean	SD	Mean	SD
Back rest angles:						
90° ($n = 12$)	1.16	4.02	−1.26	2.55	−1.51	2.27
110° ($n = 11$)	0.27	1.68	−1.07	3.53	−0.39	3.59

Changes in stature (in mm) by seat lumbar depth and back rest inclination. These values represent differences from pre- to post-trial changes in stature, expressed in mm. Positive numbers represent increases, and negative numbers represent shrinkage in height.

- the seat lumbar height did not adjust high enough to accommodate all users; this may well have differentially influenced the 90° condition, since people's lumbar height tends to lower as they lean back;
- limitations of the research protocol; for example, changes in the duration and method of lying down during the resting phase may have reduced its sensitivity.

It is also possible that the individual differences in lumbar curvature do not affect the associated user's needs in lumbar curvature. Some suggest that detrimental effects of seating may be more related to a lack of physical variation than specific characteristics of the seat (Winkel and Oxenburgh, 1990). Bendix (1993) even suggests that seats not be so comfortable as to discourage movements. Reinecke and Hazard (1993) emphasize the benefits associated with continuous changes in depth induced by passive motion of the lumbar spine.

Acknowledgment

This study was part of the Lumbar Seat Project, which was jointly conducted by Humanics ErgoSystems, Inc., the Los Angeles College of Chiropractic, and Designworks/USA. It was partially funded by the International Association of Chiropractic and the Pacific Consortium for Chiropractic Research. Our gratitude to Dr Nigel Corlett for his assistance throughout the project.

References

Andersson, G. B. J., (1980), The load on the lumbar spine in sitting postures, in Oborne, D. J. and Levis, J. A. (eds) *Human Factors in Transport Research*, New York: Academic Press.

Andersson, G. B. J., (1986), Loads on the spine during sitting, in Corlett, N., Wilson, J. and Manenica, I. (eds) *The Ergonomics of Working Postures*, London: Taylor & Francis, 109–18.

Andersson, B. J. G. and Ortengren, R., (1974), Lumbar disc pressure and myo-electric back muscle activity during sitting (Part II), *Scandinavian Journal of Rehabilitation Medicine*, **6**, 115–21.

Bendix, T., (1987), Adjustment of the seated work place – with special reference to heights and inclinations of seat and table, Master's thesis, Copenhagen: Laegeforeningens Forlag.

Bendix, T., (1994), Low back pain and seating, in Lueder, R. and Noro, K. (eds) *Hard Facts about Soft Machines: The science of seating*, London: Taylor & Francis.

Bendix, T., (1984), *Trunk posture when sitting at different inclinations of seat and different heights of seat and table*, Copenhagen: The Laboratory of Back Research, The University of Copenhagen.

Bendix, T. and Biering-Sorensen, F., (1983), Posture of the trunk when sitting on forward inclining seats, *Scandinavian Journal of Rehabilitation Medicine*, **15**, 197–203.

Bendix, T. and Hagberg, M., (1984), Trunk posture and load on the trapezius muscle whilst sitting at sloping desks, *Ergonomics*, **27**(8), 873–82.

Boudrifa, H. and Davies, B. T., (1984), The effect of back rest inclination, lumbar support and thoracic support on the intra-abdominal pressure while lifting, *Ergonomics*, **27**(4), 379–87.

Branton, P., (1984), Back shapes of seated persons – how close can the interface be designed? *Ergonomics*, **15**(2), 105–7.

Brunswic, M., (1984), Seat design in unsupported sitting, *Proceedings of the International Conference on Occupational Ergonomics*, Toronto, 294–8.

Burton, A. K., (1985), Measurement of regional sagittal lumbar mobility and posture by means of a flexible curve, in Corlett, N., Wilson, J. and Manenica, I. (eds) *The Ergonomics of Working Postures*, London: Taylor & Francis, 92–9.

Burton, A. K., (1986), Regional lumbar sagittal mobility; measurement by flexicurves, *Clinical Biomechanics*, **1**(1), 20–6.

Corlett, E. N., (1990), Evaluation of industrial seating, in Wilson, J. R. and Corlett, E. N. (eds) *Evaluation of Human Work: A practical ergonomics methodology*, London: Taylor & Francis, 510–2.

Corlett, E. N. and Bishop, R. P., (1976), A technique for assessing postural discomfort, *Ergonomics*, **19**, 175–82.

Corlett, E. N. and Eklund, J. A. E., (1986), Change of stature as an indicator of loads on the spine, in Wilson, J. R., Corlett, E. N. and Manenica, I. (eds) *The Ergonomics of Working Postures: Models, methods and cases*, London: Taylor & Francis.

Drury, C. G. and Coury, B. G., (1982), A methodology for chair evaluations, *Applied Ergonomics*, **13**, 195–202.

Eklund, J., (1984), *Industrial Seating and Spinal Loading*, Thesis submitted to the University of Nottingham, Dept. Production Engineering and Production Management, Nottingham, U.K.

Eklund, J. A. E. and Corlett, E. N., (1986*a*), Experimental and biomechanical analysis of seating, in Wilson, J. R., Corlett, E. N. and Manenica, I. (eds) *The Ergonomics of Working Postures: Models, methods and cases*, London: Taylor & Francis.

Eklund, J. A. E. and Corlett, E. N., (1986*b*), Shrinkage as a measure of the effect of load on the spine, *Spinal Load*, **9**(2), 3.2–3.7.

Grandjean, E., Boni, A. and Kretschmar, H., (1969), The development of a rest chair profile for healthy and notalgic people, in Grandjean, E. (ed) *Sitting Posture*, London: Taylor & Francis, 193–201.

Hosea, T. M., Simon, S. R., Delatizky, J., Wong, M. A. and Hsieh, C. C., (1986), Myoelectric analysis of the paraspinal musculature in relation to automobile driving, *Spine*, **11**(9), 928–36.

Nachemson, A. L., (1981), Disc pressure measurements, *Spine*, **6**(1), 93–7.

Nachemson, A. and Elfstrom, G., (1970), Intravital dynamic pressure measurements in lumbar discs, *Scandinavian Journal of Rehabilitation Medicine*, (Suppl. 1).

Reinecke, S. M. and Hazard, R. G., (1994), Continuous passive lumbar motion in seating, in Lueder, R. and Noro, K. (eds) *Hard Facts about Soft Machines: The science of seating*, London: Taylor & Francis, 157–164.

Schoberth, H., (1962), *Sitzhaltung, Sitzschaden, Sitzmoebel*, Berlin: Springer, Cited in: Branton, 1984, *op. cit.*

Schoberth, H., (1978), *Correct Sitting at the workplace* (translated from the German) Frankfurt, W. Germany: University of Frankfurt, Ostsee Clinic.

Shackel, B., Chidsey, K. D. and Shipley, P., (1969), The assessment of chair comfort, in Grandjean, E. (ed) *Sitting Posture*, London: Taylor & Francis, 155–92.

Tichauer, E. R., (1978), *The Biomechanical Basis of Ergonomics*, New York: Wiley.

Webb Assoc., (1978), *Anthropometric Source Book*, NASA Ref. No. 1023, Scientific Technical Information Service, Vol. 1, 1–24.

Winkel, J. and Oxenburgh, M., (1990), Towards optimizing physical activity in VDT/office work, in Sauter, S. L., Dainoff, M. J. and Smith, M. J. (eds) *Promoting Health and Productivity in the Computerized Office: Models of successful ergonomic interventions*, London: Taylor & Francis, 94–117.

Ydreborg, B., (1990), The Nordic Questionnaires, Cited in: Corlett, E. N., Static muscle loading and posture evaluation, in Wilson, J. R. and Corlett, E. N. (eds) *Evaluation of Human Work: A practical ergonomics methodology*, London: Taylor & Francis, 564–9.

PART VII

Seat pressure distributions

17

The biomechanical relationship of seat design to the human anatomy

Joseph A. Sember III

Force sensing array

The system developed to accurately measure pressures across the entire seat surface is called a Force Sensing Array. It comprised:

1. Mat – This is a grid of 225 force sensing resistors (FSRs) connected so as to form a series of 15 rows and 15 columns. A later development produced a system with a mat of 1064 sensors in a grid of 32 rows and 32 columns.
2. Mux – The Mat is connected to a multiplexer box (Mux) which receives the resistance readings of the FSRs, and converts them to signals that the computer can process into useful pressure information.
3. FSA Program – The software developed for this purpose massages the raw data from the Mat and displays several different output configurations.
   (a) Isobars: These display the pressures as a series of different colour levels, which change as pressures increase.
   (b) Values: The data can also be shown as a grid of numbers that correspond to the pressures being experienced by the mat, and expressed in either mm Hg (Mercury), psi or kilopascals.
   (c) Wire form: A three dimensional wire form may also be portrayed.
   (d) Finally, the data may be expressed as a series of proportional squares that vary in size with pressure.

   Other interesting types of information that are available from the system include the centre of gravity, and changes in the dynamics of the seating environment in real time.

Basic concepts

Figure 17.1 shows a representation of the human pelvis. The two bony prominences on the bottom are called the ischial tuberosities (also called ischia, or seat bones) and the two protrusions on the femurs (thigh bones) are the trochanters. When the human skeleton, and more specifically the pelvic bone, is oriented as in a seated position, the ischia are the parts of the bony structure which are closest to the seat and, therefore, experience the most pressure. We have found that when a person first assumes a seated posture (this being an inherently unnatural position) the body begins to react to gravity. This reaction begins as the fat and muscle tissue directly beneath the ischia slowly move out from under these bony prominences and allow them to 'core down' to a point approaching the skin. The longer the person remains seated, the more pronounced this condition becomes.

As this continues, the areas of the skin directly under the ischials reach a point where the pressure being experienced exceeds that at which the capillaries are forced to close. At this point the skin begins to die. When posture is constrained, between 10 and 15 minutes is usually required to reach this point.

The first sensory indication that one's seated condition is going awry, is a burning sensation under the ischials. As the soft tissue spreads out, and the contact area increases, the trochanters can also come into play, and begin to experience pressure. If no action is taken when the burning sensation occurs, the pain threshold will be reached at 8.2 psi in about 30 minutes.

Figure 17.2 is a representation of the pressure levels and gradients produced on a flat surface, immediately after sitting down. Here, it is apparent that the ischia are generating substantial pressures. In addition to pressures created on the buttocks and thighs, the seated position induces certain conditions in the lower spinal column. These are as follows:

1. When sitting upright, such as in a folding chair or church pew, the angle created between the lumbar vertebra and the sacrum (tail bone), the lumbrosacral shear angle, is 24°. In this condition, the discs of the spinal vertebrae reach pressures of 8 psi. The vertebral detail (which allows the spine to flex, but not come apart) is called the facet; pressures on these facets of the lower spine, while sitting upright, reach 6 psi. Figure 17.3 shows the pressure map of a user on a folding chair after five minutes.

2. By this time, the user's sensory warning signals should give them an overwhelming urge to move to relieve those pressures, and reinstate occluded circulation. If action is not taken when the burning sensation occurs, the pain threshold will be reached at 8.2 psi; this generally takes about 30 minutes (see Figure 17.4).

3. If the seated environment is dynamic, as with a moving wheelchair or automobile, the bouncing will magnify stresses, and reduce the time required to produce such undesirable results. Figure 17.5 shows a pressure graph resulting from sitting on a bouncing chair after only five minutes. While these pressures are momentary, the relationship of the bone to the soft tissue has shifted to the worst possible configuration. That is, most of the fat and muscle that normally provide a cushion between the bone and skin have been ploughed out of the way by the ischials. At this point, even if the bouncing stops, pain and tissue damage results.

There are several means of improving these conditions: first and most obviously, increase the bearing areas to minimize the pressure on any given area; and secondly, actively reduce the pressures under the seat bones (ischial tuberosities).

These actions will have the effect of increasing the pressures under the normally low pressure areas of fat and muscle which can afford them.

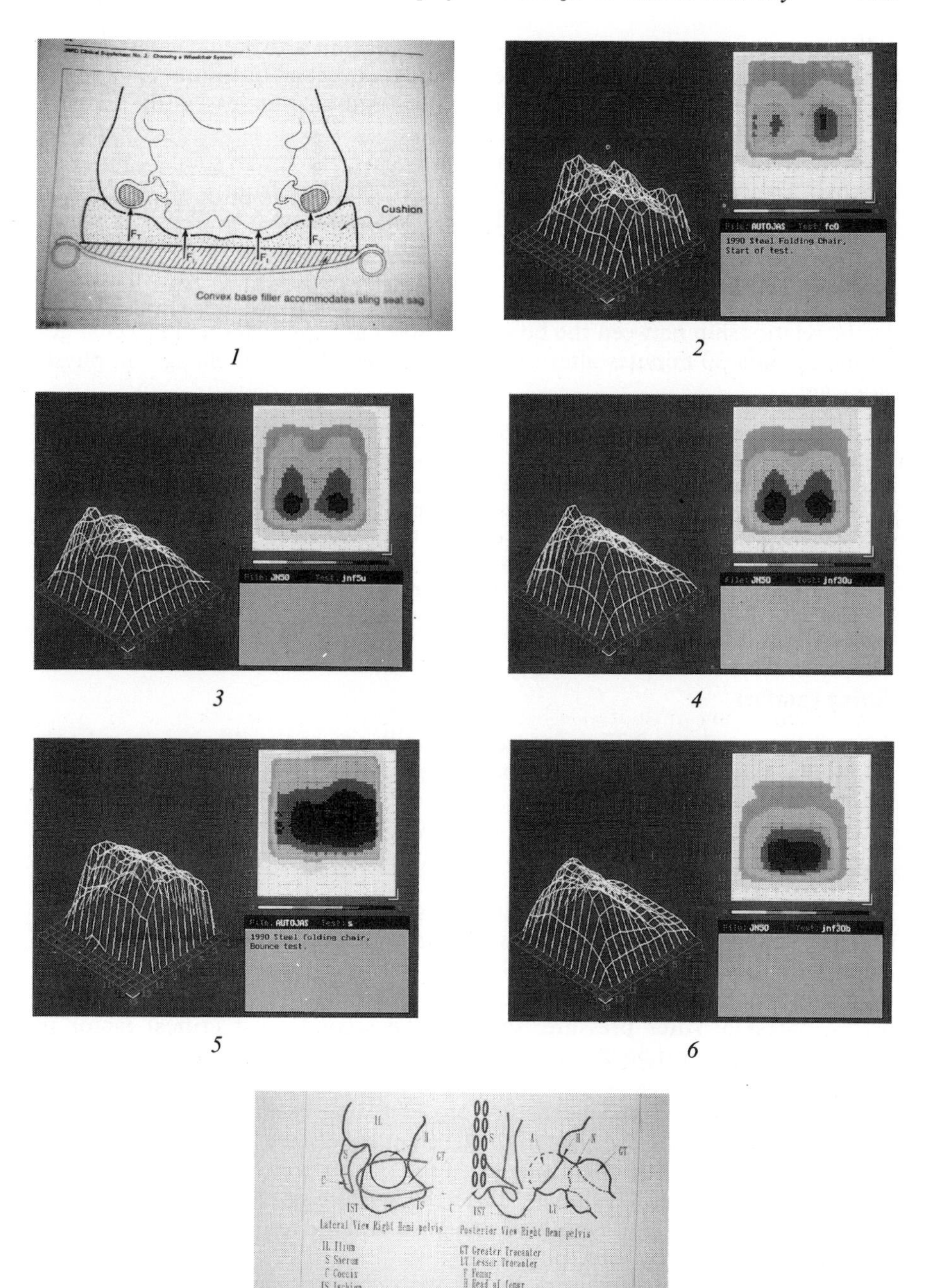

1 2 3 4 5 6 7

Figures 17.1–7

Pelvic stability can also improve seat pressures. In a dynamic environment, a side rocking motion will increase the rate of ischial pressure buildup. A front-to-back rocking motion does the same, and also increases the pressure under the coccyx (tailbone).

As the redistribution of fat and muscle occurs, the lower spinal column is affected. This increases the lumbro-sacral disc loading from 8 to 29 psi, the torque from 0 to 24 foot pounds, and facet pressures from 6 to 14 psi.

Physiological changes while seated

As the relationship between the bones and soft tissue (muscles and fat) changes during the first 30 minutes after sitting down, the additional changes in physiology are:

1. Since pressure inhibits blood flow, the tissue immediately around the blood vessels thickens, to help keep them open.
2. Lactic acid concentrates in the muscles.
3. Water builds up in subcutaneous tissue (under the skin).
4. Ischial bursae thicken to provide a cushion below the bone and, very importantly, as a result of tissue damage caused by shear, prostaglandin E2 is released, first locally, and then systemically. This is a chemical which contributes to depression, fatigue and sluggish reflexes!

Sitting comfort

Since pressures above 0.73 psi cause the capillaries in the skin to close, levels below this threshold can be tolerated indefinitely. Pressures greater than 1.7 psi lead to skin cell death (necrosis).

Pressure gradients

As can be seen in Figure 17.6, the different shaded areas represent variations in pressure level and are depicted as isobars. The width and the number of isobars in proximity to one another are referred to as gradients. A large number of narrow isobar bands in a small area indicate a high degree of shear within the skin. After pressure, shear is the second most critical factor in producing discomfort in the able-bodied and decubitus ulcers in the wheelchair bound.

Figure 17.7 depicts the pelvic region, including relative locations of the ischial tuberosities, the coccyx and the trochanters.

Unlike the flat folding chair pressure maps shown previously, in an automotive seat or reclining office chair, since the posture of the subject is slightly reclined, the pelvic region is rotated rearward and the resulting seated configuration is represented by three points.

Three automotive seats were tested with the pressure mat; Figure 17.8 depicts the first of these. This is a 1990 Toyota Corolla driver seat, after one minute of sitting. It exemplifies solution 1 (mentioned earlier); the load is distributed as much as possible to minimize local build-ups of pressure.

Figure 17.9 portrays the distribution of pressure with the same seat after 30 minutes. The skin contact area has increased and severe discomfort is induced in the green areas under the ischia.

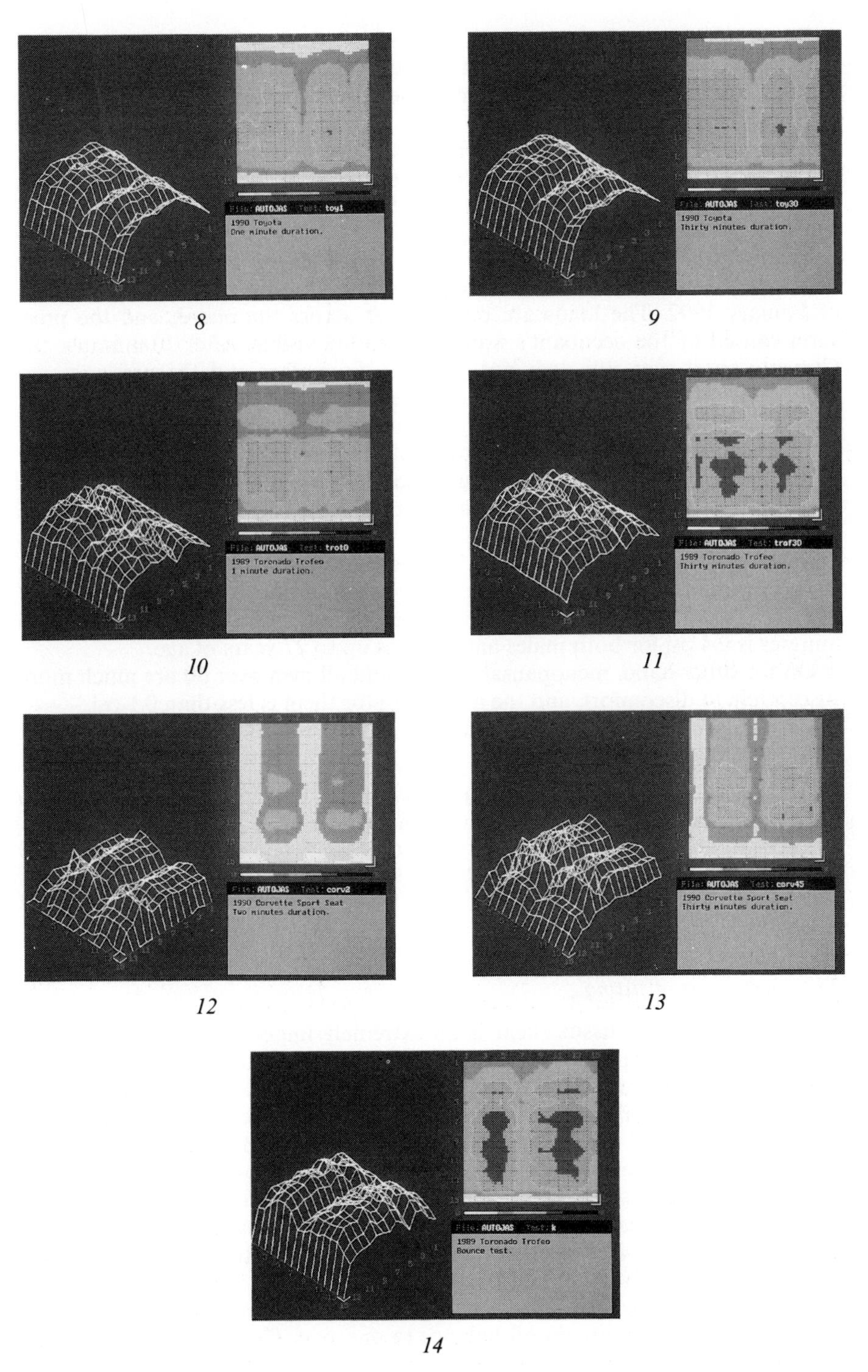

Figures 17.8–14

Figure 17.10 portrays a 1989 Oldsmobile Toronado Trofeo seat after one minute of sitting. It should be noted that the thigh and lumbar adjustments on this seat are, in this test, fully deflated. The basic pressure contours of this seat resemble those of the Toyota seat. After 30 minutes, the Trofeo exhibits the same conditions as the Toyota seat, (i.e., attempts to spread the pressures around the surface) except to a greater degree (see Figure 17.11). However, when the thigh supports are inflated, the ischial pressures improve dramatically.

Figure 17.12 is a 1990 Corvette sport seat with the lumbar support fully deflated. This is one of, if not the most, contoured production seat available as of January 1992. The loads are much lower across the board, and the pressures caused by the occupant's wallet are readily visible. After 30 minutes, the Corvette sport seat still does not produce discomfort and the pressures are more appropriately distributed (see Figure 17.13). The inflation of the lumbar support improves the curvature of the lumbar and redistributes the cushion pressures so that even these small points of pressure are eliminated.

Age and gender exert a substantial effect on the physiological results of seat pressures. All of the pressure tolerance figures described so far assume that the person is young and healthy.

As can be seen back in Figure 17.7, the sacro-coccygeal is a joint between the coccyx and the bone immediately above it (sacrum). The maximum sacro-coccygeal load which can be sustained before inducing discomfort within 15 minutes is 0.4 psi for both males and females up to 27 years of age.

On the other hand, menopausal women and all men over 50 are much more susceptible to discomfort, and the maximum for them is less than 0.1 psi.

For the ischia, the maximum loads that can be sustained without discomfort after 15 minutes are:

(a) 1.2 psi for men up to the age of 30 and women up to 40. This difference is due to the fact that female physiology provides more natural padding in the ischial areas.
(b) 0.5 to 0.9 psi for both genders over 40 years of age. It seems that after the age of 40 padding on the buttocks becomes equal.
(c) For the elderly, less than 0.3 psi is the maximum sustainable pressure without discomfort.

The energy cost of sitting

As previously noted, tissue shear is an extremely important factor due to the tissue breakdown which can result. As with other factors, the extent of this breakdown is time and age dependent. The body is capable of repairing such damage, but this ability varies with age. As a result, the maximum time that any given individual can tolerate a particular shear condition varies with that individual's ability to repair such damage. This repair is usually accomplished during sleep and, as one might expect, proceeds most rapidly in the young.

The body is constantly working to maintain balance. As a result, the buttock and leg muscles work to balance the pelvis. Activity of the back and abdominal muscles helps to maintain a centre of gravity.

Figure 17.14 shows the pressure map of the Trofeo seat during a bounce test, where the occupant was off balance to one side. The front-to-back shift in the alignment of the spine moves the centre of gravity forward and back, and induces torque. Side-to-side shifts increase rotary torque on the discs.

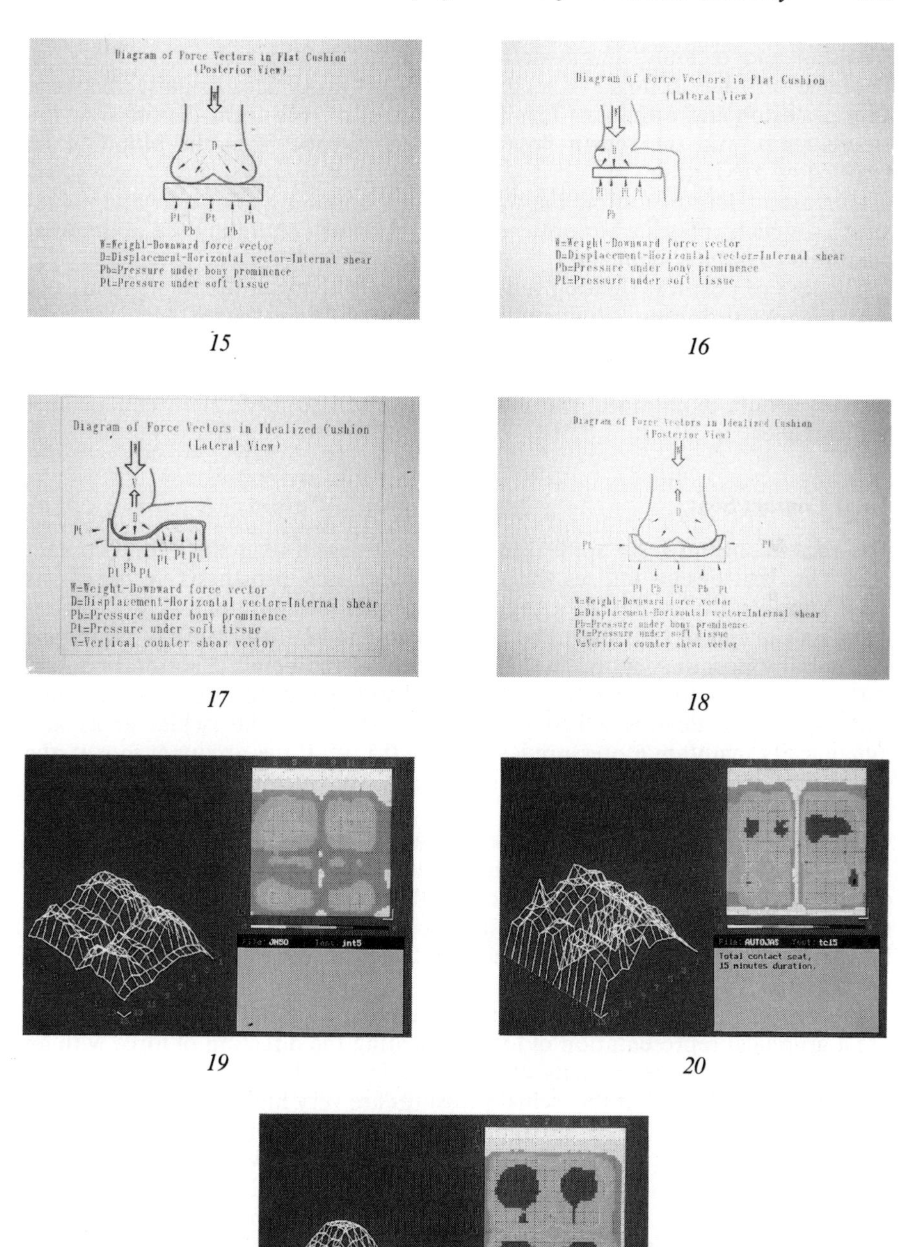

15 *16*

17 *18*

19 *20*

21

Figures 17.15–21

Tilting and reclining has a detrimental effect to the respiratory function. Thoracic tilt (due to angle of recline) of 12–34° maximizes disparity between lung perfusion and diffusion. This is the flow of oxygen from the lungs to the bloodstream, and the return flow of carbon dioxide from the blood to be expelled by exhalation.

Increased compression on the thorax shifts breathing efforts to abdominal muscles, which creates a hypnotic effect. Hypnotists commonly use abdominal breathing to induce hypnosis.

The loss of postural righting reflexes reduces activity in the cerebellum, and increases reaction time. This condition is especially important when driving, since small differences in reaction time may literally mean the difference between life and death. To the extent that centre of gravity is maintained and correct posture maintained, the seat occupant will be more alert, comfortable and capable.

Total Contact Seat

The above concepts have applied to wheelchair design to create a total contact seat. This seat design utilizes the same concept as a prosthesis. A prosthesis must support the stump of a leg, for example, without imposing excessive loads on the end of the bone. The end of the bone in a leg stump is analogous to the bony prominences (ischial tuberosities) of the pelvis. The total contact seat, in addition to providing the most advantageous contours, is equipped with pneumatic devices which sense the pressures in the ischial areas and dynamically maintain a pressure level below 0.5 psi. If the occupant moves, the seat will respond to the changed conditions and readjust the critical areas accordingly.

Another version of the total contact seat is equipped with five different inflatables, a small compressor, a ripple sequencer and a computer chip. This system not only controls pressure, in the critical areas, but sequentially changes pressures to alternate pressure points. This provides the opportunity for pressures to be applied to areas that can best handle them, yet allow periodic relief as the system cycles.

Figures 17.15 and 17.16 show the force vectors on a flat seat. Force vectors are a graphical representation of force indicating the direction of force with an arrow and the relative intensity of that force by the size of the arrow. In these figures you will note that the ischial pressures are very high.

Figures 17.17 and 17.18 are representations of the different forces acting statically on the user of the total contact seat. This shows the various forces acting on a body, and the resulting force vectors. It should be noted that unlike standard seats, its contours provide additional contact area, and correspondingly smaller force vectors.

The contours provide support in the rear, and counteract forces under the thigh areas, generating greater stability. Figure 17.19 is a pressure map of the total contact seat. It can be seen that the pressures under the ischia are very low, much lower than the 0.7 psi capillary pressures. Further, the load has been redistributed to areas of muscle and fat which can more effectively sustain the pressures.

Figure 17.20 depicts the same seat and occupant after 15 minutes. It can be seen that even after that duration, the ischial pressures in row 11 have not increased. The pressure rise in the right trochanter area is a small appointment book in the pocket of the occupant.

After 30 minutes as shown in Figure 17.21, pressures are increasing in the rear of the buttocks; yet the ischial areas remain unaffected.

Summary

The results of this research and development are very exciting to a seat designer since there is now a series of quantifiable parameters and design goals as well as a measurement tool to quantify them.

18

The biomechanical assessment and prediction of seat comfort

Clifford M. Gross, Ravindra S. Goonetilleke, Krishna K. Menon, Jose Carlos N. Banaag and Chandra M. Nair

Introduction

The widespread use of computers has been accompanied by a heightened interest in the ergonomics of office seating. In addition to improving the well-being of the worker, a comfortable seat may contribute to productivity (e.g., Dainoff and Dainoff, 1986).

Until recently, much remained unknown about how to design ergonomic seats. We have been limited by the lack of a viable operational definition for comfort and a difficulty in relating comfort to seat design parameters.

As a result, many researchers have used the notion of discomfort to understand comfort. Discomfort is caused by biomechanical stress acting on the body. One cause of stress on the body is the result of the static muscle activity required to hold the body in a near stable position.

Good seating supports the user within a minimum of cumulative stress on the back, shoulders, and legs. When the body is not properly supported, several muscle groups act together to restore stability, and may cause static muscle loading*. Branton (1969) suggested that people trade off stability with freedom of movement (e.g., crossed legs). This suggests that no single posture will provide comfort for prolonged sitting activities.

In this paper, we attempt to define the ubiquitous term 'comfort' by analysing a biomechanical correlate to sitting comfort; that is the pressure distribution between the seat surface and its occupant. A variety of approaches can be

* This does not imply that proper support to the body should be achieved by an anatomical seat pan which conforms to the shape of the body as this restricts body movement and presents sizing problems.

used to measure the pressure distribution. Traditional force measurement devices are of little value as they interfere with comfort. The steps for measuring the pressure distribution and quantifying 'comfort' requires:

1. An unobtrusive measurement interface between the seat and occupant. We developed such a system which included the supporting electronics and software necessary to drive, display and calculate the load distribution.
2. Information on how seat pressure relates to comfort. To this end, we conducted over 1100 short-term (5–10 minutes) seat-subject evaluations of 50 seats that compared load distributions with subjective comfort ratings. We found that comfort ratings can be predicted from patterns of weight distribution over a seat surface. To improve our comfort prediction capacity we constructed multiple regression equations to predict interval level comfort from seat pressure variables (r^2 greater than 90 per cent).

Although we have related seat comfort to a biomechanical variable, to be of most use this information must now be applied in the seat design process. This chapter will review the measurement of seat pressures and describe the developments for the next generation of seating—the intelligent seat.†

Seat comfort research

Studies on quantifying seat comfort have typically approached the problem in two ways:

1. Compare the dimensions and postures of seated individuals with seat measurements.
2. Relate the occupant's weight distribution on the seat to seat geometry, contour and firmness.

The approaches used for assessing the comfort in automobiles have been different from those used in office seating. A brief review of the two approaches will be presented first, followed by the common element, sitting pressures.

Anthropometric and postural studies

Industrial and office seating

Eklund (1986) has shown that when the feet support one third or more of the body weight, discomfort is experienced because continuous muscle activity is required. The force transmitted to the ground through the feet depends on many factors. One factor is the seat height, which is closely related to popliteal height.

A well-designed seat should fit all sizes and shapes of users. Size differences may be accommodated through adjustability. Many designers view sitting as a static activity and design seats based on anthropometry.

However, sitting is a dynamic activity and seat design should be based on the required changes in postural and anthropometric characteristics of the user over time. Anthropometric accommodation is necessary but not sufficient for comfort.

† Note: The *Intelligent Seat* and the transparent interface for recording pressures and calculating comfort are both BCA patented.

When changing from a standing to a sitting posture, the hip angle decreases from 180° to approximately 90°. Anatomically, this is a fairly complicated movement; about 60° of the bending takes place in the hip joint and the remaining 30°–40° is due to the flattening of the lumbar curve (i.e., the lordosis of the spine tends to flatten out because the pelvis rocks posteriorly). Most of the spinal shape changes occur between the third, fourth and fifth lumbar discs. Ergonomic seats typically attempt to help restore lumbar lordosis and minimize disc pressure with reclining seat backs as recommended by Andersson (1974) or the forward tilting of the seat pan as recommended by Mandal (1981).

Seated comfort while working has been studied primarily through the assessment of anatomical and physiological characteristics. A few ways that seats have been evaluated include the following:

1. Body height shrinkage, using a stadiometer (Corlett, 1990) to measure loads on the spine.
2. Cross-modality matching. This technique relates the seat discomfort to pressure distributions measured on the seat and the back rest (Wachsler and Learner, 1960; Habsburg and Mittendorf, 1980).
3. Spinal disc pressures and muscle loading (e.g., Andersson, 1974; Eklund and Corlett, 1984).
4. Psychophysical rating or ranking scales are the most commonly used subjective techniques for evaluating seats. The general comfort rating scale and Chair Feature Checklist (CFC) developed by Shackel *et al.* (1969) and the Body Part Discomfort Scale (BPDS) of Corlett and Bishop (1976) have been widely used.
5. Evaluation of pressure distributions.

Fleischer *et al.* (1987) adopted a different approach by studying the image patterns of weight displacements of seated subjects. They suggested that seats be designed to avoid restricting movement rather than conform to the total body. Lueder (1983) has provided a review of the many approaches used to assess seat comfort.

Car seats

The basic reference source for designing automobile seats in the United States is the SAE Handbook (1990). This handbook specifies (among other data) the dimensions, adjustability, and the configuration of automotive seating. All dimensions are determined by using a two-dimensional H-point template and a three dimensional H-point machine (Figure 18.1).

The H-point machine is one which has a seat back and seat pan representation of a deflected seat contour for adult males. This machine simulates the human torso and thigh and is mechanically articulated at the H-point. The lower leg and thigh segments can be adjusted to the 10th, 50th and 95th percentile adult male dimensions. The H-point of a seat is determined when this machine is placed in a prescribed manner with the 95th percentile male leg and thigh segments.

Earlier recommendations (Van Cott and Kincade, 1972; Rebiffe, 1969) are obsolete as the configurations of the driver's seat and the position of the eye with respect to the windscreen have changed dramatically over the past decade. Hence, only the more recent work on automotive seat comfort will be reviewed.

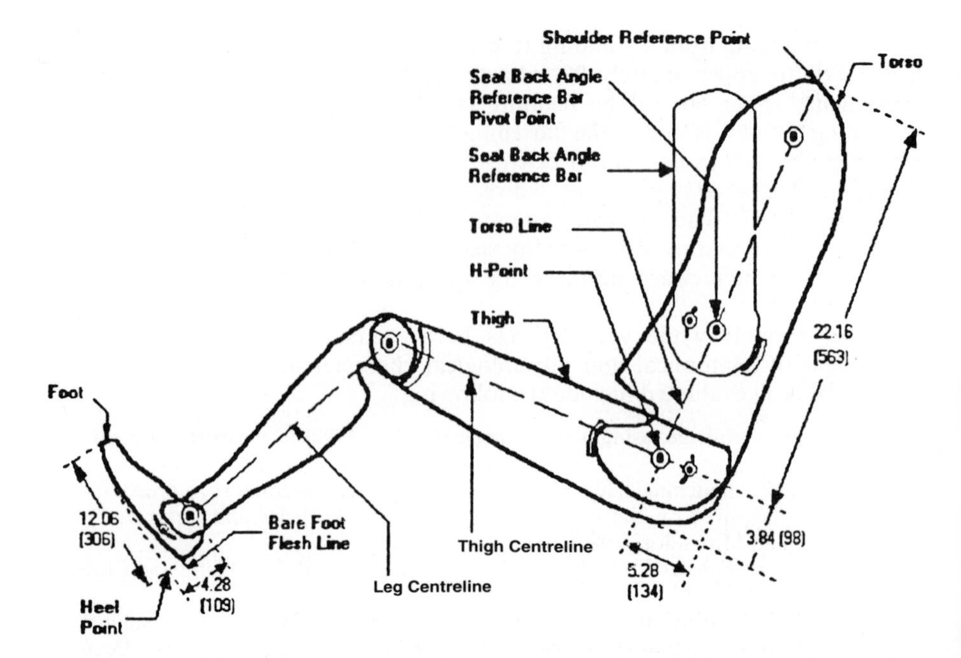

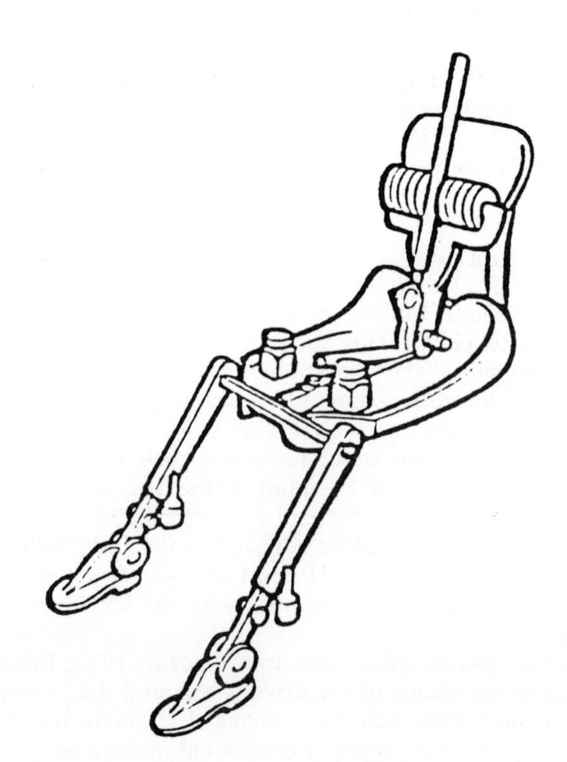

Figure 18.1 (a) *H-point template;* (b) *H-point machine*

The methodology used by a major automobile company to assess the comfort of the driver's seat is based on contour/dimension scores, subject preference scores and body pressure scores. The contour/dimension scores are based on dimensional guidelines shown in Table 18.1 for a family car, luxury car, and sports car.

Occupant preference scores were assessed by administering a seat comfort questionnaire. During the test, the subjects wore standard test clothing, sat in the driver's side of the vehicle, adjusted the seat to a comfortable driving position and rated their short-term comfort level. The subjects rated nine features of the seat, namely: fore/aft control; recliner; lumbar control; front cushion tilt; seat back wings; thigh adjustments; lower back support; seat firmness and overall comfort on a 3-point scale (poor, fair or good).

The body pressure scores were evaluated by comparing the seat pressure contours against comfortable seat back and seat cushion contours (Figure 18.2).

In another study Hubbard and Reynolds (1984) interpreted existing data on the external car body configuration and skeletal geometry to determine body positions for small women (5th percentile), average men (50th percentile), and large men (95th percentile). They note that the SAE design method of using 2-D and 3-D templates does not necessarily fit the people who will use the seat, nor predict how a person will fit a proposed seat design. They estimated the co-ordinates of 16 anthropometric landmarks for the small, average and large users, but point out the need for synthesizing basic information about seating biomechanics into a predictive model for seat design.

Seating research at Audi (Weichenrieder and Haldenwanger, 1986) led to recommendations for dimensions and postural angles to accommodate the 5th percentile female and the 95th percentile male. The postural recommendations were based on comfort levels that varied from -5 to $+5$. A comfort rating of 0 corresponds to an optimum value for passenger cars while $+5$ and -5 corresponds to the anatomical boundaries. They accounted for the differences between the sexes by considering that women on average:

1. are 10 cm smaller than men
2. weigh 10 kg less

Table 18.1 Seat dimensional guidelines

Dimensions	Family car	Luxury car	Sports car
Maximum cushion length (from H-point)	380 mm	380 mm	380 mm
Minimum cushion width	500 mm	500 mm	500 mm
Deflection at D-point	60–80 mm	80–100 mm	40–60 mm
Flat cushion surface width	150 mm	150 mm	150 mm
Centre of lumbar region (up from D-point)	230 mm	230 mm	230 mm
Radius of lumbar (plan view contour)	450 mm	800 mm	300 mm
Radius of thoracic (plan view contour)	1000 mm	1000 mm	1000 mm
Back height	510–560 mm	510–560 mm	510–560 mm

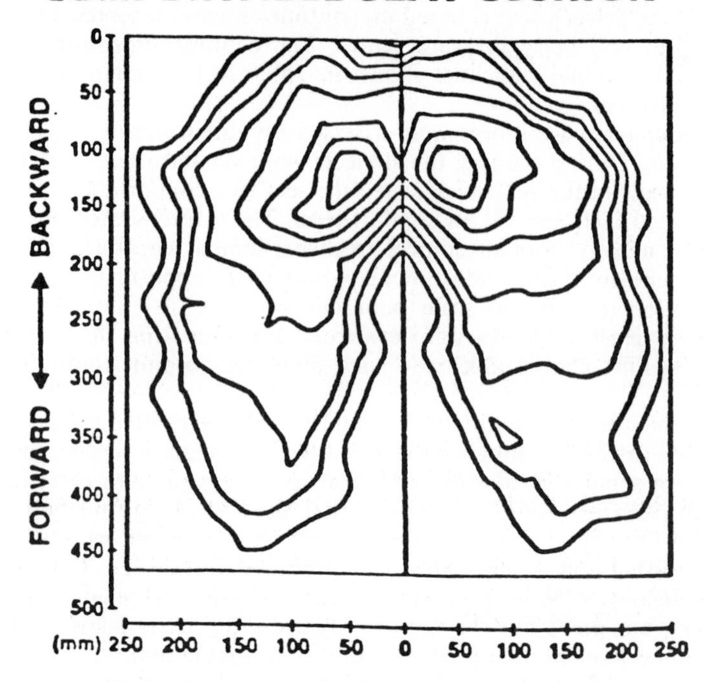

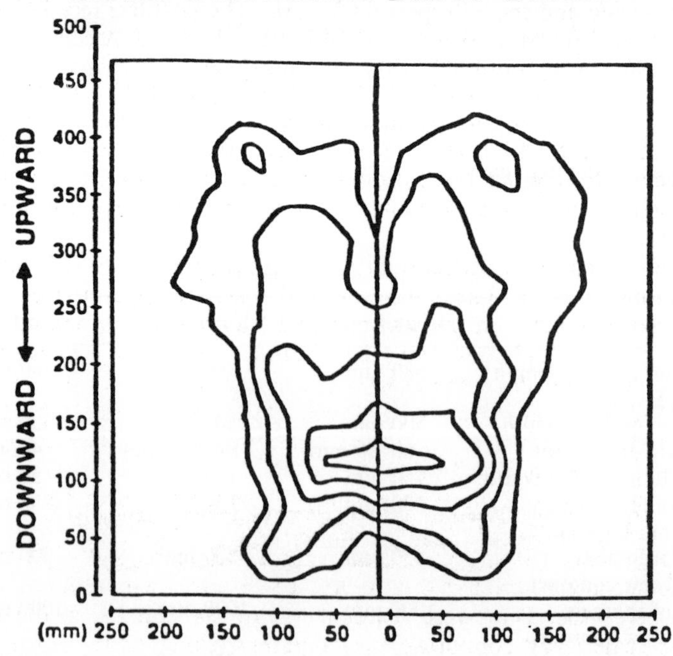

Figure 18.2 Comfortable pressure contours from automobile company study

3. have approximately two-thirds of a man's arm and leg strength
4. have more flexible limbs
5. have a thicker and more uniform layer of body fat
6. have shorter limbs turned slightly inward.

Pressure during sitting

Most of the sitting pressure is borne by the ischial tuberosities. Pressure between the seat and the body can be changed by adopting different postures, e.g., crossing one's legs or leaning forward or backward. Pressure on the body tissue restricts blood flow and may impact the nerves.

Wheelchair users who sit for long periods of time are prone to develop ischemic ulcers, usually over weight-bearing protuberances. When seated, the areas under the ischial tuberosities and the sacrum are most susceptible. Normal cellular metabolism depends on adequate circulation. Any condition which interferes with the circulation which provides nutrients and eliminates waste products, may lead to changes in the cell causing pain and discomfort. For the disabled, prolonged obstruction of the local capillary circulation eventually kills the cells. Ischemia caused by pressures greater than capillary pressure is the primary factor in the formation of ulcers (Brand, 1979; Kosiak, 1976; Landis, 1930). It has been suggested that pressures below 20–30 mm mercury are required to prevent capillary occlusion (Houle, 1969; Peterson and Adkins, 1982). When different seat cushions were used, the pressure under the ischial tuberosities ranged from 41 mm and 86 mm mercury (Mooney *et al.*, 1971; Souther *et al.*, 1974). Drummond *et al.* (1982), using a microcomputer based pressure scanner, showed that approximately 18 per cent of the body weight is distributed over each ischial tuberosity; 21 per cent over each thigh; and 5 per cent over the sacrum. These values may be modified with changes to the seat geometry and foam durometer.

Floyd and Roberts (1958) concluded that most people feel comfortable when the weight of the body is carried primarily by ischial tuberosities. Alternatively, Sanders and McCormick (1987) suggest that the weight be distributed rather evenly throughout the buttocks area, but minimized under the thighs. Such a distribution can be achieved by contouring the seat pan and varying cushion density.

Researchers may now collect point pressures from the seat back rest and the pan cushions, using a data collection and analysis apparatus consisting of pressure sensitive mats on the seat pan and back rest connected to an amplifier and analogue-to-digital converter. During testing, subjects sat on a seat with both hands on the steering wheel, the right foot on the accelerator, and the left foot on the dead pedal. The peak pressures and load distributions in each area of the seat were computed and compared to their recommended guidelines for pressure distribution (Table 18.2).

Arrowsmith (1986) performed a study in which the seat cushioning was changed in the Jaguar XJ40 seat. The study was based entirely on subjective responses to seat comfort. The subject population consisted of both sexes, a variety of social and economic groups with statures ranging from the 5th to the 95th percentile. They were required to rate the comfort of the seat on a five-point scale (very comfortable, moderately comfortable, neutral, moderately uncomfortable and very uncomfortable). Subjects rated their comfort

Table 18.2 Preferred seat pressure distribution

Region	Proportion of pressure
Lumbar	16 per cent
Shoulder	4 per cent
Back lateral	2 per cent
Ischial	54 per cent
Thigh	22 per cent
Cushion lateral	2 per cent

level after sitting for 15 minutes (showroom comfort); after 30 minutes (representing a short test drive), and at half-hour intervals for a drive of one and a half hours (representing an extended test drive). The study results were incorporated into the Series III design. Even though the changes were minor, the authors claim that they had a significant beneficial effect on passenger comfort. Some changes were the lowering of the piping across the front edge of the cushion to reduce contact with the thigh and stiffening of the rear of the cushion by reducing the size of the cavities in the foam.

Designers at Audi (Weichenrieder and Haldenwanger, 1986) used published data to distribute the weight of the driver over the entire contact area, and thus keep the overall pressure low. The highest pressure was below the ischial support points, and fell off gradually towards the boundary of the body's support area.

Seat comfort assessment

Comfort is a subjective measure and is not easily quantified. However, if comfort can be related to objective measures, then the best combination of seat adjustments may be determined. As such, distribution of sitting pressures could be directly linked to perceived comfort.

Pressure measurement

A major consideration for the seat comfort is the force exerted on the seat surface. In a manufacturing environment, force can be measured using dynamometers. However, force measures are of little use in seating research because the force acts over the entire body-seat contact surface area. Instead, pressure measurements (force per unit area) are required. When changes in posture occur, real time pressure measurements are needed.

Different techniques are used to measure force and pressure. A force sensing device will respond identically to two equal forces, regardless of the area over which the force is applied.

Alternatively, a true pressure sensor will produce an output for a given force which is inversely proportional to the area.

Sitting pressures were first measured using mechanical valves and compressed air tanks (Kosiak, 1959; Houle, 1969). More recently, optical techniques have been used based on the prototype built in 1934 by Elftman (Hertzberg, 1972; Mayo-Smith and Cochran, 1981; Treaster and Marras,

1987). Other techniques include pressure-sensitive chemicals (Frisnia *et al.*, 1970), thermographs (Trandel *et al.*, 1975), capacity sensors (Ferguson-Pell, 1976) and mechanical springs (Lindan *et al.*, 1965). The main problem with most of these techniques is the validity and the reliability of the measurements.

Recent developments in electronics technology allow for more precise measurements. Any electronic sensing device used for seat pressure measurement should have the following characteristics:

- transparent to the seated subject (i.e., cannot be felt)
- insensitive to vibration, temperature and noise
- durable
- repeatable
- optimum sensitivity and range
- low hysteresis (i.e., similar electrical characteristics during loading and unloading)
- flexible
- configure to any seat shape or size
- deformable
- linear in the pressure resistance relationship over a high range
- simple and cost-effective.

The pressure measurement system developed by BCA uses a special 'mat' comprising 225 sensors (15 × 15 matrix) on the pan (covering an area of approximately 6.3 × 6.3 cm) and 225 sensors (15 × 15 matrix) on the seat back covering an area of approximately 5.5 × 8.3 cm (Figures 18.3, 18.4, 18.5).

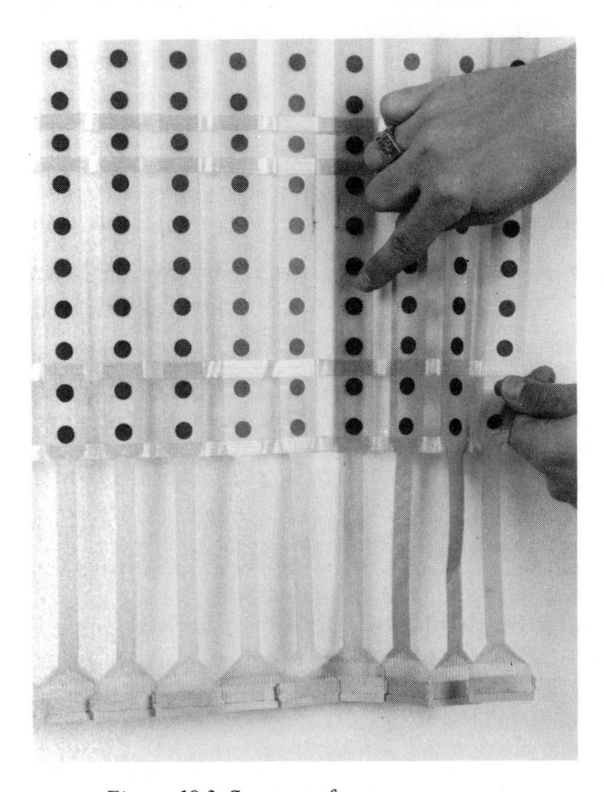

Figure 18.3 Segment of seat pressure mat

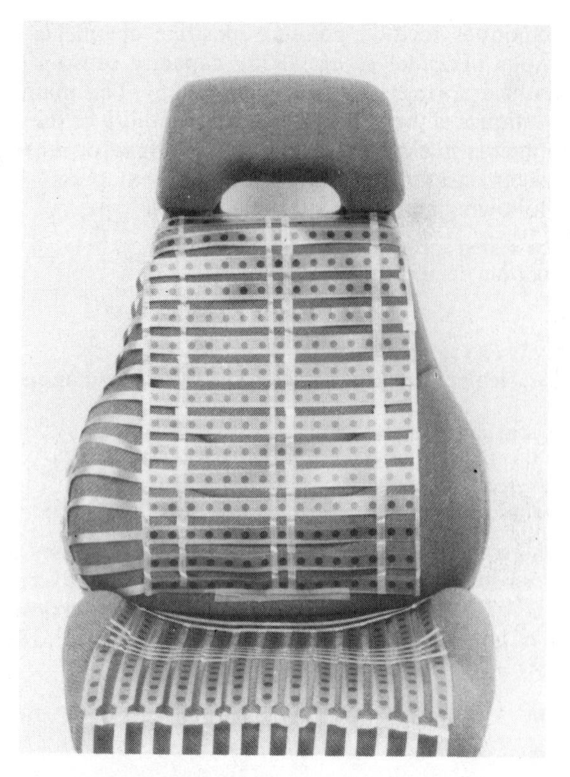

Figure 18.4 Seat pressure mat superimposed on experimental seat

At zero or very low pressures, the sensors act as an open circuit. After the pressure reaches a low threshold, increasing force rapidly reduces electrical resistance. The sensors are all connected to a 5-volt power supply through a current-limiting resistor. The voltage measured at the output end is proportional to the force as the device and the current limiting resistor form a voltage divider. To incorporate a higher current and low source impedance, operational amplifiers are used. The pressure mat is connected to a microcomputer through an analogue-to-digital data acquisition system. Proprietary BCA software is used to measure and display pressure in real time (Figure 18.6).

Modelling comfort

To quantify the factors that contribute to the perceived comfort of a car seat, the following categories of data were collected:

1. The subjective ratings of comfort for each part of the seat. Using Likert scales, the subjects rated the comfort of 12 aspects of the seat after it had been adjusted to the most comfortable position. Subjective ratings are selected on a continuous scale from one to five, where one represents very poor and five represents very good with neutral corresponding to a value of three.

Figure 18.5 Seated subject driving evaluation

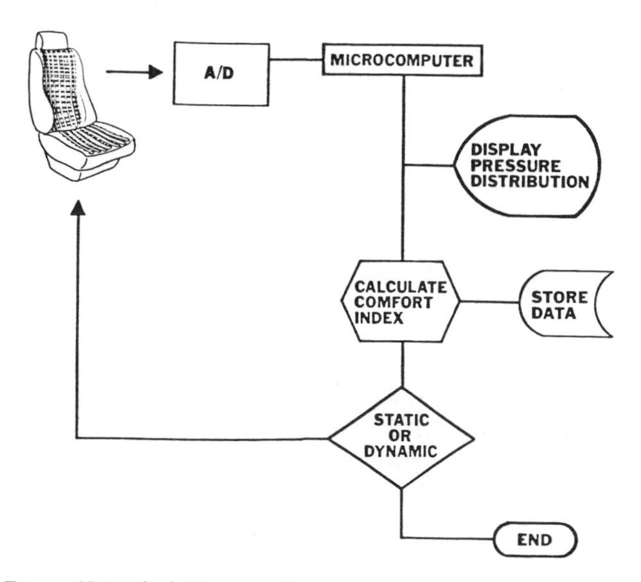

Figure 18.6 Block diagram of seat pressure measurement system

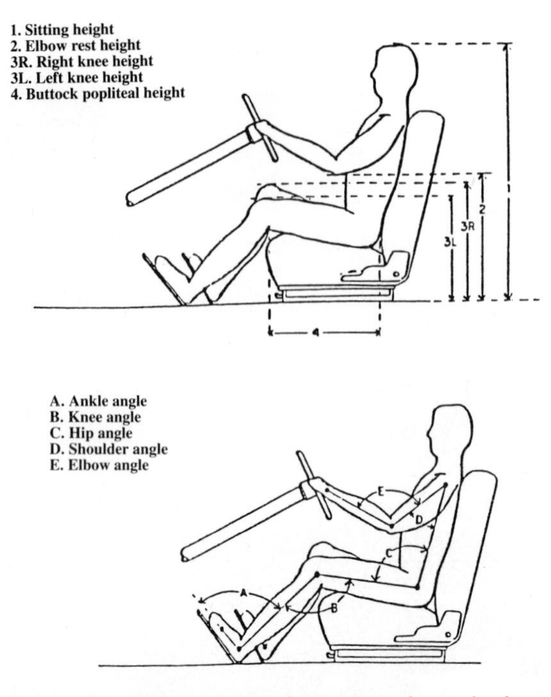

Figure 18.7 Anthropometric dimensions of seated subject

2. The anthropometric dimensions of the driver. The subjects' anthropometric characteristics were measured while seated and standing. The seated dimensions were taken with the seat adjusted to the most comfortable position. The body angles that defined the subjects' posture were measured with a goniometer. The leg and foot angles were measured separately on the left and right side as the accelerator and dead pedal were placed at different angles (Figure 18.7).
3. The dimensional and angular characteristics of the seat. Back rest height and width, seat cushion width, seat cushion length, seat cushion angle, and seat back angle were recorded (Figure 18.8).
4. The force exerted by the seated driver on the seat surface and its force distribution. The seat pan was divided into eight regions: two halves (fore and aft), two buttocks, two thighs and two bolster regions. The seat back was divided into eight regions: two halves, two lumbar, two thoracic and two bolster regions. The force measures in each of the regions were determined and analysed statistically, defining location and pressure dispersion.

More than 1100 seat-subject combinations (50 seats) were tested for short term trials (5–10 minutes). Although recordings were continuous, 20-second windows of time within these trials were used for data analysis.

The subjects were stratified by size and gender. Sample pressure plots for luxury, sports and economy seats for seat back and seat pan are shown in Figures 18.9–18.14. Pressure gradients at different points were calculated using a discrete approximation of the partial derivatives of a two-dimensional function:

$$df(x, y) = \delta f(x, y)/\delta x + \delta f(x, y)/\delta y$$

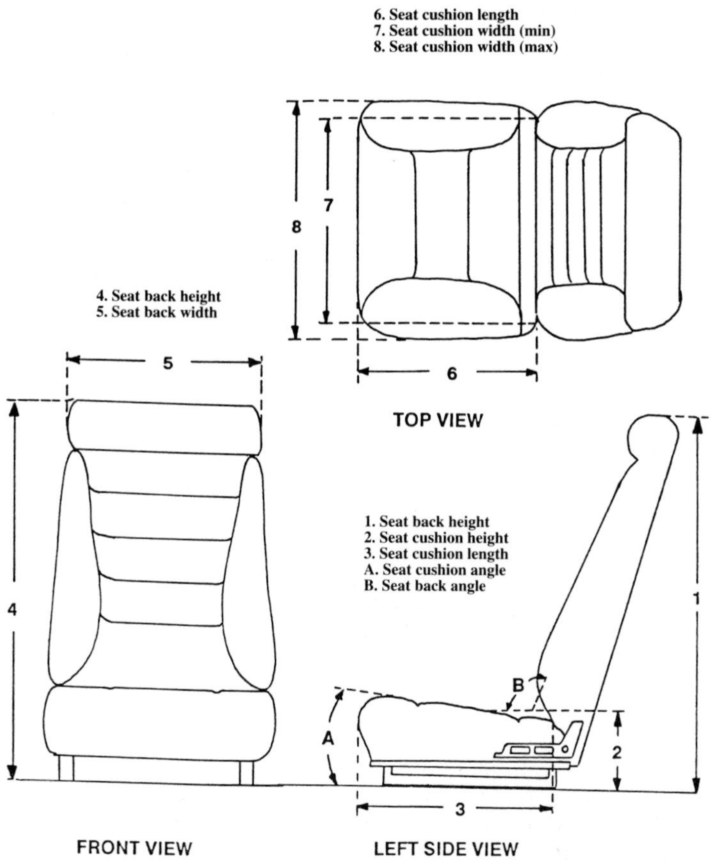

Figure 18.8 Seat dimensions and angular characteristics

The discrete approximation for the two terms in the derivative are given by:

$$\delta f(x, y)/\delta x = [f(x, y) + f(x, y+1)] - [f(x+1, y) + f(x+1, y+1)]$$

$$\delta f(x, y)/\delta y = [f(x, y) + f(x+1, y)] - [f(x, y+1) + f(x+1, y+1)]$$

Two convolution filters were used to calculate the partial derivatives. The gradient at each point is the sum of the absolute value of the derivatives. The two filters used were:

$$\begin{matrix} 1 & -1 \\ 1 & -1 \end{matrix} \qquad \begin{matrix} 1 & 1 \\ -1 & -1 \end{matrix}$$

These two operators were applied to the pressure readings at each sensor and the absolute value of the responses added to define the gradient. The maximum pressure gradient for each subject on each seat was determined by using the maximum value on the matrix that was passed through the two operators shown above.

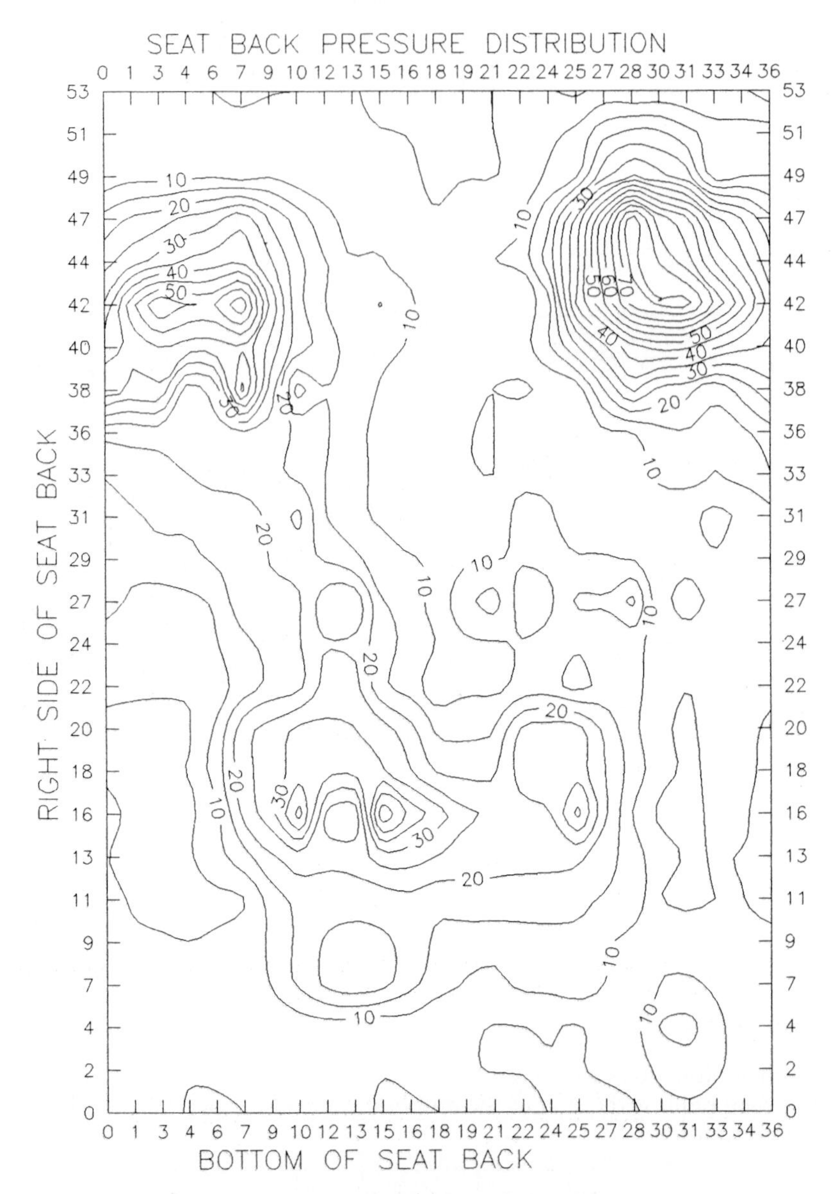

Figure 18.9 Sample pressure contours for different automotive seat classes (comfort rating is on a scale from 1 to 5). Data are normalized to peak pressure

A multivariate analysis of the data related the pressure distribution of the driver's weight on the seat surface with seat comfort (see Figures 18.15, 18.16, and 18.17 for a sample load distribution on seat pan and seat back). Statistics computed from the pressure data were strongly related to perceived comfort

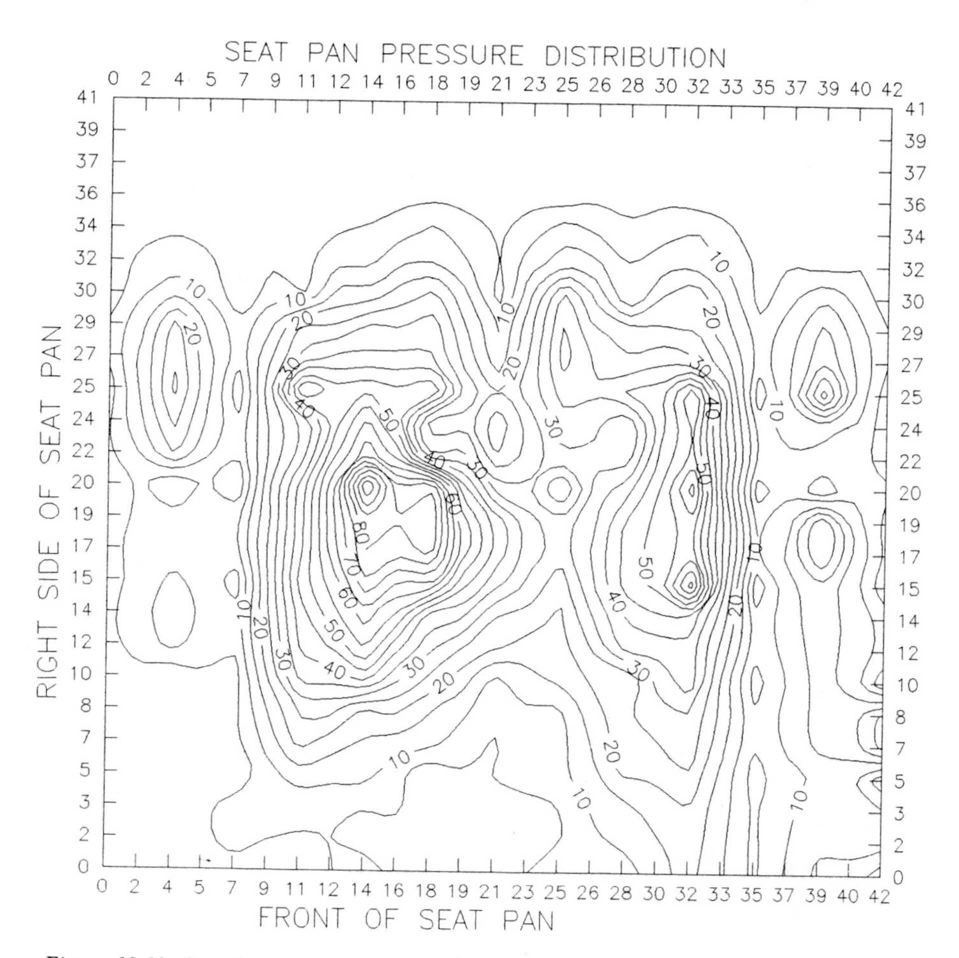

Figure 18.10. Sample pressure contours for different automotive seat classes (comfort rating is on a scale from 1 to 5). Data are normalized to peak pressure

Table 18.3 Calculation of perceived comfort

	95 per cent Confidence Interval			
Seat number	Subjective comfort	Predicted comfort	Lower limit	Upper limit
1	3.8	3.8	3.7	4.0
2	3.8	3.5	3.3	3.7
3	3.4	3.3	3.2	3.4
4	3.7	3.7	3.6	3.9
5	3.6	3.5	3.4	3.7

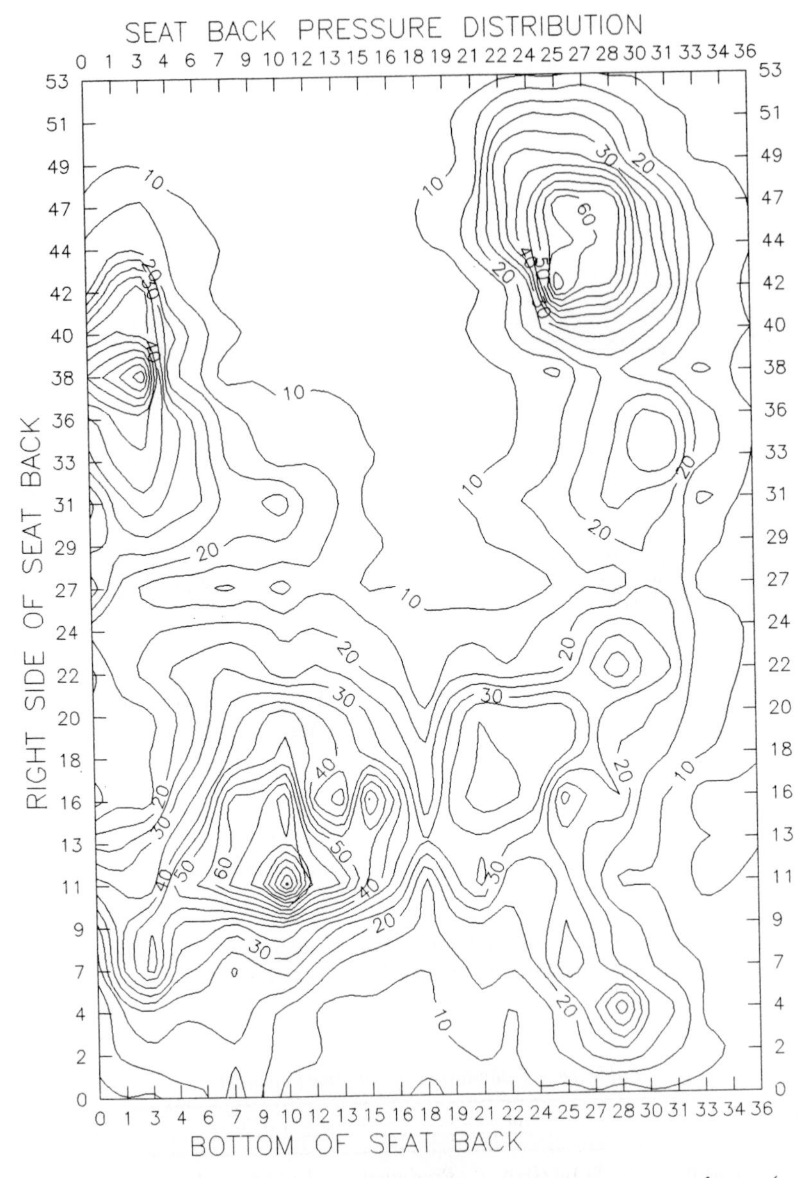

Figure 18.11 Sample pressure contours for different automotive seat classes (comfort rating is on a scale from 1 to 5). Data are normalized to peak pressure

(Table 18.3 shows sample results from 5 seats). Perceived comfort (PC) was calculated as follows:

$$PC = \sum a_i x_i \qquad \text{for all } i$$

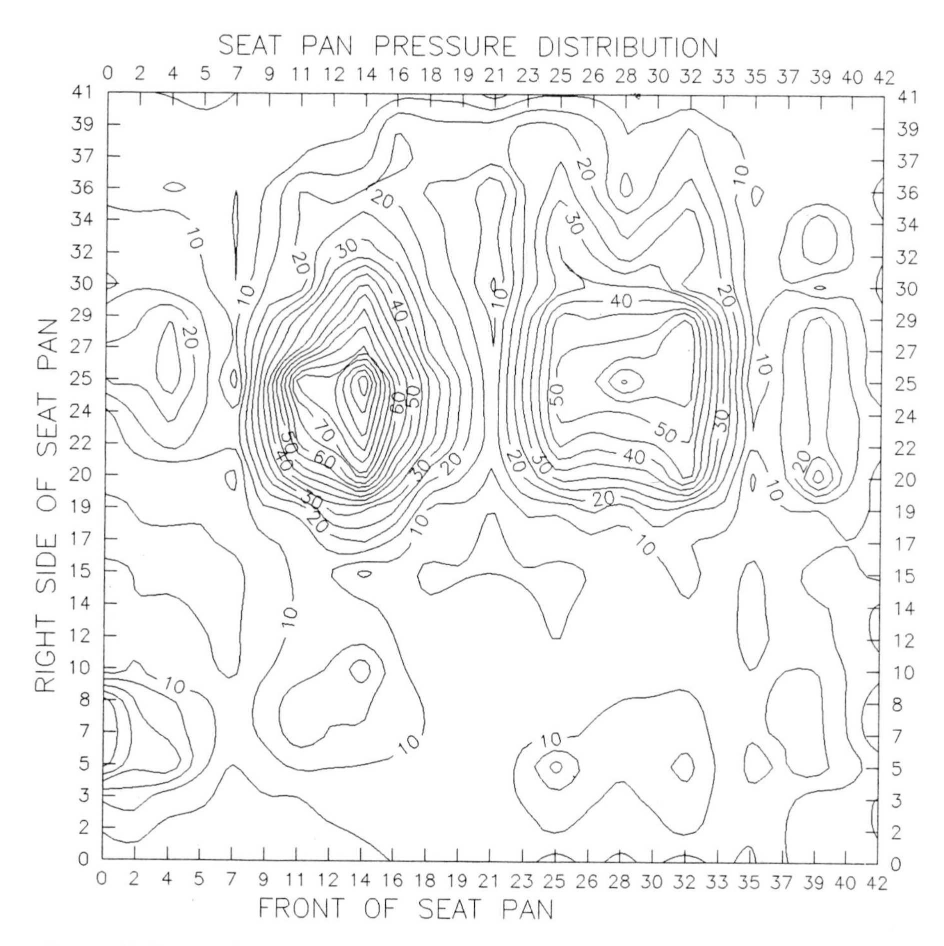

Figure 18.12 Sample pressure contours for different automotive seat classes (comfort rating is on a scale from 1 to 5). Data are normalized to peak pressure

where
a_i = regression coefficient associated with variable x_i
x_i = dependent variable of pressure.

Anthropometry and seat geometry interact in complex ways to produce seat pressures. However, statistics related to the pressure distribution were shown to predict the perceived comfort of a seat. Optimizing the comfort of prototype seats with respect to the magnitude and the pattern of the pressure distribution can significantly reduce product development time.

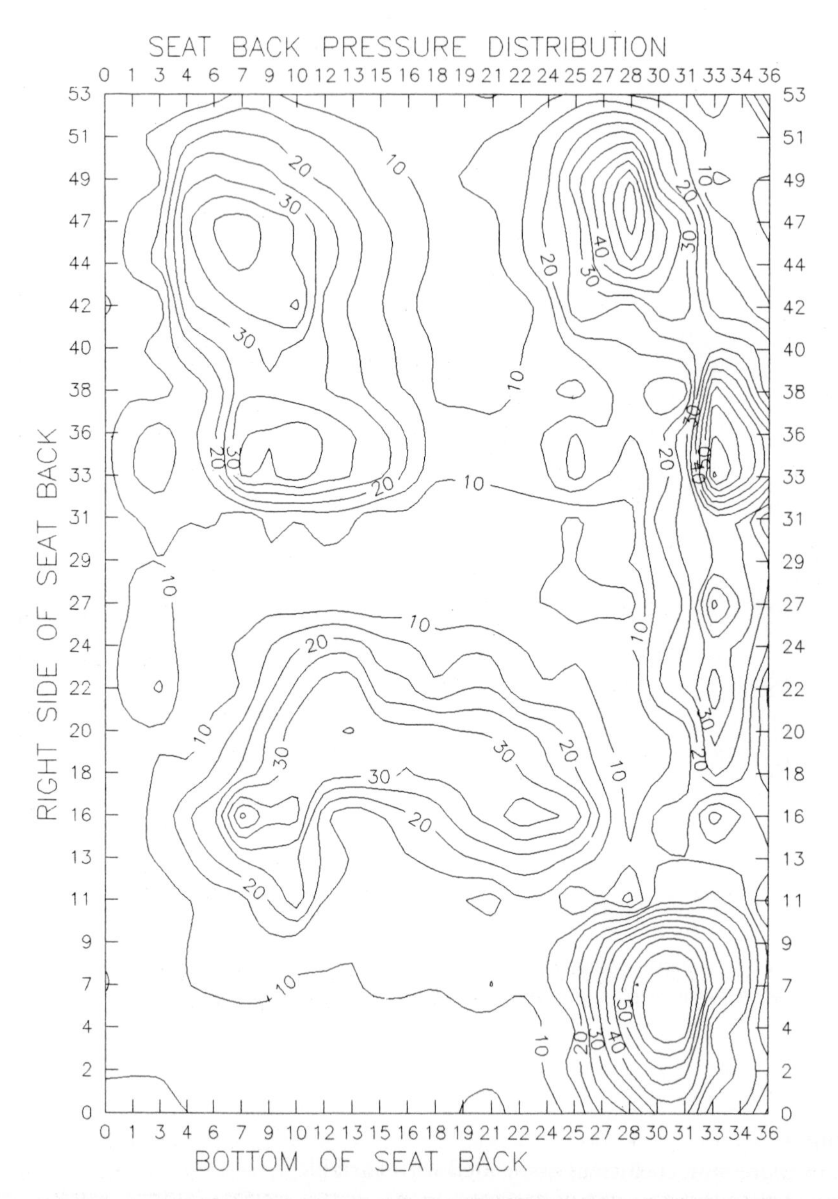

Figure 18.13 Sample pressure contours for different automotive seat classes (comfort rating is on a scale from 1 to 5). Data are normalized to peak pressure

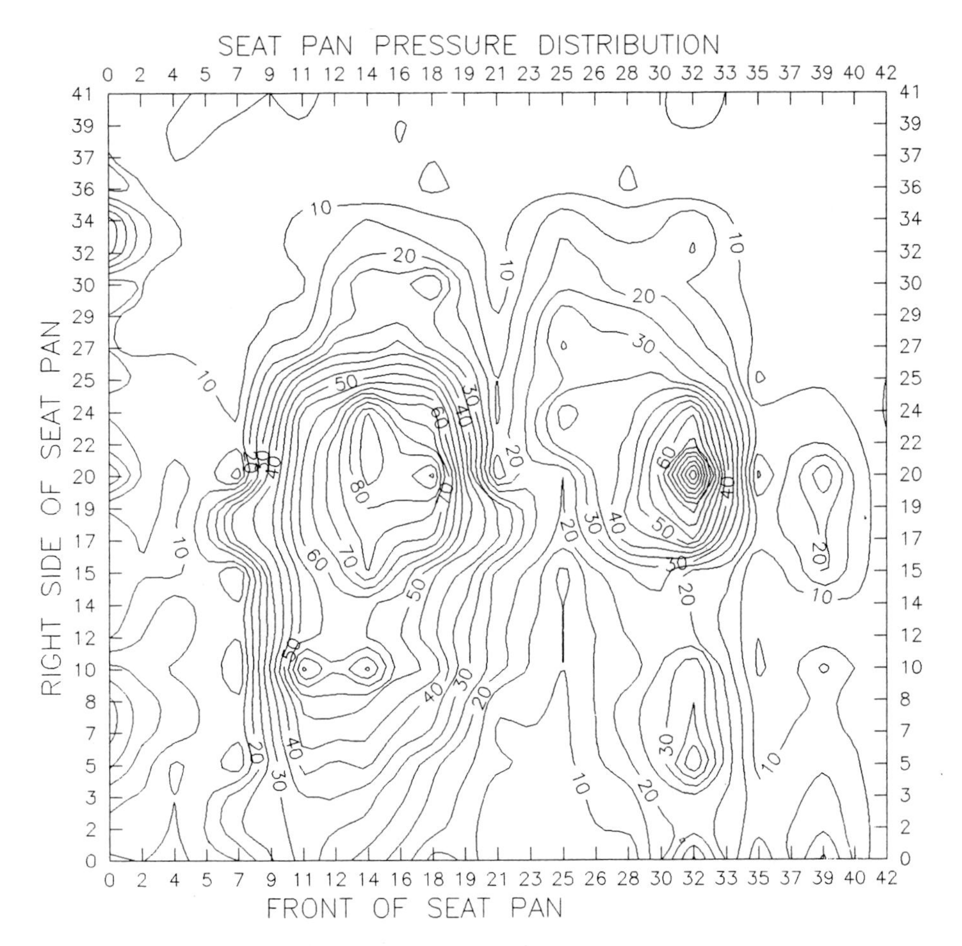

Figure 18.14 Sample pressure contours for different automotive seat classes (comfort rating is on a scale from 1 to 5). Data are normalized to peak pressure

Conclusions

The contours and firmness of seat cushions have been used in the past to redistribute pressures under the ischial tuberosities. However, in most circumstances the pressures were in excess of the capillary pressures, and not sufficiently or easily adjustable, leading to discomfort during prolonged periods of sitting.

Our research was used to develop an 'Intelligent Seat' (Figure 18.18) which avoids this by using an automatic system that can shift the pressures and reduce the pressure gradients. This device senses the pressures at the body-seat interface using sensors placed under the upholstery. Based on the pressure-comfort developed the seat will automatically adjust the elements (e.g., upper

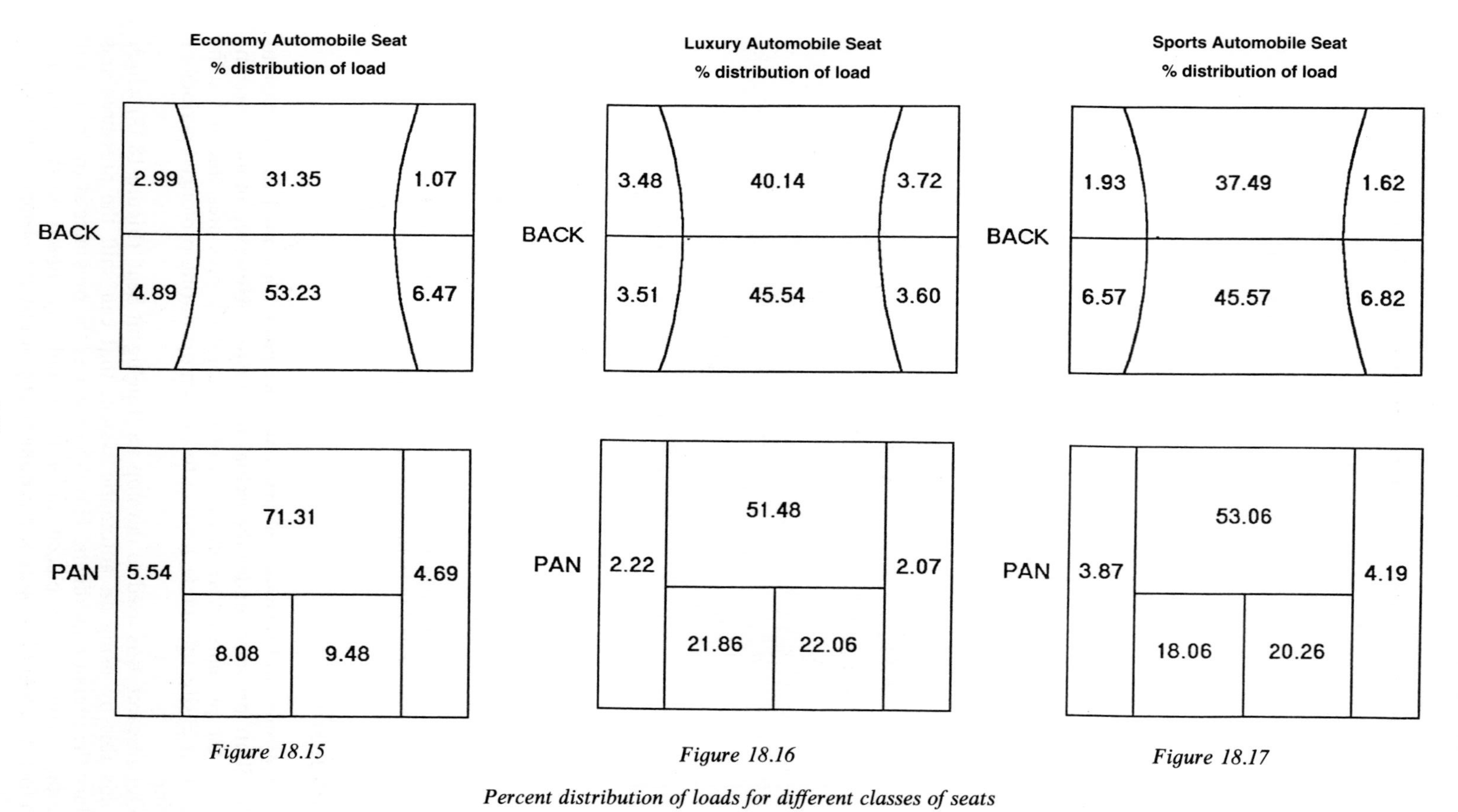

Figure 18.15

Figure 18.16

Figure 18.17

Percent distribution of loads for different classes of seats

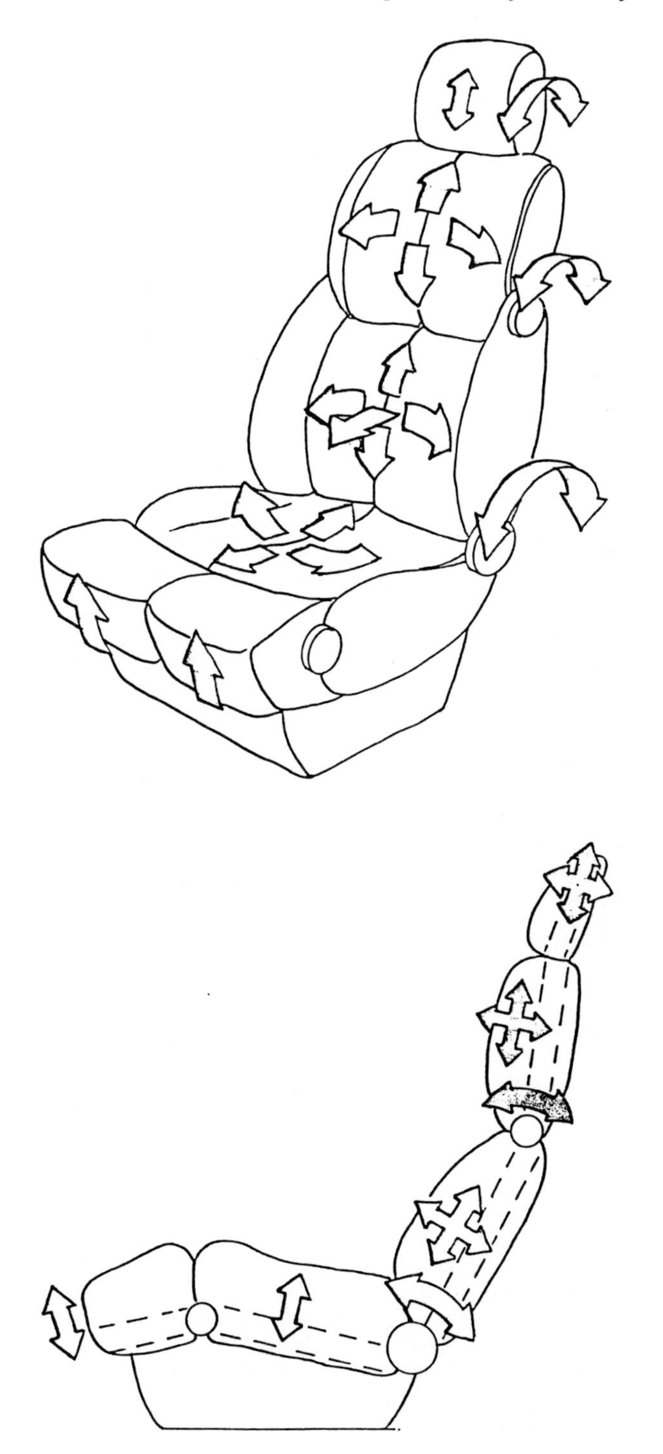

Figure 18.18 Schematic for Intelligent Seat. (*a*) *isometric view;* (*b*) *side view*

lumbar, mid lumbar, side bolster etc.) to pressure profiles of a person of any size and weight that have been found comfortable. This effectively optimizes comfort and user fit. This new seat measures the seat pressures profile of its occupants and responds to their specific 'comfort' characteristics. By bundling biomechanical 'knowledge' into the seat it can respond accordingly to its occupant.

References

Andersson, G. B. J., (1974), On myo-electric back muscle activity and lumbar disc pressure in sitting postures. Doctoral dissertation, Gotab, University of Goteborg, Goteborg, Sweden.

Arrowsmith, M. J., (1986), The design and development of the XJ40 seating system, *Proc. Instn. Mech. Engrs.*, **200**, D5, s79–s85.

Brand, P. W., (1979), Management of the insensitive limb, *Physical Therapy*, **59**, 8–12.

Branton, P., (1969), Behaviour, body mechanics and discomfort, *Ergonomics*, **12**, 316–27.

Corlett, E. N., (1990), Static muscle loading and the evaluation of posture, in *Evaluation of Human Work*. J. R. Wilson and E. N. Corlett (eds) London: Taylor & Francis.

Corlett, E. N. and Bishop, R. P., (1976), A technique for assessing postural discomfort, *Ergonomics*, **19**, 175–82.

Dainoff, M. J. and Dainoff, M. H., (1986), *People and Productivity: A Manager's Guide to Ergonomics in the Electronic Office*. Canada: Holt, Rinehart and Winston.

Drummond, D. S., Narechania, R. G., Rosenthal, A. N., Breed, A. L., Lange, T. A. and Drummond, D. K., (1982), A study of pressure distributions measured during balanced and unbalanced sitting, *The Journal of Bone and Joint Surgery*, **64**-A(7), 1034–9.

Eklund, J. A. E. and Corlett, E. N., (1984), Shrinkage as a measure of the effect of load on the spine, *Spine*, **9**, 189–94.

Eklund, J. A. E., (1986), Industrial Seating and Spinal Loading. PhD thesis. University of Nottingham. Distributed by Department of Industrial Ergonomics. University of Technology, Linkoping, Sweden.

Elftman, H., (1934), Cinematic study of distribution of pressure in human foot, *Anat Rec.*, **59**, 481–91.

Ferguson-Pell, M. W., (1976), Interface pressure sensors: Existing devices, their suitability and limitations, in *Bedsore Biomechanics*. R. M. Kenedi (ed) Baltimore: University Park Press.

Fleischer, A. G., Rademacher, U. and Windberg, H. J., (1987), Individual characteristics of sitting behaviour, *Ergonomics*, **30**(4), 703–9.

Floyd, W. F. and Roberts, D. F., (1958), Anatomical and physiological principles in chair and table design, *Ergonomics*, **2**, 1–16.

Frisnia, Warren and Lehneis, H. R., (1970), Pressure mapping: A preliminary report, *Journal of Biomechanics*, **3**, 526.

Grandjean, E., Hunting, W. and Pidermann, M., (1983), VDT work station design: Preferred settings and their effects, *Human Factors*, **25**, 161–75.

Habsburg, S. and Mittendorf, L., (1980), Calibrating comfort: Systematic studies on human responses to seating, in *Human Factors in Transport Research*. D. J. Oborne and T. A. Levis (eds) NY: Academic Press.

Hertzberg, H. T. E., (1958), Annotated bibliography of applied physical anthropology in human engineering (Report No. WADC-TR-56-30). Wright-Patterson Air Force Base, OH.

Hertzberg, H. T. E., (1972), Human buttocks in sitting: Pressure patterns and palliatives, New York Publication 72005, Society of Automotive Engineers Inc.

Houle, R. J., (1969), Evaluation of seat devices designed to prevent ischemic ulcers in paraplegic patients, *Arch. Phys. Med.*, **50**, 587–94.

Hubbard, R. P. and Reynolds, H. M., (1984), Anatomical geometry and seating, *SAE Technical Paper Series*, 840506.

Kosiak, M., (1959), Etiology and pathology of ischemic ulcers, *Arch. Phys. Med.*, **40**, 62–9.

Kosiak, M. A., (1976), Mechanical resting surface: Its effect on pressure distribution, *Arch. Phys. Med.*, **57**, 481–4.

Landis, E. M., (1930), Micro-injection studies of capillary blood pressure in humans skin, *Heart*, **15**, 209–28.

Lueder, R. K., (1983), Seat comfort: A review of the construct in the office environment, *Human Factors*, **25**(6), 701–11.

Lindan Olgierd, Greenway, R. M. and Piazza, J. M., (1965), Pressure distribution on the surface of the human body: 1. Evaluation in lying and sitting positions using a 'Bed of Springs and Nails', *Arch. Phys. Med.*, **46**, 378–85.

Mandal, A. C., (1981), The seated man (Homo Sedens), the seated work position, theory and practice, *Applied Ergonomics*, **12**, 19–26.

Mayo-Smith, W. and Cochran, G. V. B., (1981), Wheelchair cushion modification: Device for locating high-pressure regions, *Arch. Phys. Med.*, **62**, 135–6.

Mooney, V., Einbund, J. J., Rogers, J. E. and Stauffer, E. S., (1971), Comparison of pressure qualities in seat cushions, *Bull. Pros. Res.*, **10–15**, 129–43.

Peterson, M. J. and Adkins, H. V., (1982), Measurement and redistribution of excessive pressures during wheelchair sitting, *Physical Therapy*, **62**(7), 990–4.

Rebiffe, P. R., (1969), Le siège du conducteur: Son adaptation aux exigences fonctionnelles et anthropometriques, *Ergonomics*, **12**(2), 246–61.

Sanders, M. S. and McCormick, E. J., (1987), *Human Factors in Engineering Design*, NY: McGraw Hill.

Society of Automotive Engineers, (1990), *Society of Automotive Engineers Handbook*, **4**, On-Highway Vehicles and Off-Highway Machinery, Warrendale, PA.

Shackel, B., Chidsey, K. D. and Shipley, P., (1969), The assessment of chair comfort, *Ergonomics*, **12**, 269–306.

Souther, S. G., Carr, S. D. and Vistness, L., (1974), Wheelchair cushions to reduce pressure under bony prominences, *Arch. Phys. Med.*, **55**, 460–4.

Trandel, R. S., Lewis, D. W. and Verhonick, P. J., (1975), Thermographical investigation of decubitus ulcers, *Bull. Pros. Res.*, **10–24**, 137–55.

Treaster, D. and Marras, W. M., (1987), Measurement of seat pressure distributions, *Human Factors*, **29**(5), 563–75.

Van Cott, H. P. and Kincade, R. G., (1972), *Human Engineering Guide to Equipment Design*, NY: John Wiley.

Viano, D. C., Patel, M. and Ciccione, M. A., (1989), Patterns of arm position during normal driving, *Human Factors*, **31**(6), 715–20.

Wachsler, R. A. and Learner, D. B., (1960), An analysis of some factors influencing seat comfort, *Ergonomics*, **3**, 315–20.

Weichenrieder, A. and Haldenwanger, H., (1986), The best function for the seat of a passenger car, *SAE Technical Paper Series*, 850484.

PART VIII

School children

19

A fuzzy expert system for allocating chairs to elementary school children

Kageyu Noro and Takeshi Fujita

Introduction

The objective of this study was to develop a simple method for allocating chair sizes for individual elementary school pupils.

JIS S 1021 (Japanese Standards Association., 1980) specifies that Japanese elementary schools use 11 sizes of chair.

Elementary schools find it extremely time consuming to select ergonomic chairs for each pupil. Those responsible for conducting this selection process must first measure the pupils' physical dimensions. Although stature, body weight and other body dimensions are measured in regular physical checkups, popliteal height and other special body dimensions must still be ascertained. This measurement process typically requires 5–10 minutes per pupil. Because today's Japanese elementary schools have about 45 pupils per class, the process requires a half day per class. Further, accurate measurement of popliteal height requires experience and skill. Elementary school pupils' height may increase considerably in three months, underscoring the need for the development of a simplified method for allocating elementary school chairs.

JIS S 1021 includes a reference table for allocating chairs based on stature. However, this table does not account for such characteristics as obesity, which are common with many elementary school children.

An attempt was made to develop decision support software for class teachers which incorporates the expert knowledge of health teachers, who have considerable understanding of the ergonomics of seating. Such applications of fuzzy logic are designed to make inferences and judgments which approximate those of experts.

Methods

Fuzzy inference

This system applies fuzzy inference to the computer's user interface and central processing unit. The reasons for adopting the fuzzy inference is described later but summarized here.

The fuzzy inference can be constructed by using fuzzy production rules of fuzzy relations. For example, assume that the following production rules are elicited from an expert:

IF A_1, THEN B_1 IF A_2, THEN B_2

where A_1, B_2, A_2, and B_2 are described as fuzzy sets. As shown in Figure 19.1, the conclusions B_1 and B_2 are represented in terms of the statures of A_1 and A_2, respectively. Then, $(0.8 \wedge B_1)$ and $(0.2 \wedge B_2)$ are integrated by the maximum composition. Decisions associated with a particular value are based on the relative distribution of conditional and interdependent probabilities (see Figure 19.2). This distribution may also be described in terms of its central tendencies or centre of gravity.

The terms fuzzy set and membership function are characterized by ambiguous boundaries. For example, if middle age is defined as ranging between 35 and 45 years, a person aged 34 years and 11 months or a day older than 45 would not be middle aged (Figure 19.3).

Such a definition is not compatible with our more fuzzy world views. Rather, middle aged people might be more effectively characterized by a fuzzy set. Thus, a person 34 years and 11 months old is associated with the middle age group with a co-efficient of 0.9 (Figure 19.4). In this manner, non-specific expressions can be effectively represented by fuzzy sets.

System configuration and operation

As with conventional decision support software, this system includes a central processing unit, working memory, and knowledge base. Fuzzy set theory is applied to the CPU, user interface, and knowledge base.

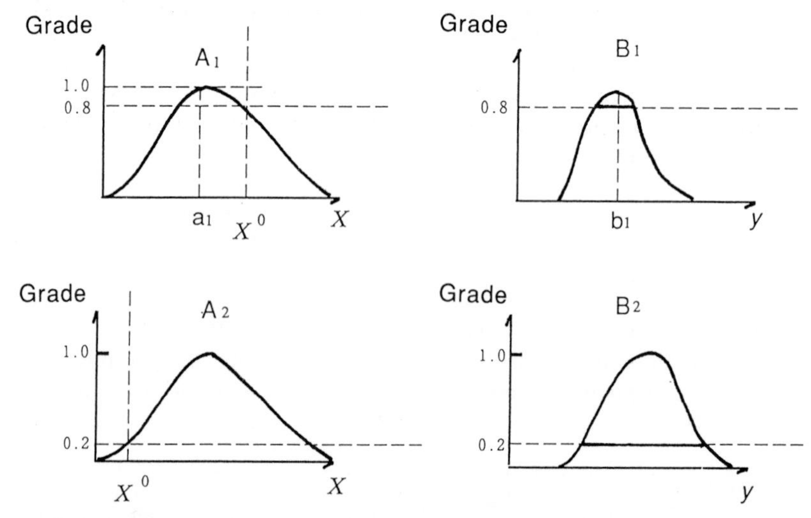

Figure 19.1 Explanation 1 of fuzzy inference. Inference is drawn using two if→ then production rules. Since input X^0 is associated with condition part A_1 by a co-efficient of 0.8 (upper left), conclusion part B_1 is set at 0.8 (upper right). Since input X^0 is associated with condition part A_2 by a co-efficient of 0.2 (lower left), conclusion part B_2 is set at 0.2 (lower right)

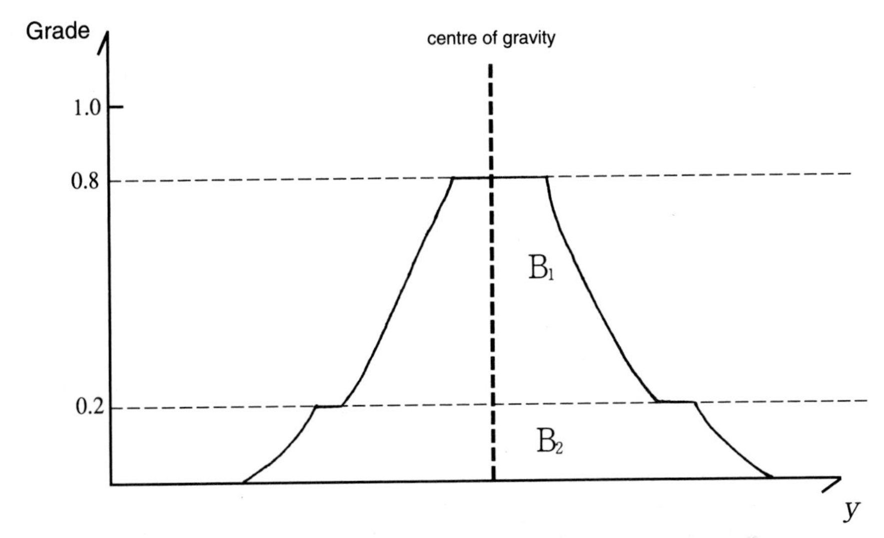

Figure 19.2 Explanation 2 of fuzzy inference process. The inferential process can be seen imposing shaded areas of conclusion production rules $A_1 \rightarrow B_1$ and $A_2 \rightarrow B_2$ in Figure 19.1. Since one conclusion must be drawn, the centre of gravity, Y^0, of the shaded area represents the final conclusion

Knowledge base

The knowledge base is composed of production rules in which features such as large and long are described using ambiguous daily language with fuzzy sets. The importance of this usage of everyday language was confirmed in a preliminary survey that found that teachers perceive the popliteal height and other body dimensions of obese and slender children very differently. The fuzzy logic production rules comprise a tree structure for defining our language according to school grade and physical image (see Figure 19.5). These production rules are listed in Table 19.1.

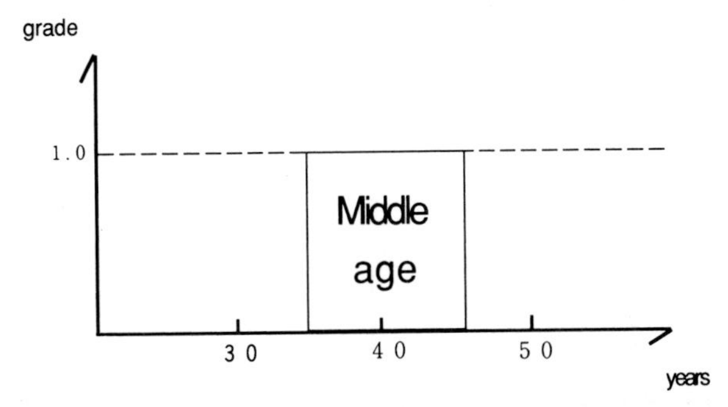

Figure 19.3 Middle age represented by a discrete, rather than fuzzy set. As evident from this graph with discrete sets, individuals who are one month younger than 35 or 1 month older than 45 are not middle aged. Such categorizations differ from our common sense perception of middle age

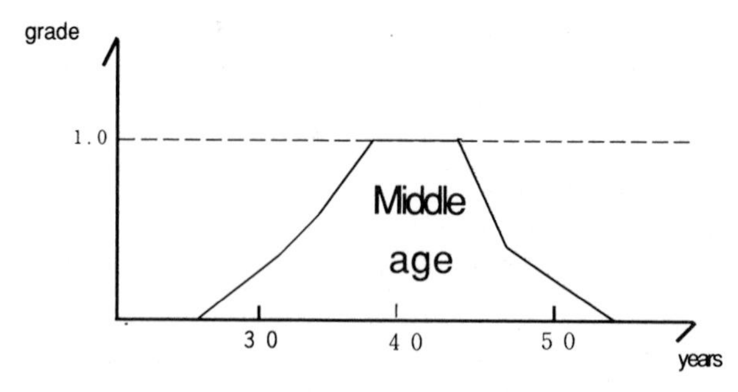

Figure 19.4 Middle age is here represented by fuzzy set, and exemplifies a membership function. A middle curve is drawn with a centre ranging between 35 and 45 years old. A person of 34 years 11 months old is 'almost', and a person of 49 years old is 'somewhat' middle aged. This approximates our common sense way of perceiving

User interface

One of the most important features of this system is that teachers do not measure each pupil's body dimensions. Fathers and teachers enter their physical image of pupils into the computer, and the system makes inferences which approximate the judgments of experts. Teachers that are accustomed to chair selection can subjectively assign an ergonomically suitable chair to each pupil by observation. Such an approach may be compared with the arts of craftsmen.

One may argue that teachers should be able to allocate chairs without fuzzy support. However, many elementary school teachers in Japan lack the knowledge or experience needed to make such ergonomics decisions. Teachers familiar with chair selection or experts were asked by questionnaire to describe how they selected chairs for different pupils. The fuzzy system makes inferences regarding the suitability of chairs for the pupils in accordance with the production rules constructed from questionnaire responses.

The fuzzy system is also equipped with a user interface that allows teachers to easily enter perceived pupil stature and production rules. The pupil's grade

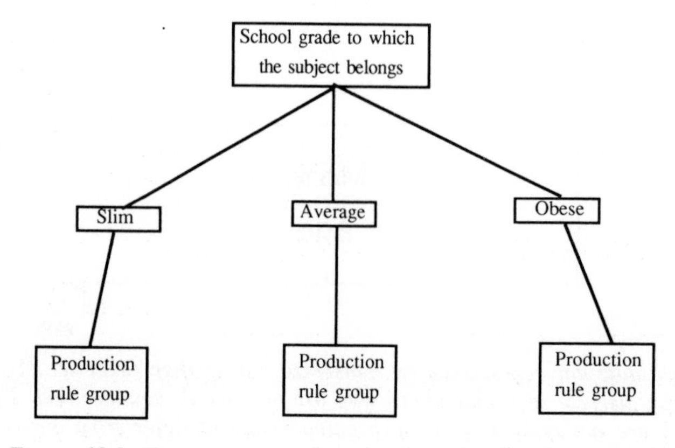

Figure 19.5 Tree structure of production rules for chair selection

Table 19.1 List of production rules

If the subject is very tall, give the subject a very large size chair.
If the subject is tall, give the subject a large size chair.
If the subject is average, give the subject an average size chair.
If the subject is short, give the subject a small size chair.
If the subject is very short, give the subject a very small size chair.
If the subject has a very large popliteal height, give the subject a very large size chair.
If the subject has a large popliteal height, give the subject a large size chair.
If the subject has an average popliteal height, give the subject an average size chair.
If the subject has a small popliteal height, give the subject a small size chair.
If the subject has a very small popliteal height, give the subject a very small size chair.

Condition and conclusion parts are underlined. Note that these production rules are not expressed in discrete numbers (in 00 cm to 00 cm), but rather described in our standard ambiguous daily language.

and perceived image of physique and body dimensions are encoded with a mouse, and displayed on the computer monitor (Figure 19.6).

Limitations of the system exist as well. The construction of the production rules are shown in Figure 19.7. Even slight changes of the plotted position may alter conclusions greatly. Users are not always consistent in how they plot the physical image of a given pupil. Great changes in output caused by slight changes in plotted position (input) are problematic.

To address such limitations, the features of the production rules are made to overlap so that (*see* Figure 19.8) membership function and production rules are made conditional and interdependent to increase reliability of the image recorded by the users (*see* Figure 19.8). This use of fuzzy inference incorporates information from two production rules for each position drawn on the

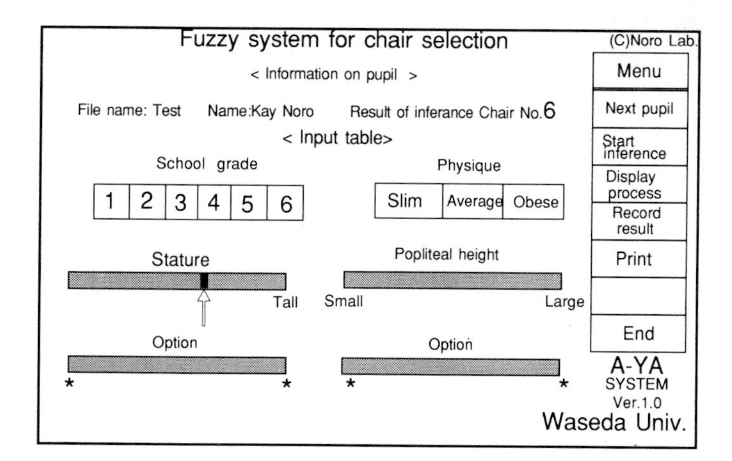

Figure 19.6 A sample page of input of fuzzy system for selecting chairs is displayed in the lower half of the screen. Ratings of the subject are plotted with a finger-shaped mouse cursor (lower left), and entered into the system. Commands are invoked by clicking the menu bar at right with a mouse. Little keyboard use is required

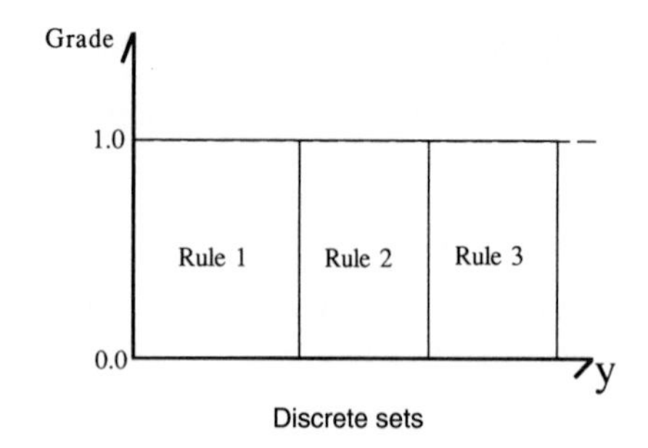

Figure 19.7 Unlike fuzzy inferences, discrete inferences provide clear cut distinctions regarding which rule applies to a plotted point

monitor and thereby increases the applicability of the decision support. Consequently, different interpretations of plotted positions do not change the conclusion substantially.

Response validity is increased by drawing inference from two overlapping adjacent production rules.

Supposed users

Proper use of the fuzzy system characteristics of elementary school pupils. For this reason, the use of the fuzzy system is limited to elementary school teachers.

A description of the fuzzy system

Fuzzy inference involves performing a statistical mini-max analysis of the data, and a determination of the centre of gravity of the distribution of probabilities for a given conclusion. For example, the fuzzy system incorporates the computer. This is achieved by selecting display process from the input menu. This inference process display page is shown in Figure 19.9.

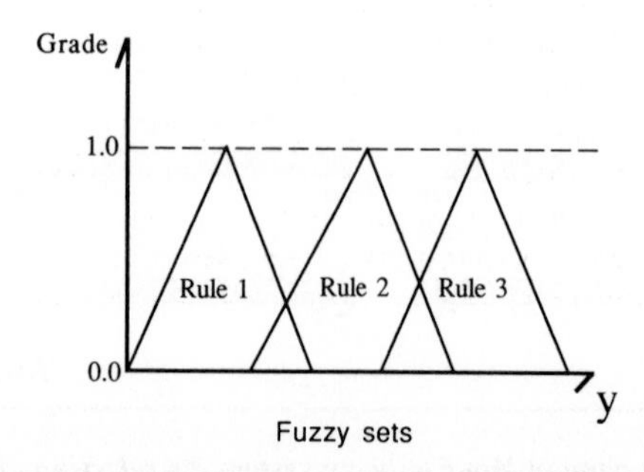

Figure 19.8 Fuzzy inference. If plotted point is located at a boundary between rules 1 and 2, it may apply equally to each. Unless the plotted point evidences a dramatic shift toward rule 1 or 3, an association is maintained with both

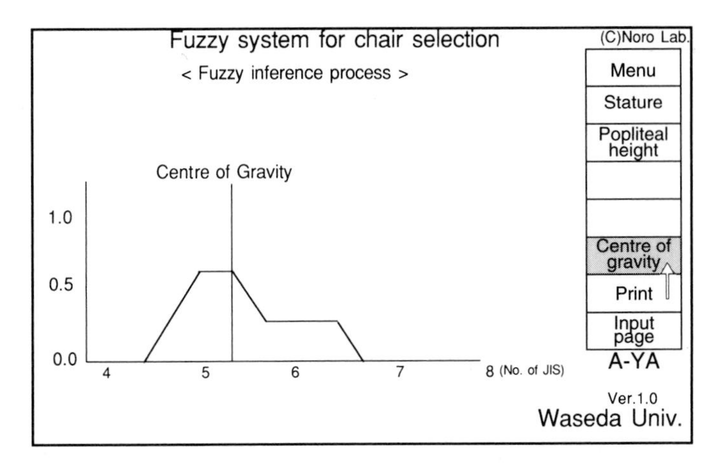

Figure 19.9 This graph of the fuzzy inference process displays how the distribution of the data is searched to reach an inference

Investigation

Collection of production rules

Survey data used in the development of production rules were collected by an if-then questionnaire (*see* Figure 19.10).

Identification of membership function

A prime characteristic in fuzzy inference is the identification of membership function.

Teachers are asked to observe each of their pupils. If, as an example, a production rule says: If the pupil is tall, provide a large chair, the teacher is asked to determine whether 'tall student' can be applied to pupil A with a grade of about 0.75 and to pupil B with a grade of about 0.25. The teacher is then asked to write the grade of each pupil on the questionnaire. Values on lines increase incrementally as 0, 0.25, 0.5, 0.75, and 1 for ease of entry. If the teacher regards a 135 cm student as tall, they assign a co-efficient of about 0.75. This survey is conducted for each production rule and pupil. The stature and grade are plotted along the x-axis and y-axis, respectively, to construct a membership function which describes the values associated with each production rule. An example of membership function is shown in Figure 19.11. The plotted points are connected, and represent initial values which are then refined by incorporating data derived from the pilot study.

Field experiment

The production rules (described as IF . . . THEN) and the membership functions which ascribe values to the production rules were prepared as described above, and incorporated into the knowledge base. The resultant prototype was used to select chairs at an elementary school. Results of the experiment follow.

Question

Observe each pupil, determine the degree to which the conditional part of the rule applies to the pupil, and indicate your answer by circling 0%, 25%, 75%, 100% on the line. The number at left of each line denotes the pupil's class.

Rule

1	0%	25%	50%	75%	100%
2	0%	25%	50%	75%	100%
3	0%	25%	50%	75%	100%
4	0%	25%	50%	75%	100%
5	0%	25%	50%	75%	100%
6	0%	25%	50%	75%	100%
7	0%	25%	50%	75%	100%
8	0%	25%	50%	75%	100%
9	0%	25%	50%	75%	100%
10	0%	25%	50%	75%	100%
11	0%	25%	50%	75%	100%
12	0%	25%	50%	75%	100%
13	0%	25%	50%	75%	100%
14	0%	25%	50%	75%	100%
15	0%	25%	50%	75%	100%

Figure 19.10 Questionnaire items for establishing values (membership functions) for production rules are listed at the top of the questionnaire. The degree of association between student characteristics is marked on a line. Numbers located next to each of these lines is used to assign the subject rank order

Research site

Kifune Elementary school, Kitakyusyu, Fukuoka Prefecture included elementary school teachers who had responded to the questionnaire used to develop the fuzzy system. Four staff members from Noro Laboratory (Waseda University) and five health teachers from Kitakyusyu elementary schools participated in the field experiment. Subjects were 26 third graders (14 boys, 12 girls); 54 fourth graders (29 boys, 25 girls) and 33 fifth graders (17 boys, 16 girls) at the Kifune Elementary School. The experimental site is depicted in Figure 19.12. A subject testing is shown in Figure 19.13.

The prototype system consisted of an NEC PC-9801 VM21 personal computer, a 14-inch diameter high-resolution colour monitor, and a 200-counts/inch bus mouse.

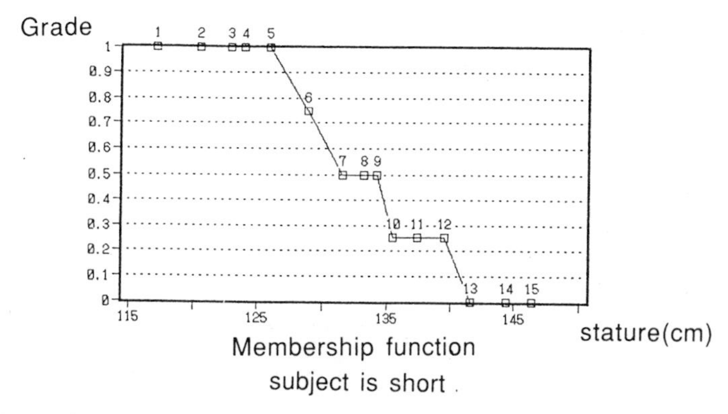

Figure 19.11 This example of a membership function ascribes a value (membership function) to the statement 'subject is short' on a production rule. The degree of subject shortness is plotted along the vertical axis, and subject stature is plotted along the horizontal axis. The numbers plotted in the graph represent subject numbers in class. This membership function was established from a preliminary survey, and refined before it was entered into the knowledge base

Experimental procedure

The experimental procedure consisted of the following steps.

1. Male subjects wore gymnastic pants and training shoes. Female subjects wore T-shirts and bloomers.

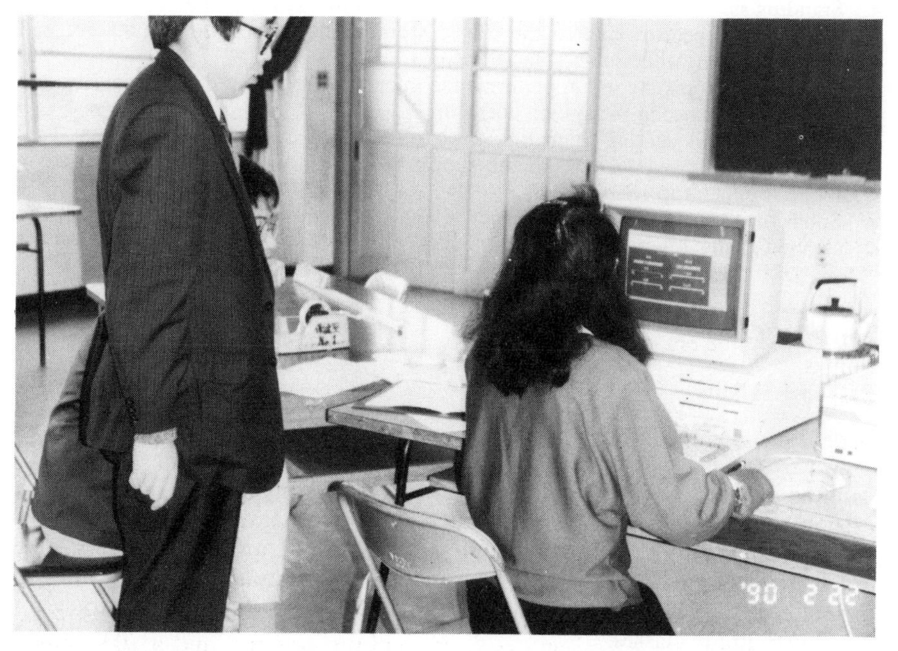

Figure 19.12 View of experimental site. The means of fuzzy system operation is shown

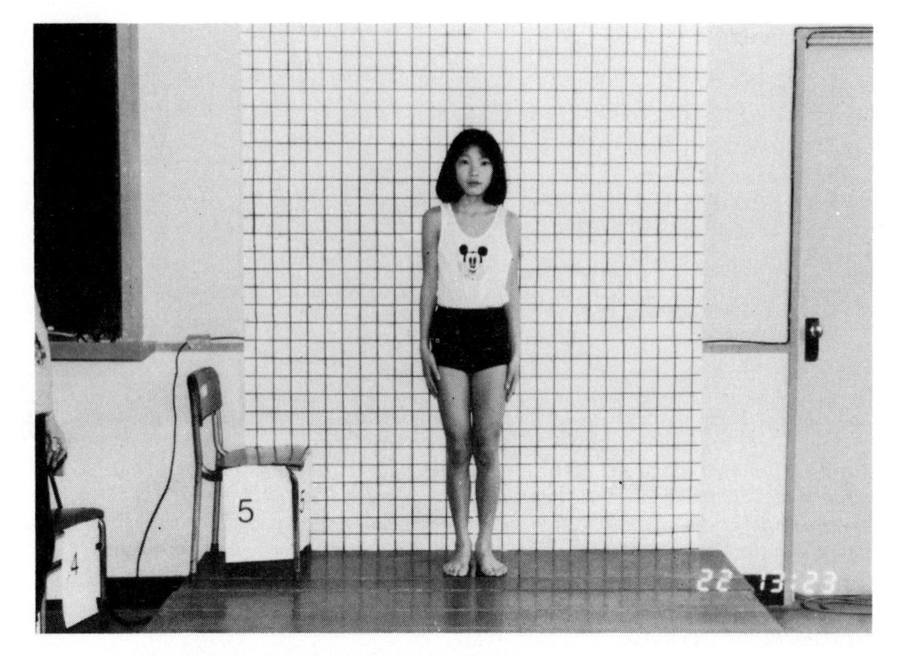

Figure 19.13 The fuzzy system is selecting a suitable chair for this female subject. These chairs can be selected with pupils standing to save time

2. Subject stature, body weight, sitting height, and popliteal length were measured.
3. Subjects stood on a platform with their backs against a mesh screen.
4. Standing subjects were photographed from the front and right side.
5. The system operator entered the subject's grade and rated physique, stature and popliteal height while subjects faced them.
6. The system stores the input, performs the inference process, and communicates the resulting inference to the system operator.
7. Health teachers serving as seating experts selected a chair for each subject. Teachers were allowed to observe the subject both standing and seated when they found it difficult to select a chair.
8. Subjects were photographed from the front and right side while seated on chairs selected in step (6).

Findings

Response validity

Expert seat selections were presumed correct, and compared with inferences drawn by the fuzzy system. The objective of the field experiment was to validate the production rules (IF . . . THEN) and to refine the membership functions which ascribe values to these rules. This system can currently only be generalized for allocating chairs to fourth graders. Third and fifth graders were tested in a subsequent series of experiments. The fuzzy system was found to predict seats selected by experts with 48 percent and 70 percent accuracy for two fourth grade classes.

Effect of training

These differences in accuracy rates of plotted positions between classes were considered unacceptable. Accordingly, the system operator was instructed to exaggerate her plotted image of the subject. This instruction improved the system operator's accuracy rate for the second group of fourth-graders.

For the first group, the system operator was preoccupied with small differences and did not deviate the cursor much from each real line.

Possibility of utilizing neural network

The membership function (which ascribes value) represents the mathematical mechanism for inferring conclusions from a fuzzy set model. To this end, the Tanaka Laboratory of the University of Osaka Prefecture has developed a very powerful tool which identifies membership function with neural networks (Ishibuchi *et al.*, 1989). The fuzzy system described in this chapter was tested to refine membership functions and undergo subsequent field evaluations.

Apparent benefits of this fuzzy system include:

1. Chair selections for each pupil can be achieved in about 15 seconds.
2. The process does not require sitting pupils on various chairs.
3. Familiarity with a personal computer is not required because almost all operations can be made with menus and a mouse.

A proposed new user interface

The findings described in this chapter suggest that the user interface of the fuzzy system allows teachers to enter their images of the pupils into the computer. It is anticipated that this tool will be used to promote Kansei engineering. Kansei is a Japanese term that may be roughly translated as sensibility.

Acknowledgments

The authors are grateful to the health teachers in the Kokura Minami and Kita Ward Health Teacher Committee of Kitakyusyu for providing the production rules for the fuzzy system; the teachers and pupils of Jyono and Kifune Elementary Schools of Kitakyusyu for co-operating in the field experiment; and the Izumidai Elementary School of Kitakyusyu for lending a personal computer for field research.

References

Ishibuchi, H., Tanaka, H., Tamura, R. and Fujioka, R., (1989), Identification of membership functions by neural network. Transactions of the Institute of Electronics and Communication Engineers of Japan. Section D-II, T73-D-II, 1227-1237, (in Japanese).

JIS S 1021, (1980), School Furniture (Desks and Chairs for Classroom) (Japanese Standards Association, Tokyo).

20

The prevention of back pain in school children

A. C. Mandal

There has until recently been world-wide unanimity with respect to 'correct' sitting posture, namely that the body should be upright with a 90° flexion or bending of the hip joint and with preserved lordosis of the lumbar region. However, nobody is able to sit in this posture while working.

There has been no real explanation as to why this right-angled position should be better than any other posture. Nevertheless, it had uncritically been accepted by experts all over the world as the only correct one. Some had considered the lumbar support to be the means to improve seated posture (Akerblom, 1948), but the back rest only carries about 5 per cent of trunk weight – and only in a reclined position of 15° when the arms are supported (Branton, 1969).

In comparison little interest had been attached to the seat pan itself which carries about 80 per cent of the trunk weight. Its influence on the posture of the trunk must therefore be much more important (*see* Figure 20.1).

The sketches in Figure 20.1, however, have absolutely no scientific background. They are nice-looking, but entirely based on the wishful thinking, aesthetics, morals and discipline from the days of Chancellor Bismarck and Queen Victoria. They are based on the false assumption that you are able to sit and work with a 90° flexion in the hip joint and a preserved lordosis of the lumbar region. No one has apparently bothered to check the consequences of these recommendations. Nevertheless, they constitute the basis of:

1. International standardization of school furniture (ISO);
2. European standarization of office furniture (CEN);
3. education of furniture designers (USA) (Diffrient, 1974);
4. anthropometry (AUS), (Oxford, 1969); and
5. instructions for 'correct' sitting posture (DK) (Snorrason, 1968).

The Danish sketch by Snorrason (DK) is perhaps the most interesting of them all, as it explains the origin of this type of instruction. This is simply a drawing of a skeleton sitting on a chair. But sitting problems of a skeleton

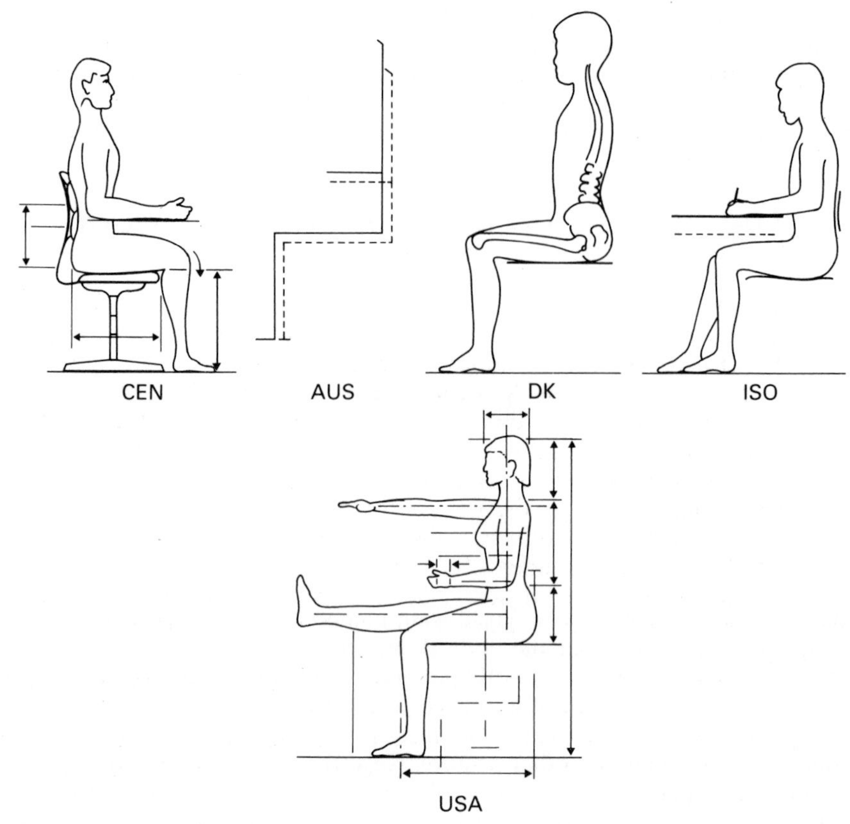

Figure 20.1 Schematic models representing 'correct right-angled posture' from various countries

have nothing in common with the sitting problems of a working person. This is a good example of the low priority the medical profession gives to prevention in the work place.

In Scandinavia enormous efforts have been made to teach people to sit in 'right angled sitting postures' in schools, offices and factories, hoping this

Figure 20.2 The results of intensive posture training with traditional school furniture. Table height 72 cm. All the pupils have unacceptable posture

would reduce the great number of back pain sufferers. In one local school authority (Gentofte) the pupils were given 90 lessons of posture training during 5 years. The poor results of these efforts can be seen in Figure 20.2.

The photographs were taken with an automatic camera at 24-minute intervals during a four-hour examination. All pupils are sitting in postures very harmful to the back.

The anatomy of the seated man

Most doctors, furniture designers, and physiotherapists have little knowledge of the anatomy of the seated person but the German orthopaedic surgeon, Hanns Schoberth, has carried out some excellent research on the problems of sitting posture (Schoberth, 1962). Figure 20.3 is taken from his book.

When standing (A) there is almost a vertical axis through the thigh and the pelvis, and a concavity, or lordosis, is present in the small of the back. When a person is seated (B) the thigh is horizontal, the hip joint is flexed by about 60° and the pelvis has a sloping axis. The lumber region then exhibits a convexity, or kyphosis. Schoberth found from X-ray examinations of 25 people sitting upright that there was an average of 60° flexion in the hip joint and an average of 30° flexion in the lumbar region. When leaning forward over the desk, the increased flexion mainly took place in the lumbar region. This explains why people normally sit with a kyphotic back.

This finding of Schoberth has been confirmed by Akerblom (1948) and Keegan (1953). No scientific investigation has found a 90° flexion in the hip joint combined with a lordosis in the seated work position. The sketches from Figure 20.1 should consequently be abandoned in serious ergonomics literature in the future.

The lumbar support will only have beneficial effect if the seat is sloping backwards so that the point of gravity of the body is forced backwards, behind the supporting areas. This, however, leads to increased pressure under the knees and will tend to tilt the axis of the pelvis backwards. In this way, the

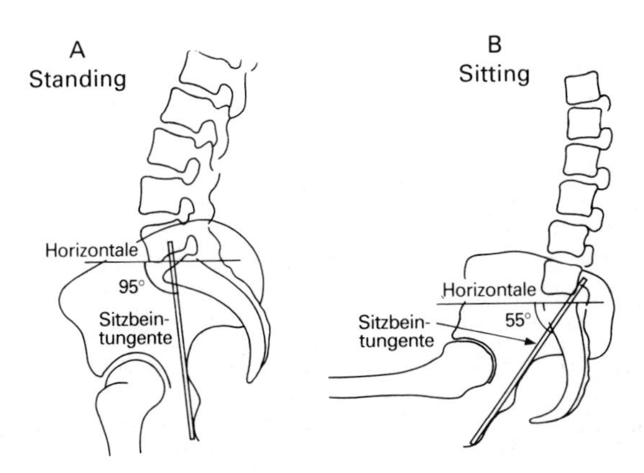

Figure 20.3 The normal anatomy of the lumbar region, standing, A, and upright sitting, B (Schoberth, 1962)

Figure 20.4. *Figure 20.5*

Sitting with sloping thighs can preserve a straight back

effect of the lumbar support is counteracted. To avoid knee-pressure the chairs and tables have become lower and lower.

In the standing position you will automatically obtain a lordosis, a concavity, of the small back. This also happens when sitting on horse-back or on the edge of a table (Figure 20.4). All children know that it is more comfortable to tilt forward on the front legs of a chair when working at a table (Figure 20.5). In all these ways you can sit with thighs sloping about 30° and this reduces the lumbar flexion to a similar degree, resulting in an upright balanced position, which needs no lumbar support.

What is the ideal height of furniture?

During the 20th century the average height of man has increased about 10 cm and during the same period furniture height has decreased (!) about 10–20 cm. As the visual distance has remained 20–40 cm, this will tend to give more constrained postures when working at a desk. Some years ago I investigated which furniture height a group of 80 people preferred when reading and writing.

All preferred to have furniture 15–20 cm higher than the standard, provided the seat and the desk were sloping towards one another. At the preferred height most of them were able to sit in a balanced position with a straight back (Mandal, 1982).

To evaluate the influence of furniture height on flexion of the back a 171 cm tall person was asked to sit reading/writing for a period of 20 minutes at:

1. Traditional furniture with 5° backward sloping seat and a 72 cm high horizontal desktop, i.e. ISO-standard.
2. 20 cm higher furniture with seat and desktop sloping towards one another, Figure 20.6.

This experiment was repeated over a period of 10 days. With an automatic camera photographs were taken at 4-minute intervals. To control the flexion of the various parts of the body, well defined anatomical points were marked on the skin:

1. Knee-joint (capitulum fibulae)
2. Hip-joint (trochantor major)
3. Fourth lumbar disc (midway between anterior and posterior iliac spine)
4. Shoulder-joint (acromion).

At the end of the experiment 50 photographs of the two situations were available. The skin marks were connected with lines on the photos. The resulting angles between these lines were measured:

1. Desk height 72 cm (ISO) showed an average lumbar flexion of 42°.
2. Desk height 92 cm showed an average lumbar flexion of 10°.

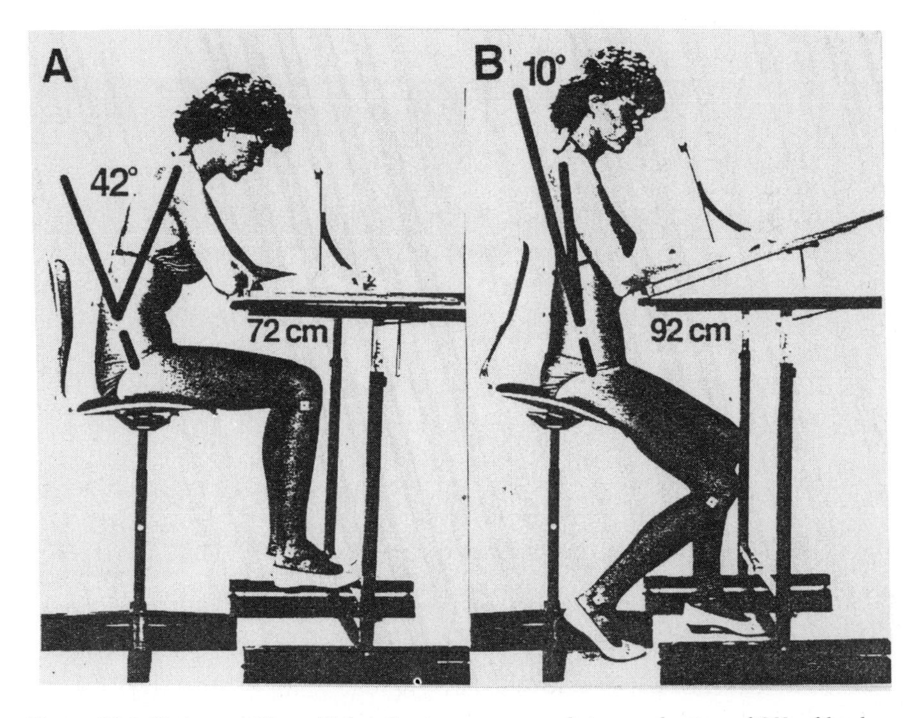

Figure 20.6 Sitting on 20 cm higher furniture may result in a reduction of 32° of lumbar flexion and 20° of neck flexion. The girl definitely preferred the higher position

This means a reduction of lumbar flexion of 32°! The difference was highly significant ($p < 0.0001$).

Also, the flexion in the hip joint was reduced by 15° when changing from 72 cm to 92 cm. Finally, the neck angle was measured and a reduction of 20° was found. In all, this meant a reduction of the total back flexion by 67°; and the girl definitely preferred the higher position. It is obvious that the disc pressure is very low in the position, as it is half standing with preserved lordosis.

New Scandinavian school furniture

The mentioned experiment was made on slightly modified Danish school furniture, which for the past 10 years has proved an effective means of reducing and preventing back pain. About 25 000 are now in use in Denmark. The only problem was a rather complicated and expensive construction. A few years ago we succeeded in making a much simpler version which has proved to be just as effective. The first Danish classroom was equipped with the new type of furniture in 1987 and is shown in Figure 20.7. When the pupils first got the new furniture they had a 10 minute lesson on correct use. Three months later the photograph was taken. As you will notice, all the pupils are sitting much more upright while reading and writing. The furniture is 15–20 cm higher than recommended by ISO. The profile of the seats is part of a circle with a marked

Figure 20.7 The first Danish classroom with the new furniture. The pupils are all sitting much more upright while writing, than on the old ISO furniture in Figure 20.2

forward slope. To hinder sliding forward, the seat is treated with friction varnish. The desk top is sloping.

After a test period of a few months the school decided to replace all the old ISO furniture with the new type.

One year after the installation of the new furniture a questionnaire was given to the pupils. Two hundred and eighty four pupils answered, which represented virtually all those who had used the furniture for one year. 48 per cent of the pupils reported that they had had back and neck pain problems on the old ISO furniture. Out of these, 86 per cent reported a reduction in pain with the new furniture (Figure 20.8), but 91 per cent of the pupils using the 'preferred' height reported improvement. The main complaint, that the furniture was too low, was probably because the pupils had grown considerably during the year.

Of course any form of new furniture will apparently reduce back problems when using a questionnaire. On the other hand, a good many of the pupils had already at this time forgotten the back problems they had one year earlier.

Each pupil was given written instructions about correct height and how to use the furniture (Figure 20.9). This means that the majority of the pupils were able to find the correct height – which is the same as the most comfortable height – and gradually they would learn that it is also comfortable to sit in the front half of the seat with their feet on the floor while reading and writing. When the thighs are sloping 30°–40° they will normally sit with a straight back.

PERCENTAGE OF PUPILS WITH REDUCED BACK PAIN ON NEW FURNITURE

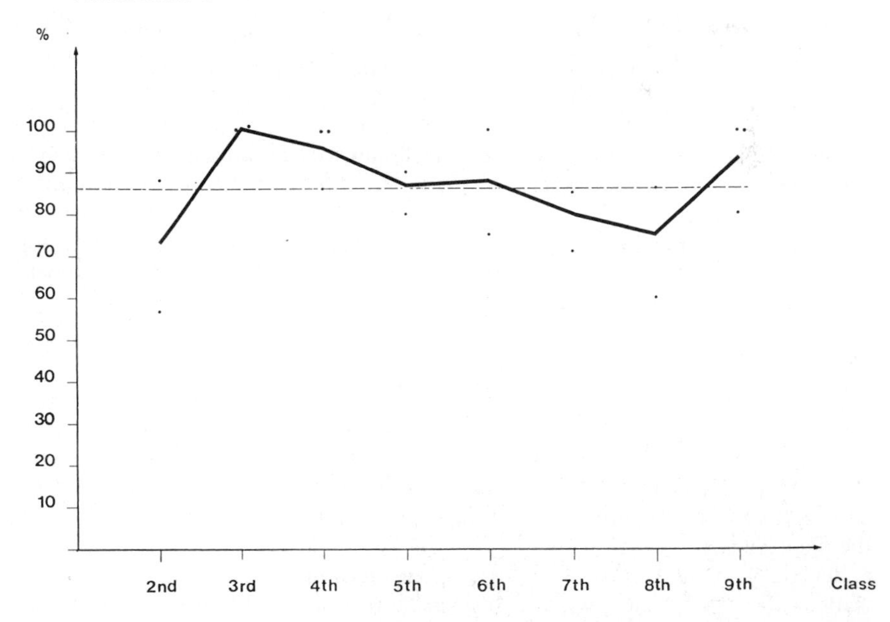

Figure 20.8 An average of 86 per cent of the pupils with back-neck complaints on the old ISO furniture reported reduced pain with the new furniture. Each dot indicates the average of one class

INSTRUCTION:

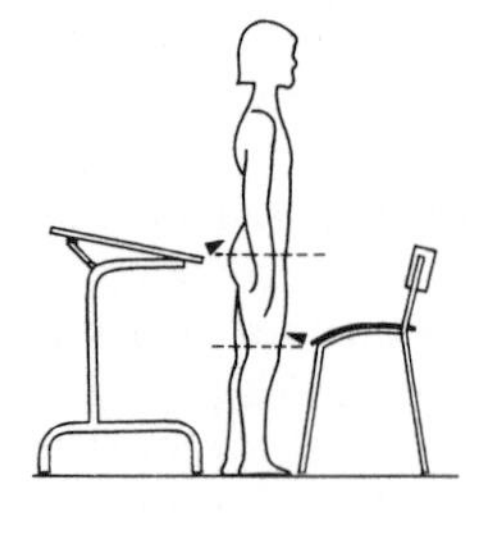

A correct height of furniture is important

Edge of desk at level with your buttock
Front edge of chair 2–4cm above knee

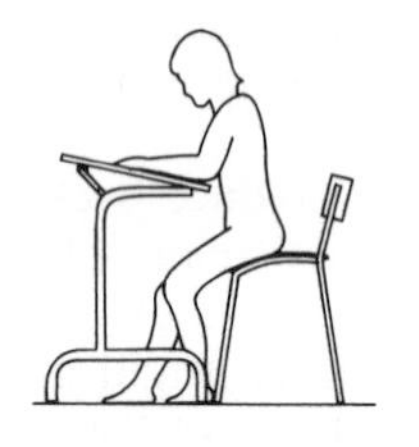

Reading/writing position

Use front half of the seat
Both feet on the floor
Desktop sloping

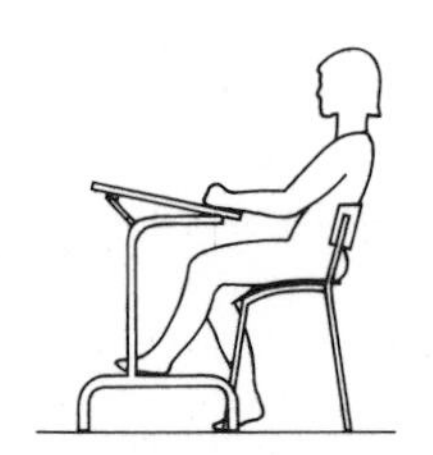

Resting position

Move backwards on the seat
Use the lumbar support
Feet on footrest

Figure 20.9 All pupils are given this written instruction about correct height and how to sit while working and resting

The new furniture has already been accepted to such an extent that most of the new furniture purchased now by Danish and Swedish schools is of the new type. The teachers have also benefited from the tall desks as they avoid the worst bending while assisting the pupils. Several municipal school authorities have, after a short test period, decided to replace all the old ISO furniture with the new type in the course of the coming years.

Conclusion

In the industrialized world more and more work time is spent in a seated position. Very often people have to sit bent over a desk for many hours every day with constant repetitive work. This places a very high degree of strain on the back.

Several Danish French and Swiss investigations have shown that 60 per cent of 15–16 year old pupils have back and neck complaints after using the ISO standard furniture. Even the most intensive training will not enable them to sit with acceptable posture while reading and writing. This is clearly demonstrated in Figure 20.2 and indicates that the ISO furniture is the most likely to cause the large number of back problems in schoolchildren. Furthermore, ISO may well be responsible for a good deal of the back problems in the grown-up population.

At many congresses around the world during the last decade I have asked if there was an explanation or excuse for the principles of the ISO school furniture. There was none. Most school furniture in the industrialized world is still made according to ISO standards. No one has ever been able to show me a class with acceptable working postures on standard furniture.

I have during the last 20 years tried to improve school furniture and I can only advocate that similar experiments are started in other countries.

It is obvious from anatomical and laboratory studies, as well as experiences with a very large number of Scandinavian pupils, that the ISO standards are harmful and ought to be changed.

References

Akerblom, B., (1948), *Standing and Sitting Posture*, (Stockholm: Nordiske Bokhandeln).

Branton, P., (1969), *Ergonomics*, **12**, 316–27.

CEN (Comité Europeén de Normalisation), (1982), Pr En, (Afnor, Tour Europe CEDEX 7, 920-80 Paris, La Defense).

Diffrient, N., (1974), *Human scale*, (Cambridge, Mass: MIT Press).

ISO (International Organisation of Standardisation), (1978), TC 136/SC7, Cologne.

Keegan, J. J., (1953), *Journal of Bone and Joint Surgery*, **35**A, 3.

Mandal, A. C., (1982), Correct height of school furniture, *Human Factors*, **24**, 257–69.

Oxford, H. W., (1969), Anthropometric data for educational chairs, *Ergonomics*, **12**, 140–61.

Schoberth, H., (1962), *Sitzschaden, Sitzmöbel*, (Berlin: Springer-Verlag).

Snorrason, E., (1968), *Tidsskrift for Danske Sygehuse*, Copenhagen.

21

Fitting the chair to the school child in Korea

Am Cho

An understanding of the relationship between school-children's posture and their school seating can promote their health and education. The objective of this research is to evaluate existing school chairs and suggest alternative designs.

One hundred and twenty five teachers were randomly selected to participate in questionnaire research. The anthropometric dimensions and postures of fourth graders from Seoul (175) and Pusan (109) and the dimensions of their chairs and desks were measured, and photographed.

The following problems were identified:

1. Chair allocations were based on student stature.
2. The teachers are not satisfied with available chairs and the method of chair allocation.
3. Pupils complain of low seat pan heights, narrow seats, hard back rests, and other design inadequacies.
4. Pupils typically select chairs that are one size larger than suggested by their size.
5. Poor postures are associated with inappropriate seat heights and sasyaku* (sitting height in cm $\times \frac{1}{3} - 1$) of their chairs, their writing posture and lack of teacher's instruction.

This leads to the following recommendations:

1. develop conversion tables to help specify chairs for each pupil;
2. allocate a chair and desk to each pupil;
3. individually adjust the seat pan height and desk height for each pupil; and
4. teach both teachers and pupils how to sit properly.

To these ends, new values of sasyaku and desk height (not specified in Korean Industrial Standards (KS)) are proposed.

---

* Sasyaku is the distance between chair and desk height.

Introduction

School-children sit for long periods of time. It is important to understand how their chairs affect posture to promote their health and education. Primary school-children grow and change faster than at any other age (Mandal, 1982). Their growth and physical well being may be hindered if the school chairs do not fit their users.

Objectives

The objective of this research is to analyze the following:

1. Methods for determining chair and desk size.
2. The chair and desk sizes selected by school-children who are dissatisfied with their present chair and desk sizes.
3. Differences between existing and selected furniture to develop recommendations for the future.
4. The causes of poor seated postures.
5. The 'fit' between chair and desk heights and anthropometric dimensions of users.
6. New values of sasyaku and desk heights and the appropriateness of those provided in Korean Industrial Standards (KS).

The methods

This research was conducted from November 1988 to June 1989. Subjects were:

1. 125 primary school teachers in Seoul and Pusan randomly selected and sent questionnaires.
2. 175 fourth graders in Seoul, and 109 fourth graders in Pusan. Table 21.1 depicts variables investigated.

Table 21.1 Experimental factors

- Factors associated with present usage:
  - (1) Method for determining chair size
  - (2) Design features used in establishing the pupil's chair size
  - (3) Complaints of school chairs and desks
  - (4) Causes of incorrect seating posture
  - (5) Problems associated with seating, plus 3 other items in questionnaire to 125 primary school teachers
- Additional measures:
  - (1) School-children dimensions
  - (2) Present usage of chairs
  - (3) Dimensions of existing chairs and desks

* Signifies that only the Seoul population was used for this evaluation.

Measurements included:

1. physical dimensions of fourth grade school-children;
2. dimensions of school chairs and desks;
3. photography of seated student postures while studying;
4. video recordings of postures of students attending lectures; and
5. questionnaire responses regarding school chairs and desks.

School chair and desk usage

The school chair and desk height could adjust independently. Each table was used by two students, and each chair was used by one pupil. Figure 21.1 shows the numbers of each size and type of chair currently used in the fourth grade.

Chair types 4 and 5 represent 71 per cent (124/175) of existing furniture (*see* Figure 21.2). School seats were moved every 1–2 weeks, so that students sat at different chairs and desks.

Questionnaire responses

The questionnaire identified the following problems:

1. Each class teacher developed his/her own criteria for selecting chair size. 94.4 per cent of these decisions were based on children's stature (118/125) and 5.6 per cent on lower leg height (7/125). Teachers also used the sasyaku and the viewing distance to allocate chairs.
2. Chair size should be determined by sasyaku. Some teachers reported that overweight children were not able to fit their belongings under the school desk.
3. School-children's chairs are selected by the class teacher. Pupils did not complain unless the chair was broken or their possessions did not fit under the desk.

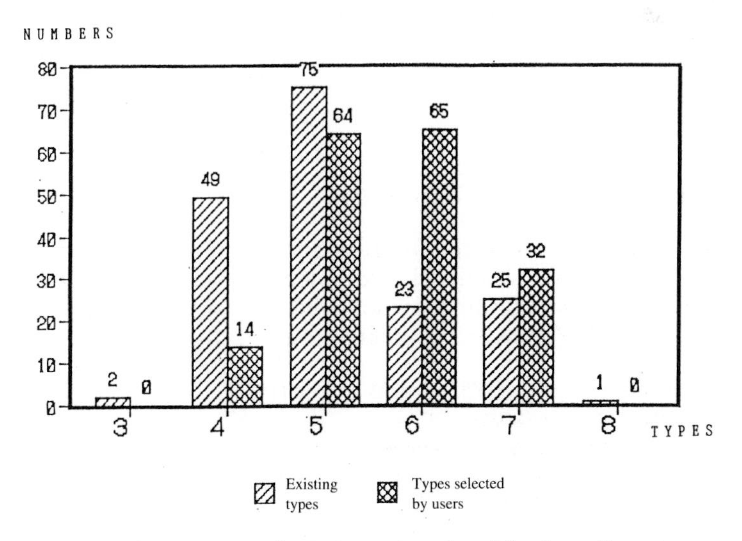

Figure 21.1 Frequency of existing and preferred furniture dimensions

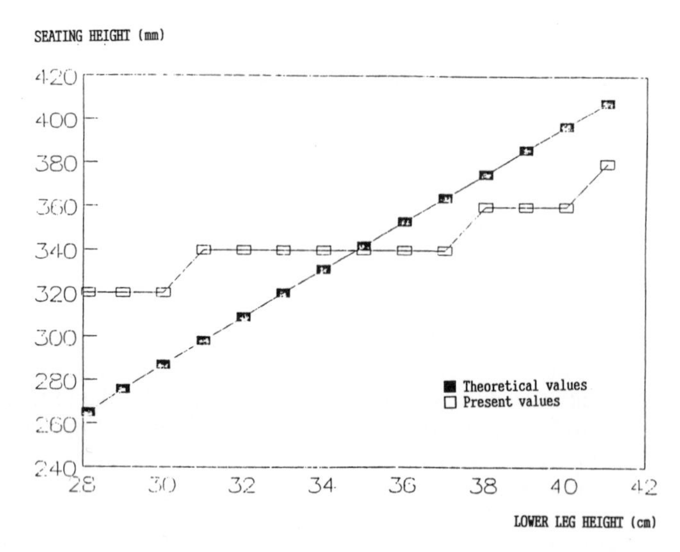

Figure 21.2 Existing and recommended seat heights vs. lower leg heights

4. Poor student postures are associated with inappropriate seat heights and sasyaku (sitting height in cm × $\frac{1}{3}$ – 1); the narrow width and hard surface of seats while writing, and to lack of instruction. Their unnatural writing posture is due to the low desk height, which forces children to hunch over their desk.
5. Other students complained that seat pans were too low and narrow, and back rests too hard.
6. Teachers complained that existing chairs and desks did not accommodate the spectrum of user sizes.
7. Teachers expressed concern because poor seats may hinder a pupil's ability to concentrate and adversely affect their physical growth.
8. Teachers indicate that there were not enough types of chair and desk. Consequently, chairs cannot accommodate the spectrum of users.

Questionnaires and site reviews indicate that the most prevalent problem was associated with the relationship between chair and desk heights; that is, sasyaku.

The height of chair and sasyaku

Figure 21.1 indicates that in the fourth grade, the most commonly available chair sizes are type 4–6. Table 21.2 compares Korean Industrial Standards with present findings regarding school-children's anthropometry and chair size. Figure 21.1 and Figure 21.2 compare preferred chair height;

1. Sitting heights by Keegan (1953) and Akerblom (1954)
2. Dimensions of existing chairs
3. User preferences (the height of chair selected by pupils).

Figures 21.1 and 21.3 indicate that pupils selected chairs one size larger than is generally suggested for their body size. This result is in agreement with that of Dr Mandal (1985).

Table 21.2 Experimental values of sasyaku and anthropometric data for each type of desk and chair

Stature (cm)	Types (size)	Desk height (mm)	Drawer thickness (mm)	Actual value of sasyaku (mm)	Sitting height (cm) 0.55 × stature	Proposed value of sasyaku (cm) sitting height/3^{-1}
109–116	1	440	90	90	63.3	20.1
116–123	2	470	90	100	67.1	21.4
123–130	3	500	110	90	72.0	22.7
130–137	4	530	110	100	74.8	23.9
137–144	5	560	120	100	78.7	25.2
144–151	6	590	120	110	82.5	26.5
151–158	7	620	120	120	86.4	27.8
158–165	8	650	120	130	90.2	29.1
165–172	9	680	120	140	94.1	30.4
172–179	10	710	120	150	97.9	31.6
179–	11	740	120	160	101.8	32.9

The tolerance limit ± 2.0 mm[9].

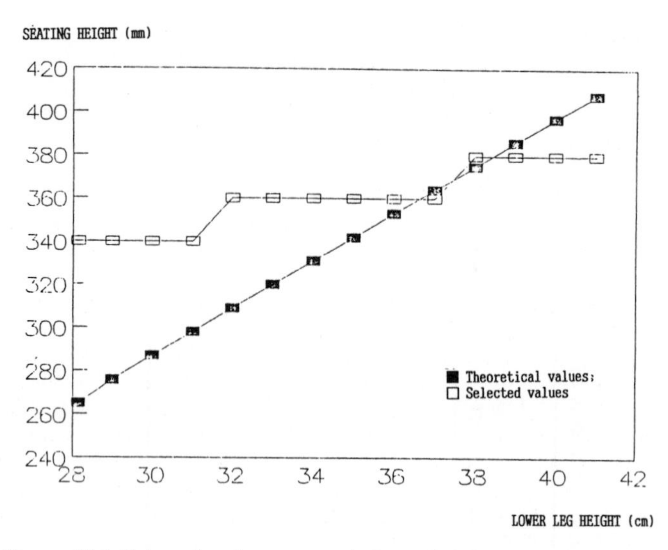

Figure 21.3 Selected and recommended seat height vs. lower leg heights

High seats pans were selected to:

1. increase lower leg mobility;
2. increase back rest comfort;
3. improve visibility of the front of the class; and
4. increase comfort associated with wider seat pans.

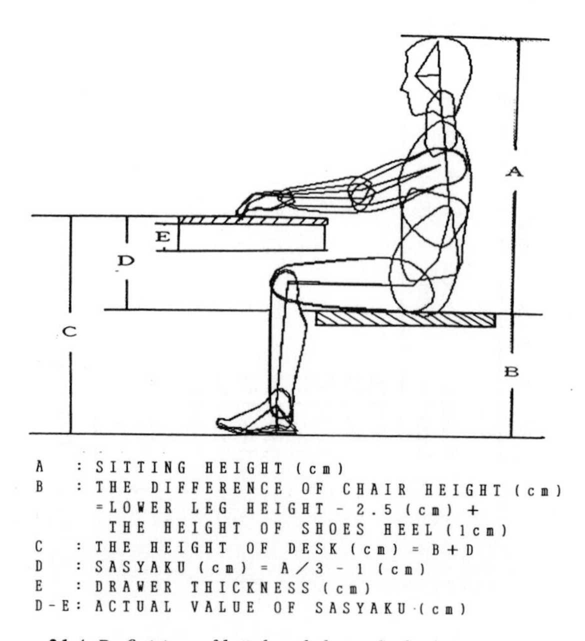

A : SITTING HEIGHT (cm)
B : THE DIFFERENCE OF CHAIR HEIGHT (cm) = LOWER LEG HEIGHT - 2.5 (cm) + THE HEIGHT OF SHOES HEEL (1cm)
C : THE HEIGHT OF DESK (cm) = B + D
D : SASYAKU (cm) = A / 3 - 1 (cm)
E : DRAWER THICKNESS (cm)
D-E : ACTUAL VALUE OF SASYAKU (cm)

Figure 21.4 Definition of height of chair, desk, drawer and sasyaku

Differences between the existing and selected chair heights were minor. Where such differences existed, a simple seat adjustment was able to accommodate pupils.

Korean Industrial Standards (KS) address design of school chairs and desks and specify values for sasyaku. This research indicates that provision of sasyaku does not ensure that fourth grade students are accommodated.

* Sasyaku; 210 mm, Thickness of desk drawer; 110 mm

Average thigh depths were 100 mm (standard deviation; 14 mm). Consequently, it is difficult to accommodate the lower leg under the desk, as when sitting at the desk while writing. The theoretical values of sasyaku were calculated to establish minimum distances between the school chair and desk. Seat heights were calculated as 0.55 multi stature. These calculations are depicted in Table 21.2.

Table 21.2 shows the exponential values of sasyaku which were without a drawer from present sasyaku value. Based on sasyaku, the height of the desk should be 21 mm to 49 m higher than the present desk height. Optimum desk heights were calculated as chair height (KS) plus proposed sasyaku values. That is, present desk heights should be increased by 21 mm to 49 mm to avoid poor writing postures associated with low desk heights. Figure 21.4 defines the height of chair and desk, sasyaku, and drawer thickness. Figure 21.5 compares the values of present, actual and proposed (*see* Figure 21.4). Figure 21.6 compares the present with proposed desk height.

Discussion

1. 90.4 per cent of teachers and pupils are not satisfied with the chairs and the method of allocating them.
2. Pupils complain of low seat height, narrow seats, hard back rests, and other design limitations.

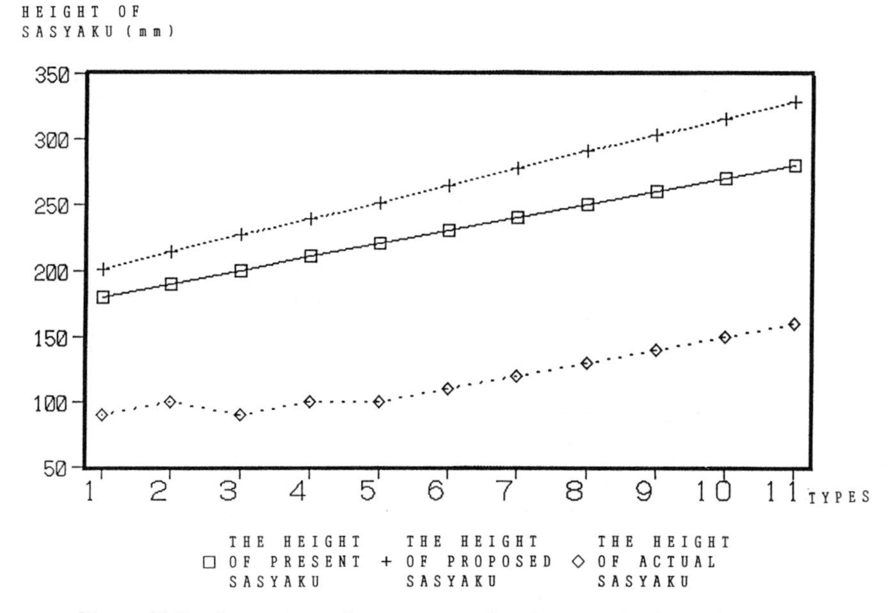

Figure 21.5 Comparison of present, actual and proposed values of sasyaku

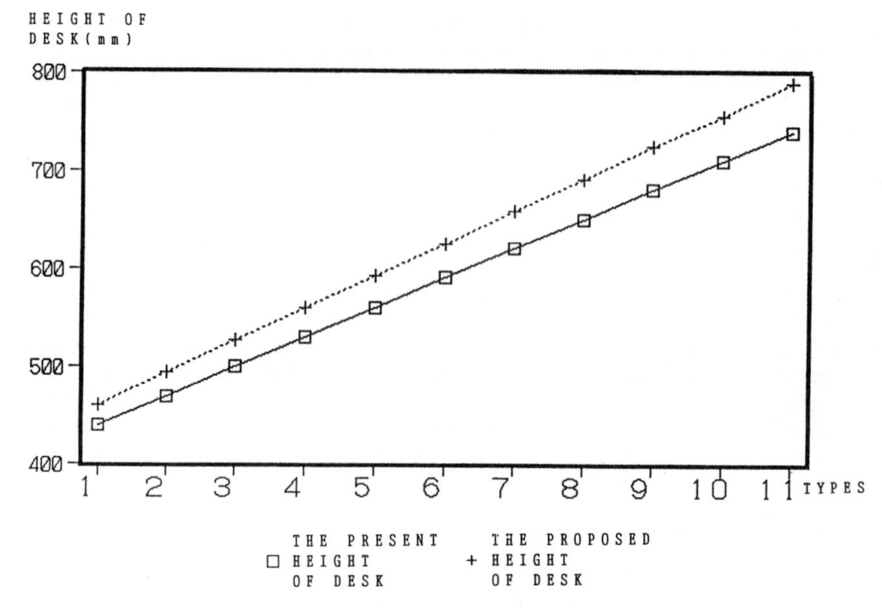

Figure 21.6 Comparison of present and proposed desk heights

3. Chairs are typically allocated to the pupils based on their stature. The author recommends that popliteal height be used to determine chair and desk height.
4. Teachers expressed frustration with the inappropriate sizes of chairs and desks. Because chairs and desks provided are adjustable, desks must by adjusted appropriately to accommodate individual differences in user sizes.

Table 21.3 Proposed guidelines for desk heights and sasyaku for different size chairs

		values of sasyaku (mm) (sitting height/3 – 1)		height of desk (mm)	
types (size)	sitting height (cm) (0.55 × stature)	present values	proposed values	present values	proposed values
1	63.3	180	201	440	461
2	67.1	190	214	470	494
3	71.0	200	227	500	527
4	74.8	210	239	530	559
5	78.7	220	252	560	592
6	82.5	230	265	590	625
7	86.4	240	278	620	658
8	90.2	250	291	650	691
9	94.1	260	304	680	724
10	97.9	270	316	710	756
11	97.9	280	328	740	789

The tolerance limit ±2.0 mm[9].

5. Poor student posture is associated with inappropriate seat heights and sasyaku (sitting height $\times \frac{1}{3} - 1$), their habitual writing postures, and lack of teacher instruction.

Both teachers and pupils need training regarding chairs and seated postures. New values of sasyaku and height of desk here proposed are not specified in Korean Industrial Standards (KS) for the design of school desks and chairs.

Conclusion

The following problems were identified as a result of the investigation:

1. Poor student postures are associated with the seat height and sasyaku (sitting height $\times \frac{1}{3} - 1$) of their chairs, their writing postures and lack of teacher instruction.
2. Both teachers and pupils lack information regarding how to adjust and use their chairs.
3. Sasyaku is not specified in Korean Industrial Standards (KS).

The following may resolve these problem:

1. Adjust seat and desk height to accommodate each pupil.
2. Train both teachers and pupils on correct seated postures.
3. Specify desks individually and adjust to accommodate the user.
4. Allocate chairs and desks individually.
5. Seat heights should serve as the foundation for determining the height of the desk.
6. Research and develop conversion tables to specify chairs for individual pupils.

New values of sasyaku and desk height, not specified in Korean Industrial Standards (KS), are proposed in Table 21.2.

Acknowledgments

The authors would like to thank the reviewers for helpful comments, Professor K. Noro of Waseda University, and also thank Mr N. Hirasawa, Ms K. W. Kang, and Mrs R. S. Lee in Korea, for their assistance with experimental research.

References

Akerblom, B., (1954), Chairs and sitting, in W. F. Floyd, and A. T. Welford (eds), *Symposium on Human Factors in Equipment Design*, Proceedings of the Ergonomics Research Society Vol. II (pp. 29–39) UK: The Ergonomics Society.

Chung, B. Y. and Park, K. S., (1986), *An Ergonomics Study of Standard Sizes of Educational Chairs and Desks*. Journal of the Human Engineering Society of Korea, Vol **5**(1) June, 29–41, 1986.

Evans, W. A., Courtney, A. J. and Fork, K. F., (1988), The design of school furniture for Hong Kong school-children, *Applied Ergonomics*, **19**(2), 122–34.

JIS S1021 – 1980, 1–5, (1985).

Keegan, J. J., (1953), Alteration of the lumbar curve related to posture and seating, *Journal of Bone and Joint Surgery*, **35A**, 583–603.

Kikuzaa, Y., (1983), Height adjustment of the student's desk and chairs for home study, *Journal of Home Economics of Japan*, **34**(8), 488–97.

Mandal, A. C., (1982), The correct height of school furniture, *Human Factors*, **24**(3), 259–69.
Mandal, A. C., (1985), *The seated man* (*Homo sedens*), Dafnia Publications, Denmark, 45–6.
Ouchi, K., Wakai, S. and Kato, M., (1975), A field survey on the use of school furniture, *Japanese Journal of Ergonomics*, **11**(2–3), 63–7.
UDC 645. 444/411: 371. 63 KS (Korean Standard) G 2010-1-2.

22

A procedure for allocating chairs to school children

Tomoko Hibaru and Toshiko Watanabe

Introduction

Health disorders suffered by elementary and junior high school children were first noticed in Japan in the early 1970s. These disorders included diabetes, hypertension, and gastric ulcer, which are generally considered adult diseases. Simultaneous evidence indicated that although Japanese school children are not ill, they are also not healthy.

Chairs are considered essential for promoting good postures. Class and health teachers in Japan observed that poor postures were common in elementary and junior high school children. Many class teachers found that children often sat in a slumped posture during class. About 9000 elementary and junior high school children in the Kokura Minami and Kita Wards of Kitakyushu were investigated to establish optimum seat dimensions. The results are depicted in Figure 22.1. The differences between the range indicated by the straight lines and the range indicated by the histograms demonstrate that chair selections for school children from 6–15 years old should not be based on stature, as specified in JIS S 1021 – School Furniture (Desks and Chairs for Class Room) (Japanese Standards Association, 1980).

Rather, chair selections should be based on user postures and preferences. Since children are still growing, however, this practice has limited applicability for schools. Correspondingly, this study was designed to clarify the relationship between student anthropometric dimensions and chair size requirements in order to facilitate the allocation of chairs to individual school-children.

Methods

It was assumed that with a few exceptions such as obesity, desks are compatible with chairs. Consequently, the objective of this investigation was to evaluate school seating.

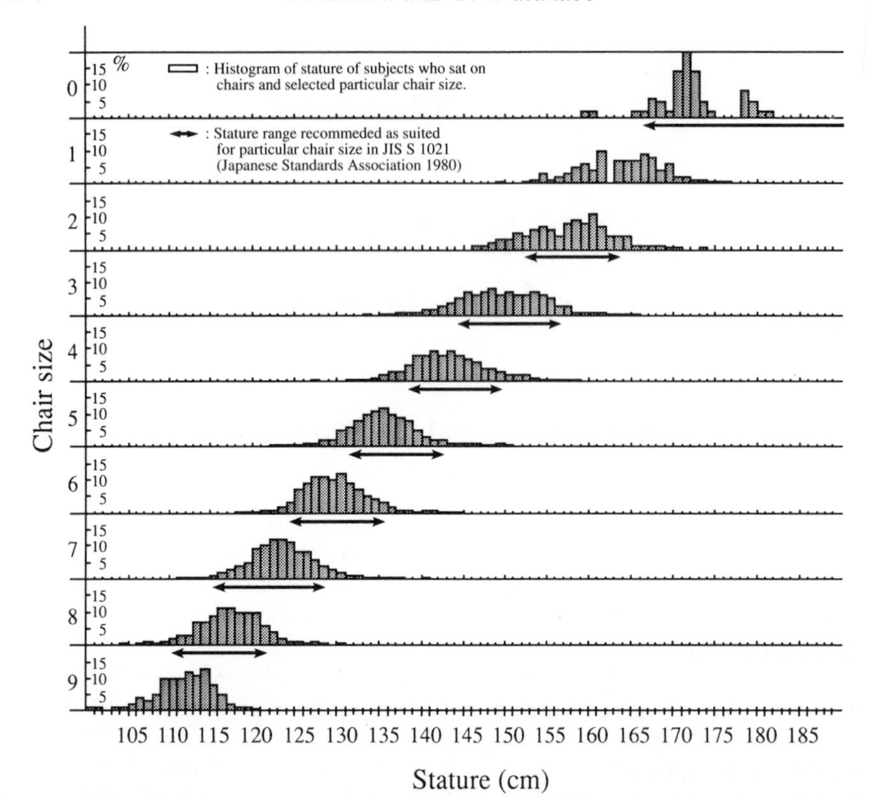

Figure 22.1 Line representation of stature range conforming to chair size recommended by JIS S 1021 (Japanese Standards Association, 1980) and histograms of chair sizes used according to stature by 9000 elementary and junior high school-children in Kitakyushu

Figure 22.1 shows that the chair size that is appropriate for children of the same stature will vary in accordance with body shape, secondary sex characteristics, and other individual factors. The objective of this research was to develop a simple method for allocating chairs in accordance with the child's stage of growth.

In 1984, 124 fourth graders participated in the study. Preferences for school seating were compared with eight body dimensions in the standing posture and nine body dimensions while seated. It was found that:

1. The chair size selections were strongly correlated with the lower leg length;
2. Two or more chair sizes accommodated one body size (referred to as overlap).

The subsequent phase of the evaluation examined why the chair size selections overlapped. The following problems were identified with school chairs:

1. Chair sizes supplied to the schools did not match the distribution of body sizes of school-children.
2. More small (size 9 and 10) chairs and desks were required for elementary schools.
3. More large (size 3) chairs were badly needed for junior high schools.
4. Chairs used at elementary and junior high schools were too high.

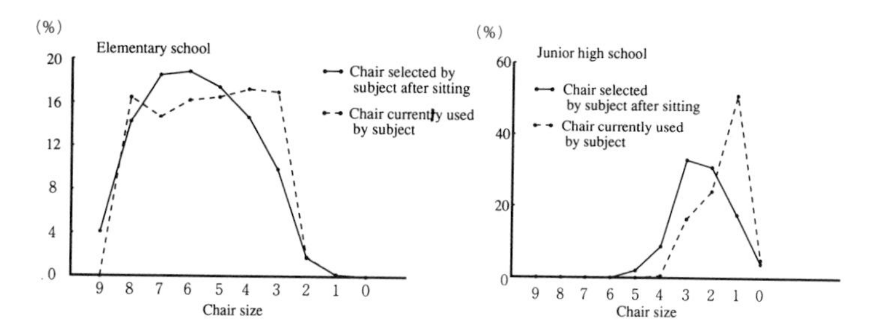

Figure 22.2 Comparison between number of chair sizes currently used and number of chair sizes selected after sitting by elementary and junior high school-children

In 1986, an additional survey was conducted to investigate how teachers and children selected chair sizes. Two fourth grade classes with a total of 90 subjects had their stature, weight, sitting height, upper leg length, and lower leg length measured. Prior to sitting trials, teachers developed postural criteria for evaluating chairs.

Suitability check points were established. Subsequently, subjects sat on chairs, and teachers and subjects selected chairs together.

The teachers and subjects were questioned regarding how they used these criteria to guide their selection process. These responses were correlated with chair characteristics and subject gender. The dimensions of elementary school chairs were measured at 17 points (*see* Figure 22.3). These dimensions are listed in Table 22.1.

Data collection and analysis

Data collection

Subject dimensions are provided in Table 22.2. These values were comparable to national averages.

Figure 22.4 depicts subject chair size selections. The interview responses of the teachers and subjects are summarized in Table 22.3. About half of the subjects found it difficult to select a chair size. When asked why they selected

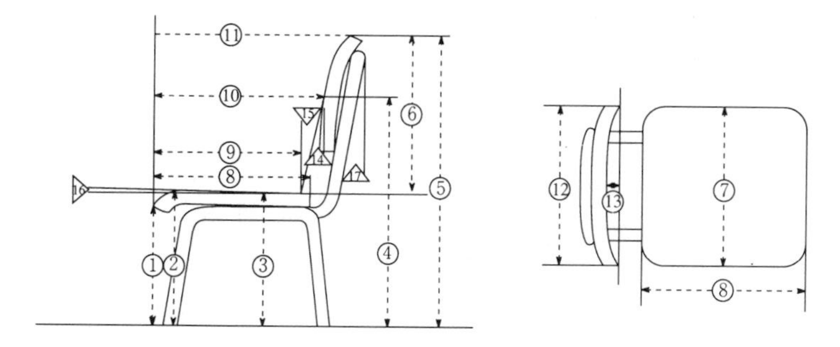

Figure 22.3 Seventeen chair dimensions measured

Table 22.1 Primary seat dimensions

Chair manufacturer	Chair size	Chair dimension (mm) 1	2	3	4	5	6	7	8	9	10	11	12	13	14	15	16	17
B	1	44.1	45.9	44.3	70.3	77.4	33.2	35.8	36.0	38.7	39.2	40.3	36.1	2.5	88.5	93.0	3.5	93.0
B	2	41.1	43.0	42.4	68.9	75.7	34.2	36.0	36.0	37.8	39.5	40.1	36.2	2.7	90.0	94.5	4.0	91.0
B	3	39.6	41.6	40.3	63.9	70.9	30.5	35.7	32.1	36.2	35.7	36.7	36.0	2.5	89.5	92.5	3.0	92.5
B	4	38.0	39.4	38.1	61.0	68.4	30.5	35.9	32.7	34.8	35.9	36.8	36.0	2.5	91.5	93.0	3.5	93.0
B	5	36.4	37.8	36.4	57.9	64.6	28.9	35.9	29.4	31.6	32.4	33.7	36.0	2.5	89.5	93.5	2.5	93.5
C	6	33.4	34.6	33.6	53.3	60.0	26.7	34.5	30.1	34.6	35.7	36.7	31.5	1.7	90.0	93.0	2.5	94.0
B	7	31.1	32.2	32.3	52.4	59.5	27.8	34.0	25.3	28.2	28.4	28.8	34.0	2.2	88.0	93.0	2.5	93.0
B	8	29.5	30.7	30.7	49.3	56.7	26.7	34.0	25.1	28.9	28.0	30.4	34.0	2.2	90.0	93.0	3.5	93.0
B	9	27.2	28.0	27.9	45.3	53.6	26.2	34.0	25.2	29.4	24.8	25.0	34.0	2.2	89.0	92.0	3.0	90.5

For chair dimension numbers, see Figure 22.3.

Table 22.2 Means and standard deviations of body measurements of fourth graders in elementary school.

		Stature (cm)	Weight (kg)	Sitting height (cm)	Lower leg length (cm)	Upper leg length (cm)
Boys	Mean	135.9	32.8	73.5	36.0	35.2
	Standard deviation	5.068	5.982	2.517	1.805	1.613
Girls	Mean	136.8	32.4	74.1	35.9	36.2
	Standard deviation	5.781	5.421	3.319	1.851	2.219

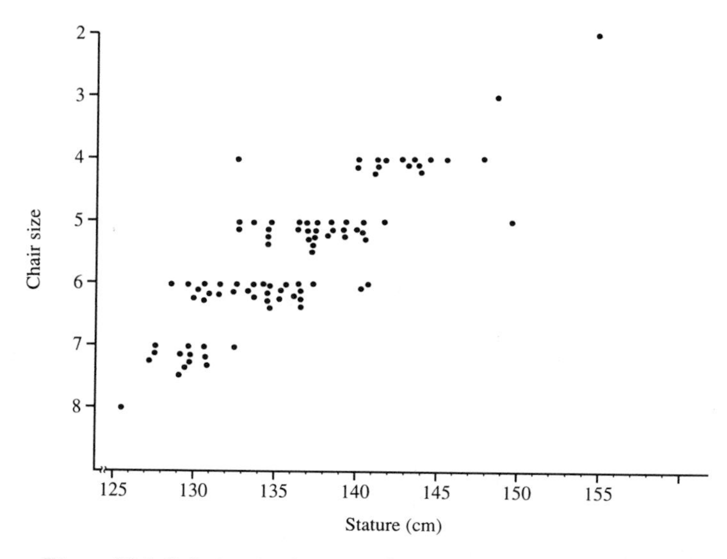

Figure 22.4 Relationship between chair size selections and stature

the particular chair size, many subjects answered that their decision was based on personal preference and the fact that their teacher judged it suitable. When the subjects were asked why they liked the chair size, many cited fit and comfort. The thighs of obese subjects exceeded the largest seat pan width; this problem was more pronounced for females.

Data analysis

The correlation between chair size selections and lower leg length is shown in Figure 22.5. Despite this high correlation, children with similar lower leg lengths selected different chair sizes. This finding is in agreement with research of 9000 school-children (discussed in previous sections). This overlap was investigated further.

Children who selected high chairs were compared with children with similar lower leg lengths who selected lower chairs. Since the lower leg length is often considered to correspond to seat height and upper leg length to seat depth, these dimensions were compared with seats selected.

Children who selected small chairs were compared with those with similar lower leg lengths who selected larger chairs. Generally, children who selected high chairs had long upper leg lengths. Children with long legs (upper and lower) tended to select higher chairs with large seat pan depths.

Obese children tend to select higher chairs. To clarify why they did so, their seated posture was examined. Figure 22.6 shows a typical example. Since this girl has an obese back and hip, her ischial tuberosities displace forward, causing her to appear unsupported by her chair. Deep chairs presumably provide more support for obese children, as long as they can maintain feet on the floor. Obese children do not necessarily have long leg lengths, but have excessive back and hip breadths. Consequently, they select high chairs with deep seats. Children with 36 cm upper leg lengths selected size 5 and 6 chairs.

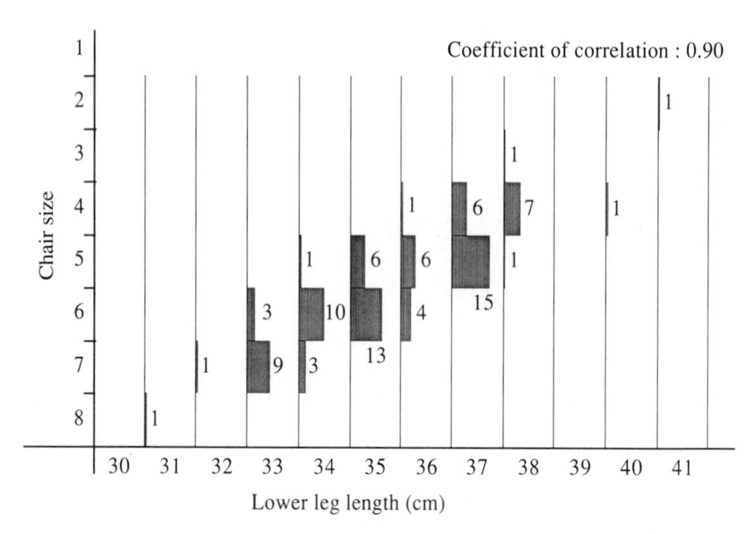

Figure 22.5 Relationship between chair size and lower leg length, based on survey results. Each histogram shows the number of children by chair size

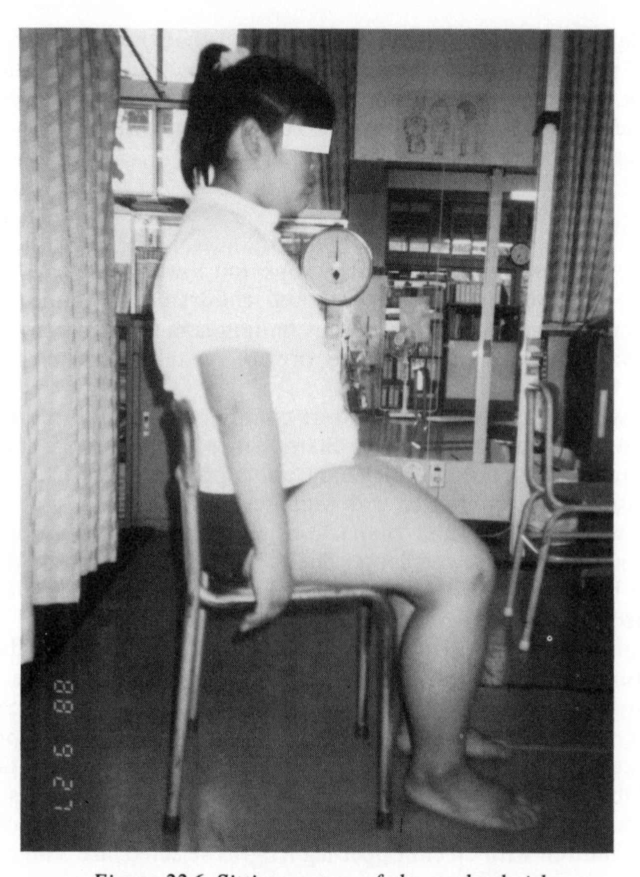

Figure 22.6 Sitting posture of obese school girl

Lower leg length, upper leg length, and body weight each affects chair selections. Body weight is closely related with selected seat depths.

Proposed method for allocating chairs

The above research was used to develop a method to facilitate the allocation of school chairs. It is proposed that chair sizes be selected based on the child's lower leg length. This dimension was most strongly correlated with selected chair sizes. In Figure 22.7, two chair sizes are shown for a given range of lower leg lengths.

It is recommended that children with long upper leg lengths or large body weights receive high chairs, and other children receive lower chairs. This approach was tested on 90 school-children. Of the 90 children, 54 with relatively long upper leg lengths were selected for the evaluation. After sitting, 43 of the 54 (79.6 per cent) selected chair sizes which agreed with the proposed method. This percentage exceeded those obtained from the method specified in JIS S 1021 (Japanese Standards Association, 1980).

Summary

Physical dimensions of elementary and junior high school-children were compared with their chair selections after sitting. This comparison was used to develop a more effective method of allocating chairs to elementary and junior high school-children with different body shapes.

Once or twice per year at the city of Kitakyushu schools, children sat, and had their chairs evaluated. Chairs are reselected as often as feasible. The children are then taught how to sit. The city recently bought size 9 and 10 chairs. The schools now exchange chairs among themselves. The city's board of education periodically adjusts desk heights to accommodate each child.

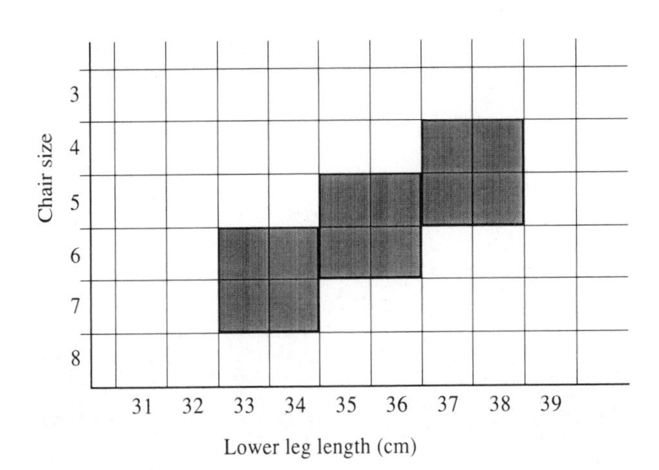

Figure 22.7 Each shaded square shows two chair sizes recommended for each lower leg length range by proposed method

The ratios of seat pan depth to seat pan height of school seating are inappropriate. This causes users to give priority to one dimension over the other during the selection process. Consequently, the chair selected is often inappropriate, and affects their posture adversely. This procedure cannot accommodate users with long upper leg lengths and large body weights, which frequently characterize today's school children.

This study was conducted by S. Saito, T. Hibaru, T. Watanabe, T. Takata, N. Furumoto, and Y. Iwashita, members of school health teacher groups in the Kokura Kita and Minami Wards of Kitakyushu, Japan.

Acknowledgments

The authors would like to thank Professor Kageyu Noro and Doctor Narumi Hirao of Hyogo Hospital for guidance and advice in the study, and the Board of Education of Kitakyushu City as well as the teachers and children of elementary and junior high schools in the Kokura Kita and Minami Wards of the city for participating in the study.

References

Japanese Standards Association, (1980), JIS S 1021 – School Furniture (Desks and Chairs for Class Room) (Japanese Standards Association, Tokyo), (in Japanese).

Kikuzawa, Y., (1983), Accommodation of home study desks and chairs to postures of children from standpoint of height adjustment, *Journal of Home Economics of Japan*, **34**, 488–97, (in Japanese).

Komatsu, Y., (1981), Putting ourselves in the place of children, *Shinano Kyoiku Kaiho*, **347**, (in Japanese).

Komatsu, Y., (1985), Building good postures, *Shinano Kyoiku Kaiho*, **1187**, 23–7, (in Japanese).

Mandal, A. C., (1982), The correct height of school furniture, *Human Factors*, **24**, 257–69.

Ministry of Education, (1981), *Guide to School Furniture* (Gyosei, Tokyo), (in Japanese).

Ouchi, K., Wakai, S. and Kato, M., (1975), Survey on actual state of usage of school furniture, *Japanese Journal of Ergonomics*, **11**, 63–7, (in Japanese).

PART IX

Users with special needs

23

The classification of physical disabilities in the design of products and spaces

Tadao Koga

Environmental and human factors in wheel-chair design

Wheelchairs provide mobility to people with spinal injury and those with functional deterioration of walking caused by paralysis, fracture or leg amputation. Wheelchairs also assist those unable to use walking sticks and other aids because of insufficient strength or co-ordination of the upper limbs, heart disease, cerebral palsy, or muscular dystrophy.

The design of wheelchairs involves the following considerations.

1. The human factors of the disabled. Many of the physically handicapped have experienced spinal injury, cerebral palsy, cerebrovascular disorder, progressive muscular dystrophy, or rheumatoid arthritis. The mobility of these disabled people varies in accordance with the type and degree of their disorder. Individuals with the same type of disability may differ in their physical strength and co-ordination. No single wheelchair can accommodate all disabled. Conversely, people with different types of disability may exhibit similar motor characteristics. Human factors is an essential consideration in the design and planning of living environments for the physically handicapped.
2. The shape and size of wheelchairs have implications for the design and planning of living and exterior spaces, facilities, public transportation systems, and private transport vehicles.
3. In Japan, guidelines for hand-propelled and motorized wheelchairs are provided in JIS T 9201 and JIS T 9203, respectively (Japan Standard Association, 1971). These Japanese Industrial Standards (JISs) do not, however, address the variety of wheelchair applications and the physical characteristics and requirements of wheelchair users. Revisions of the JIS Standards are now being considered to meet these inadequacies as well as to conform with the International Standards developed by the Technical Committee TC173 of the International Standards Organization (ISO). JIS Standards must address both whether wheelchairs can support the physical and functional requirements of the physically handicapped and the extent that various types of disability are accommodated in the design of wheelchair components. Such standards must specify the performance, construction, size, and angle of wheelchair components.

4. Because many physically handicapped people require assistance, wheelchair design must make the mechanism, shape and size of the wheelchair easy for helpers to operate.

The 3M · E system approach to design is based on the analysis of the machine (M), physically handicapped person (Man 1), helper (Man 2), and environment (E). This approach facilitates the evaluation of the design of wheelchairs and other appliances used by the physically handicapped (Noro and Koga, *et al.* 1990a).

A human factors classification of physical disabilities

The functional requirements of the physically handicapped depend on the type and degree of their disabilities. These specific differences must be understood before they can be translated into design criteria. Various classifications for disability are described below.

Building Design Standards Committee, Architectural Institute of Japan

The Building Design Standards Committee for the Physically Handicapped, Architectural Institute of Japan, has classified how buildings obstruct the physically handicapped. These handicaps include (a) informational handicaps associated with perceiving space, (b) movement handicaps associated with locomotion, and (c) skill handicaps associated with the utilization of space (Japan Architect Meeting, 1982).

US Building Code Standards

The US Building Code Standards classify handicaps as relating to (a) functional ability to walk, (b) walking impairment, (c) vision, and (d) motor control.

British Building Code

The British Building Code classifies handicaps as either motor or perceptual. The physically handicapped are categorized as those (a) who routinely use wheelchairs, (b) who find it difficult to walk, (c) visually impaired, and (d) the hearing disabled.

ICTA of Sweden

The International Council of Technical Aid (ICTA) of Sweden classifies the physically handicapped as those (a) who have no walking disability or who can almost walk, (b) whose ability to walk is slightly impaired, and (c) who are unable to walk or require assistance.

World Health Organization

The World Health Organization (WHO) classifies the physically handicapped as those with (a) impairment, (b) disability, and (c) handicap. Impairment is defined as an anatomical, physiological or psychological disturbance. Disability is defined as a limitation or loss of abilities that most people possess which results from the impairment. Handicap is defined as a social disadvantage incurred as a consequence of the impairment or disability which limits or interferes with their ability to perform a specific role (given age, sex, societal, and cultural factors).

Ministry of Health and Welfare of Japan

The Disabled Persons Welfare Law (Ministry of Health and Welfare, 1985), classifies disabilities as pertaining to (a) vision, (b) hearing or sense of balance, (c) voice and speech, (d) limb or trunk, and (e) the internal organs. These disabilities are rated from Class 1 (most severe) to Class 7 (least severe). Physical disabilities are classified further by the type and extent of impairment. As such, a person whose lower limbs are paralyzed by injury to the thoracic region of the spinal cord has no upper limb disorder and is designated Class 1. Similarly, quadriplegics who must depend on helpers for living activities due to injury to the cervical region of the spinal cord are Class 1. Those with systemic disabilities due to athetotic cerebral palsy who must depend on a helper in living activities are designated Class 4 or 3 if they possess some ability to walk and move upper limbs. This disability classification is unreliable, and has many contradictions in designing for the physically handicapped.

Koga's six-by-six matrix

The above classifications may facilitate the design and planning for the physically disabled. However, observation indicates that individuals with different disabilities may display the same motion characteristics. Consequently, disabilities are here classified as relating to the upper limb and trunk, and to the lower limb; these factors are common to both groups (Koga, 1986). The subjects had spinal cord injuries or other types of physical disability. Subject classifications were based on those used in rehabilitation medicine. The resultant classification is suggested as a standard for the design of products and spaces for the physically disabled.

Disability degree classification matrix

In the disability degree classfication matrix shown in Figure 23.1, the disability of the upper limb function is divided into Classes I to VI, and the disability of the trunk and lower limb function is divided into Classes 1 to 6. Ratings of Classes I and 1 are serious, and Classes VI and 6 approach normal functioning.

Classification of upper limb dysfunctions

Dysfunctions of the upper limbs may be classified as follows:

Class I: Complete loss of function of both upper limbs.

Class II: The function of both upper limbs is either severely impaired, or there is complete loss of function of one upper limb, and the other upper limb is severely impaired. Severe impairment signifies that subjects can perform

	I	II	III	IV	V	VI
1						
2						
3						
4						
5						
6						

Figure 23.1 Six-by-six matrix for classification of physical disabilities

gross upper body movement, but lack the co-ordination required for precise limb positioning. Such subjects cannot hold objects with their fingers, control their elbows, and move upper limbs significantly.

Class III: The functioning of both upper limbs is moderately impaired, or one upper limb is severely impaired while the other functions moderately. In this case, 'moderate functionality' indicates that there is some voluntary control over the shoulders and elbows, but co-ordination and stamina is absent. Subjects' fingers are capable of grasping large objects, but not precise movements.

Class IV: Functionality of both upper limbs is slightly impaired, or one upper limb is moderately impaired, with the other slightly impaired. The term 'slight impairment' signifies that there is satisfactory voluntary control over the shoulders and elbows, but these efforts cannot be sustained. Fingers can grip objects, but are unable to perform essential acts of co-ordination.

Class V: Either the functioning of both upper limbs is slightly impaired; or, one upper limb is impaired, while the other is slightly impaired.

Class VI: Functioning of both upper limbs may be virtually normal; or, one upper limb shows very slight impairments in function while the other functions satisfactorily; or, the functioning of the dominant hand is slightly impaired while the non-dominant hand functions normally; or, the functioning of the non-dominant hand is moderately or very slightly impaired and the dominant hand functions normally. Fingers can manipulate objects.

Classification of dysfunction of the trunk and lower limbs

The dysfunction of the trunk and lower limb dysfunctions may be classified as follows.

Class 1: It is difficult or impossible to achieve a seated posture without support.

Class 2: Subjects can maintain the seated posture, but cannot sit upright, or move while seated.

Class 3: Subjects can sit upright, maintain the seated posture and move or shift seat positions. However, they cannot, or find it difficult, to stand up.

Class 4: Subjects can stand up, but find it difficult to walk in a sustained or stable fashion.

Class 5: Subjects can walk, but require support or assistance to walk up and down steps or a stairway, find it difficult to avoid obstacles, and sustain movement.

Class 6: Subjects are able to walk independently and effectively step, walk up and down stairways, and avoid obstacles.

The above classifications of upper limb dysfunctions are based on the extent of voluntary control over the upper limbs (e.g., reach) and ability of the fingers to grip for mobility and work activities. Trunk and lower limb dysfunctions are classified by the ability to sustain seated postures and perform movements.

Classification of physical disabilities

Spinal cord injuries are classified by the extent of disability

The spinal cord may be injured at work, in traffic accidents, or during sports. These injuries may cause fracture, dislocation or inflammation, which in turn injure the spinal cord.

The spinal cord consists of eight cervical vertebrae at the neck (C_1 to C_8), twelve thoracic vertebrae at the chest (T_1 to T_{12}), five lumbar vertebrae of the lower spine (L_1 to L_5), five sacral vertebrae (S_1 to S_5), and the coccyx.

The spinal nerves exit the anterior and posterior horns on the left and right sides and are united on either side of the spinal cord. Spinal nerves that emerge from the front of the spinal cord are anchored at the ventral root; spinal nerves that emerge from the back of the spinal cord are anchored at the dorsal root. Most of the nerves that pass through the ventral roots represent efferent motor nerves; when these nerves are cut, motor paralysis is induced. Nerves passing through the dorsal roots are afferent sensory nerves; injury to these nerves induces sensory paralysis.

For example, C_3 governs the sternocleidomastoid muscle, trapezius muscle, and diaphragm. Any injury to C_3 causes quadriplegia; that is, motor and sensory paralysis in the associated regions, including difficulty in breathing.

The C_4 site corresponds to the trapezius muscle and diaphragm; C_5 to the deltoid muscle, biceps muscle of the arm, brachial muscle, greater pectoral, and supinator muscle; C_6 to the carpal extensor, ulnar extensor, biceps of the arm, and greater pectoral muscle; C_7 to the triceps of the arm and carpal extensor group; C_8 to the digital flexor group; and T_1 to the digital abductor and adductor (Imai, 1981).

A matrix classification of cervical, thoracic or lumbar nerve injuries to the spinal cord is shown in Figure 23.2.

Injury to C_2 and C_3 is associated with impairment of respiration. Since the diaphragm is governed by C_4, any injury to this site results in quadriplegia. The quadriplegic finds it difficult to breathe and requires artificial respiration. Since this disability is more severe than Class I-1, it is not included in this classification.

If C_4 is left intact, functions of the trapezius and the levator scapulae remain. Products used by people with C_2 and C_4 injuries must be actuated and operated by head, shoulder, jaw or eye movements, respiration, or the voice. These individuals belong to class I-1, since they can maintain a seated posture with a back rest or the aid of a helper.

If C_5 is left intact, some functions of the deltoid and biceps muscles of the arm remain. Such individuals can abduct, adduct and extend the shoulders

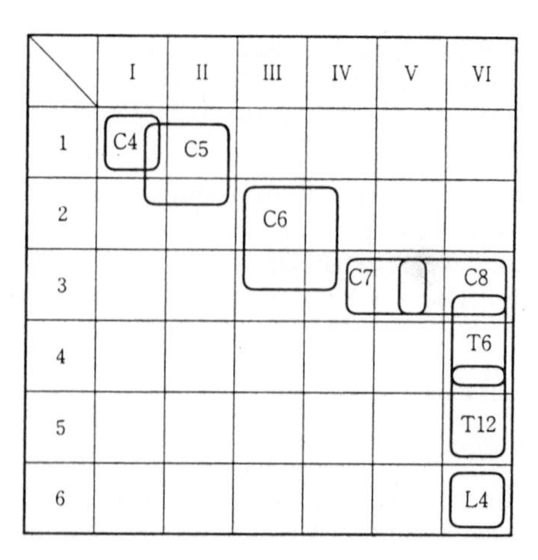

Figure 23.2 Six-by-six matrix classification of people with spinal cord injuries

and flex the elbows. Although wrist joints and finger muscles do not function, some daily life activities can be performed with simple self-help devices that attach to splints which extend the wrist joints. They cannot wear such devices or operate the wheel-chair by themselves. Their upper limb disabilities may be of Class I or II, and the trunk and lower limb disabilities may belong to Class 1 or 2.

If C_6 is left intact, the peri-articular muscle (of the shoulder), biceps muscle (of the arm), and carpal extensor (of the wrist) can function. These individuals can move shoulders, flex elbows, supinate forearms and the joints of the arm. Since finger dorsiflexion is impaired, objects are gripped by holding together the outer portion (hypothenar eminences) of both hands, held by dorsiflexion of the wrist, picked up with the palmar sides of the fingertips, and caught with the palmar sides of the forearms and forefingers. The upper limbs supplement these functions. Individuals can sit up and maintain the seated posture and operate the wheel-chair on a flat surface by flexing elbows, hands, and shoulder muscles. Such disabilities belong to Class III or IV for the upper limbs and to Class 1 or 3 for the trunk and lower limbs.

If the C_7 is intact, the largest muscle of the back, triceps muscle of the arm, flexor muscle of the hand, and extensor muscle of the fingers remain functional. Grip force is weak, but strong enough to hold objects. The upper limbs supplement these actions. Since the largest muscle of the back and triceps muscle of the arm function, subjects can sit upright, maintain the seated posture, and crawl. Such disabilities belong to Class IV to V for the upper limbs, and to Class 3 for the trunk and lower limbs.

If the C_8 is left intact, all muscles except the palmar muscle group are functional. The upper limbs can therefore be used independently. The functions of the trunk and lower limbs correspond to those described with C_7. The disability belongs to Class V to VI for the upper limbs and to Class 3 for the trunk and lower limbs.

If thoracic segments T_1 to T_6 remain intact, the upper limbs can be manipulated independently, and use of the trunk and lower limbs is somewhat improved. As the site of injury on the spine lowers, individuals can sit upright, shift postures, and stand up using lower limb orthoses (leg braces). However, since such braces are often impractical, wheelchair use is more prevalent, such disabilities belong to Class VI for the upper limbs and Class 3 or 4 for the trunk and lower limbs.

If the thoracic segments T_7 to T_{12} remain intact, the muscle groups of the trunk (particularly the abdominal muscles) are unaffected. Upper limbs function as well or better than that of the hands and upper limbs of able-bodied people. The pelvic ligament can be lifted, but walking is impaired. Although knee-ankle-foot orthoses (long leg braces) assist in walking, wheelchairs are commonly used. This disability belongs to Class VI for the upper limbs and Class 4 or 5 for the trunk and lower limbs.

If the lumbar segment L_1 is intact, the iliopsoas muscle functions slightly, the hip is capable of moderate flexing by rotating the pelvis, but the lower limbs are completely paralysed. Use of the upper limbs compares with those of able-bodied people. Subjects can walk with sticks or knee-ankle-foot orthoses for the trunk and lower limbs (long leg braces). This disability belongs to Class VI for the upper limbs and Class 6 for the trunk and lower limbs.

If the lumbar segment L_2 is intact, the iliopsoas muscle functions slightly, and the hip has some flexion. Use of the upper limbs, trunk and lower limbs is similar to that described for the lumbar segment L_1. The disorder belongs to Class VI for the upper limbs and Class 6 for the trunk and lower limbs.

If the lumbar segment L_3 is intact, the iliopsoas muscle, adductor muscle group, and quadriceps muscle of the thigh function. Such individuals can walk using a stick and wearing ankle-foot orthoses. The disability belongs to Class VI for the upper limbs and Class 6 for the trunk and lower limbs.

If the lumbar segment L_4 is severed, the quadriceps muscle of the thigh and the anterior tibia (of the leg muscles) continue to function. The subject can walk, wearing specially-designed boots. This disability belongs to Class VI for the upper limbs and Class 6 for the trunk and lower limbs.

When the coccyx muscle is injured, the lower limbs are slightly impaired. Such individuals can walk independently and effectively, corresponding to normal functioning.

The cervical, thoracic, and lumbar segments generally correspond to the upper limbs, trunk, and lower limbs, respectively. Injury to the cervical vertebrae results in quadriplegia; injury to the lower vertebrae results in paraplegia. Consequently, the site of injury must be understood to evaluate the nature of the disability.

Disability classification of cerebral palsy, muscular dystrophy, cerebrovascular disease, and rheumatoid arthritis

As described above, people suffering from spinal cord injuries may be classified in a matrix, as shown in Figure 23.3.

Cerebral palsy is a motor disorder caused by injury to the brain during pregnancy, at birth or in infancy. Cerebral palsy is also said to involve many other disorders. Since cerebral palsy exhibits the same symptoms in adults, classifications for cerebral palsy-disabled adults correspond to those of children with cerebral palsy.

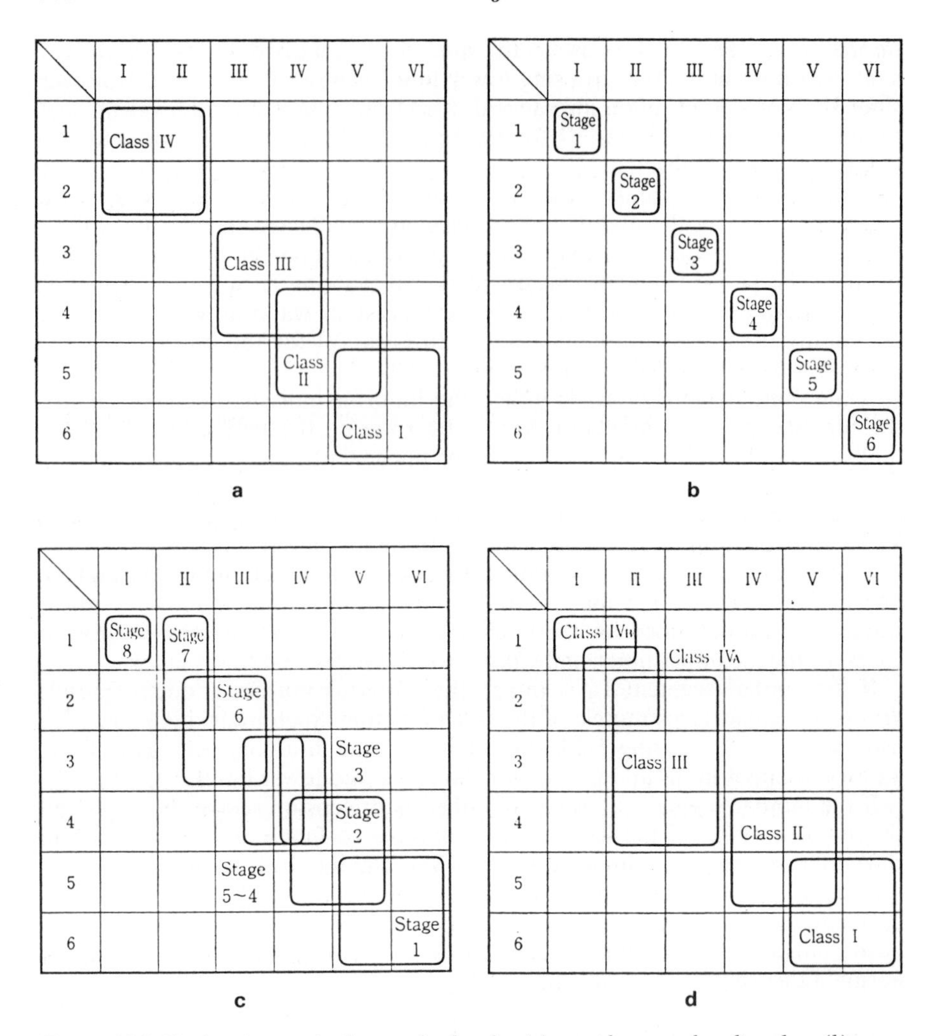

Figure 23.3 Six-by-six matrix for cerebral palsy (a); cerebrovascular disorders (b); progressive muscular dystrophy (c); rheumatoid arthritis (d)

The American Academy for Cerebral Palsy has developed physiological classifications for cerebral palsy as spastic, athetoid, rigid, ataxic, tremor, atonic, and mixed, with other types which defy classifications. The spastic and athetoid types may account for 90 per cent of cerebral palsy cases (Koike, 1974).

The American Academy for Cerebral Palsy has classified these disabilities as follows.

Class 1: The cerebral palsy child displays little loss of co-ordination and does not require care.

Class 2: The cerebral palsy child has slight or moderate curtailment of activities, needs to wear some orthoses (braces), and requires some care.

Class 3: Activities of the cerebral palsy child are seriously limited; orthoses must be worn, and their use requires medical instruction.
Class 4: The cerebral palsy child cannot perform any useful activities and requires long periods of hospitalization and care.

Muscular dystrophy begins to progress at 3–4 years for the Duchenne type, 10–18 years for the facio-scapulohumeral type, and 10–40 years for the limb-girdle type.

The severity of progressive muscular dystrophy is classified into eight stages by the Central Rehabilitation Service of the University of Tokyo Hospital (Ueda, 1968).

Stage 1: The lumbar spine is flexed, walk is slightly unstable, and one can go up and down stairs without assistance.
Stage 2: The lumbar spine is flexed, walk is very unstable, and assistance is needed to go up and down stairs.
Stage 3: There is severe lumbar flexure, walk is very unstable, and one cannot go up and down stairs. One can walk on a flat ground and rise from a chair.
Stage 4: The lumbar spine evidences extremely severe flexion, walk is extremely unstable, and one cannot rise from a chair.
Stage 5: One can operate a wheelchair and can independently ambulate in the wheelchair. One cannot walk but can crawl on hands and knees.
Stage 6: A wheelchair can be operated but assistance is needed in daily life. One can crawl on knees, but not on hands and knees.
Stage 7: Subjects cannot crawl but can independently maintain seated posture.
Stage 8: Subjects are bedridden, cannot shift positions, and are completely dependent on assistance for survival.

Cerebrovascular disorder is commonly referred to as apoplexia. This disorder includes intracranial hemorrhage, encephalomalacia, and brain infarction. Hemiplegia accounts for 80 per cent of cerebrovascular disorders. Paralysis of either the upper or lower limbs or of both sides of the body (displegia) are said to also occur. The cerebrovascular disorder is typically accompanied by many other dysfunctions.

Brunstrom and Sakuma classify cerebrovascular disorders in six stages, according to the severity (Mishima, 1975).

Stage 1: The subject has muscle paralysis of the upper limbs, trunk and lower limbs.
Stage 2: Subjects exhibit slight voluntary movements (autokinesis) and associative movements (synkenesis) of the upper limbs. They have little or no capacity to perform all or part of finger flexion, shoulder elevation, retraction, abduction, lateral rotation and extension, and elbow flexion and supination. They have little control over voluntary movements of the trunk and lower limbs.
Stage 3: Subjects can perform voluntary associative movements of the joints, can hold or hook objects by gripping with all fingers, but they cannot release objects. They can extend the fingers by reflex, but not with volition. They can flex or initiate associative movements with the hip, knee, and foot joints.
Stage 4: Subjects cannot move the upper limbs in conjunction but they can move the upper limbs separately. They can grip objects laterally with fingers and release objects by moving the thumb. They can flex the knees to angles exceeding 90° while seated, move the feet backward on the floor, and voluntarily dorsiflex the feet while seated without lifting the heels from the floor.
Stage 5: Subjects have greater control over voluntary movements and can combine more complicated opposite assistive movements than is possible in stage 4. They can hold objects with opposing fingers and grip cylinders and balls, but

finger motions lack skill. They can flex the knees with little hip movement and can straighten their feet by moving the affected limb slightly forward keeping knees extended.

Stage 6: Subjects can move the upper limbs independently, rapidly and with high levels of co-ordination. They can make all types of grip, use the upper limbs skilfully, and can extend them throughout the entire range of movement. They can abduct the hip while standing and rotate the legs laterally and medially by the reciprocal contraction of the medial and lateral knee flexor muscles.

The causes of rheumatoid arthritis are not well understood, although possible contributors include bacterial allergies, metabolic disorders, poor constitution, and hormonal imbalance. Almost all joints are affected with rheumatoid arthritis. The incidence of rheumatoid arthritis is highest among people aged 20–30 years. Middle-aged women are particularly susceptible to this disorder. It is widely believed that there is no real cure for this disease (Sasaki *et al.*, 1974).

Steinbrocker (1949) divided rheumatoid arthritics into four classes according to the severity of the disability:

Class 1: These individuals can live and work without any inconvenience.

Class 2: Rheumatoid arthritics have some joint discomfort and motor handicap, but can live normally.

Class 3: These rheumatoid arthritics cannot work or take care of themselves.

Class 4: This class is subdivided into Class 4A and Class 4B. The Class 4A rheumatoid arthritic can use a wheelchair, while the Class 4B rheumatoid arthritic remains bedridden.

The types and degrees of several predominant limb and trunk disabilities have been summarized above. It is evident that one disability encompasses a large spectrum of dysfunctions, ranging from those who function normally to others who are severely disabled.

The design and planning of housing and the living environment for the physically handicapped requires that their specific physical dysfunctions are understood and accommodated.

Correlation of spinal cord disabilities with other physical disabilities

The classifications of severity of physical disabilities are presented in a matrix (*see* Figure 23.4). The physically disabled are equally distributed above and below the diagonal line connecting Group I · 1 with Group VI · 6. It can be seen that the degree of disability varies from slight to severe.

The disability of the upper limbs is not correlated with that of the trunk and lower limbs in Groups V · VI-1.2 and I · II-5.6. The physically disabled in Group V · VI-1.2 resemble able-bodied people in the upper limbs and have the thighs amputated, hip transected, or other related dysfunctions. The physically disabled in Group I · II-5.6 are mainly upper limb amputees or thalidomides, and resemble the able-bodied in the lower limbs.

Attempts were made to classify the physically disabled using a matrix according to disability evaluation criteria commonly used in rehabilitation medicine, neurology, or those which reflect approaches used to evaluate motion characteristics of the physically disabled. These classifications show that the spectrum of characteristics that may be found in one disability cannot

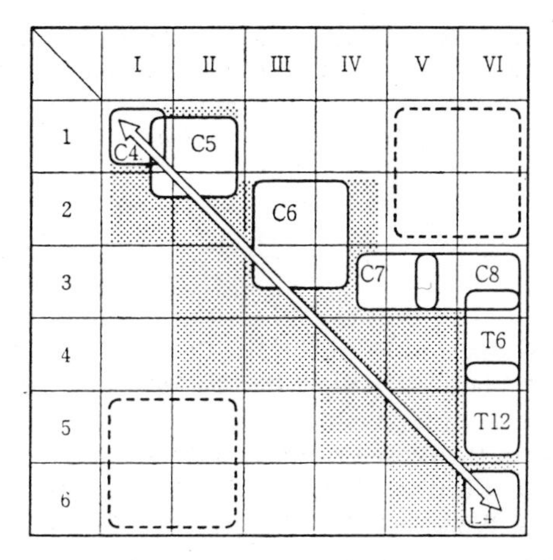

Figure 23.4 Superimposition of classifications of physical disabilities on a six-by-six matrix

be fully described in a single matrix. Some disabilities might be categorized by any of several matrix elements, depending on its type or site of injury. As a result, the classification of physical disabilities depicted in Figure 23.5 represents one approach for designing and evaluating products and housing. In this classification, the upper limbs, trunk and lower limbs are divided into three levels of disability.

People in Group A cannot move their upper limbs independently, maintain a seated posture, or walk. They are typically bedridden. Those in Group B cannot move their upper limbs, cannot walk, and depend on assistants to perform life functions. For those in Group D, their only independent upper limb functions consist of grasping and catching large objects. They depend on assistants to perform basic life functions, such as those in Group A.

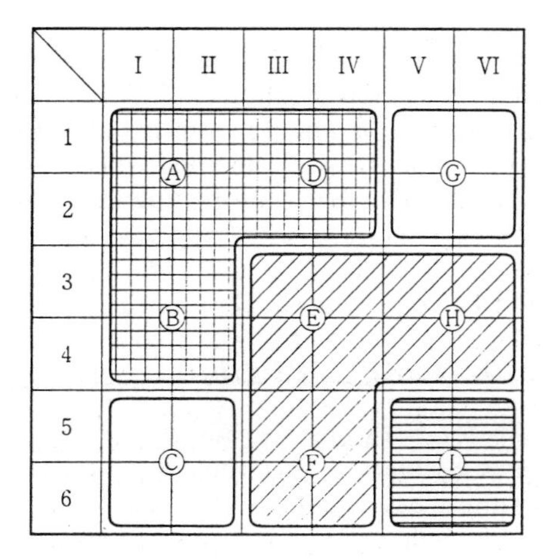

Figure 23.5 Method of classification

The upper limb disabilities of those in Group E resemble those of Group D, while their trunk and lower limb disabilities correspond to Group B. The people in Group E need both direct and indirect assistance to perform daily life activities. The upper limb function of people in Group F is the same as that of those in Groups D and E. Individuals in Group F have some motor control of the trunk and lower limbs; they are effectively able to walk, but require the direct and indirect aid of helpers in their daily life activities. The people in Group H can perform coarse finger movements and can manipulate the fingers with skill and co-ordination. Group H resembles Groups B and E in functional use of trunk and lower limbs; these people need the indirect aid of helpers to perform daily activities. People in groups E, F, and H have a moderately severe degree of physical disability.

The people in Group I can maintain some degree of function over the upper limbs and the trunk and lower limbs. They are considered severely disabled in the Disabled Persons Welfare Law, but are able to perform basic daily life activities.

People in Group C cannot use their upper limbs, but are able to maintain some or all function of the trunk and lower limbs. Group C is similar to Groups A and B in that total help is required for functional use of the upper limbs.

Groups G, H and I have similar use of the upper limbs. However Groups G, A and B are all severely disabled in the trunk and lower limbs. Individuals in these groups resemble able-bodied people in the upper limbs and can walk with the help of prosthetic devices. Their ability to perform daily life activities is slightly impaired.

The above discussion underscores that individuals with different disabilities may have similar motion characteristics. The physically disabled can be classified by degree of severity (moderately severe/severe/extremely severe). They can also be characterized by their dependence on outside assistance for performing daily life activities (total/partial/no dependence).

Appliances and buildings for the physically handicapped must accommodate their specific disabilities. An effective design evaluation must include information regarding characteristics and capabilities of the physically disabled involved.

Classifications of the physically disabled can be used in the design process to formulate basic wheelchair specifications. That is, motorized wheelchairs are appropriate for Groups A, B and D; motorized wheelchairs with voice recognition devices for Groups A and C; manual wheelchairs with the aid of a helper for Groups E, F and H; and self-operable wheelchairs for Groups I and G.

Ergonomics of wheelchair design

Wheelchair design calls for both engineering and ergonomics studies of performance. An ergonomic evaluation must include a variety of techniques from physiological to behavioural assessment.

Necessary data must include body dimensions, physical workload, functional demands, posture, subjective evaluation and safety. Movements associated

Relationship
●; High
○; Middle
△; Low

			Case of Attendant Operation							Case of Self Operation										
			Motion							Motion										
			Loading	Folding	Braking	Assisting	Wheel Chair Drive		Transfer	Loading	Folding	Braking	Sitting			Wheel Chair Drive				Transfer
						Step Scaling	Slope Climbing	Locomotion(Running)					Relax	Exercise	Holding	Wheeling	Step Scaling	Slope Climbing	Locomotion(Running)	
Anthropometry	Static	Body Dimension	●	●	●	○	○	○	○	○	○	○	○	○	○	○	○	○	○	○
		Body Mass	\	\	○	○	○	○	○	\	\	○	○	○	○	○	○	○	○	○
	Dynamic	Movable Range of Joints	○	○	○	○	○	○	○	●	○	●	○	○	○	●	●	●	●	○
		Work Space	○	○	●	○	○	○	○	○	○	○	▷	●	▷	○	○	○	○	○
Physiological Cost		Heart rate	○	▷	▷	○	○	○	○	○	▷	▷	○	●	○	○	○	●	●	○
		Oxygen Consumption	○	▷	▷	○	○	○	○	○	▷	▷	○	○	○	○	○	●	●	○
		Respiratory Frequency	○	▷	▷	○	○	○	○	○	▷	▷	○	○	○	○	○	○	○	○
		Blood Pressure	○	▷	▷	○	○	○	○	▷	▷	▷	○	○	○	○	○	○	○	○
		Biochemical Measuring	○	▷	▷	○	○	○	○	▷	▷	▷	○	○	▷	○	○	○	○	○
Controls		Muscle Strength	●	●	●	●	●	●	●	●	○	●	▷	○	\	●	●	●	●	●
		Driving Power	●	●	●	●	●	●	●	○	○	●	▷	○	\	●	●	●	●	○
		Torque	\	\	\	○	○	○	\	\	\	▷	\	\	\	●	●	●	●	\
		EMG	○	○	○	○	○	○	○	○	○	▷	○	○	○	○	○	○	○	○
		Motion Analysis	●	●	○	●	●	●	●	○	○	▷	○	○	○	○	○	○	○	●
Posture		Working Posture Analysis	●	●	○	●	●	●	●	○	○	○	●	●	●	○	○	○	○	●
		Centre of gravity of Men	▷	\	▷	●	○	○	○	○	▷	▷	○	○	●	●	●	○	○	●
		Pressure of Sitting	\	\	\	\	\	\	\	○	▷	▷	●	●	●	○	○	○	○	\
Psychophysical Method			○	○	○	○	○	○	○	▷	▷	▷	○	○	○	\	\	\	\	▷
Environmental Physiology		Mean Skin Temperature	\	\	\	\	\	\	\	▷	▷	▷	○	○	○	\	\	▷	▷	▷
		Rectal Temperature	\	\	\	\	\	\	\	▷	▷	▷	○	○	○	\	\	▷	▷	▷
		Optimum Temperature	\	\	\	\	\	\	\	▷	▷	▷	○	○	○	\	\	▷	▷	▷
Safety		Muscle Strength	●	●	●	●	●	●	●	●	○	●	▷	○	\	●	●	●	●	●
		Driving Power	●	●	●	●	●	●	●	○	○	●	▷	○	\	●	●	●	●	○
		Skill	▷	▷	○	●	○	○	○	○	○	●	\	▷	\	○	○	○	○	▷
		Balance Centre of gravity	▷	\	▷	●	○	○	○	○	▷	▷	○	○	○	○	○	○	○	●

Figure 23.6 Ergonomic approach to wheelchair design

with transfer, driving, sitting, braking, folding, and loading must be considered. When a helper is needed to use a wheelchair, the workload of this second user is as relevant to wheelchair design as that of the physically handicapped person. For example, seat evaluations must include study of posture, centre of gravity, buttock pressure distribution, balance for maintaining the seated posture, range of motion, heart rate, posture, and buttock pressure distribution for work. These are portrayed by the open circles in Figure 23.6 (Noro and Koga, *et al.* 1990b).

Design of an ergonomic compact wheelchair

Because standard wheelchairs have 24-inch wheels, making them difficult to load into a car, a folding model was designed with smaller wheels (*see* Figure 23.7).

The design of the compact wheelchair was based on research on the effect of wheel diameters on the physical workload of the physically handicapped driving wheelchairs.

Ten students participated as subjects in the experiment. Each subject was asked to operate a 20-inch and 24-inch wheelchair on a treadmill. The correlation between the heart rate and oxygen uptake associated with use of the two wheelchairs at specific levels of oxygen uptake was investigated. The experimental apparatus is shown in Figure 23.8 (Koga, 1977).

The subject operated each wheelchair at a speed of 10, 35 or 60 m/min on the treadmill. The oxygen uptake and heart rate data are shown in Figure 23.9.

The oxygen uptake and heart rate data are plotted along the x-axis and y-axis, respectively. The solid and dotted lines represent oxygen uptake-heart rate regression lines for the 20-inch and 24-inch wheelchairs at speeds of 10, 35 and 60 m/min, respectively. Heart rate increases are proportionate to oxygen uptake for both wheelchairs. The oxygen uptake and heart rate for the two

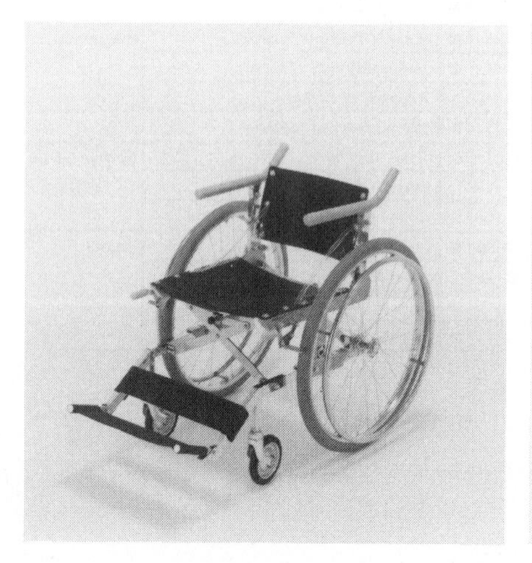
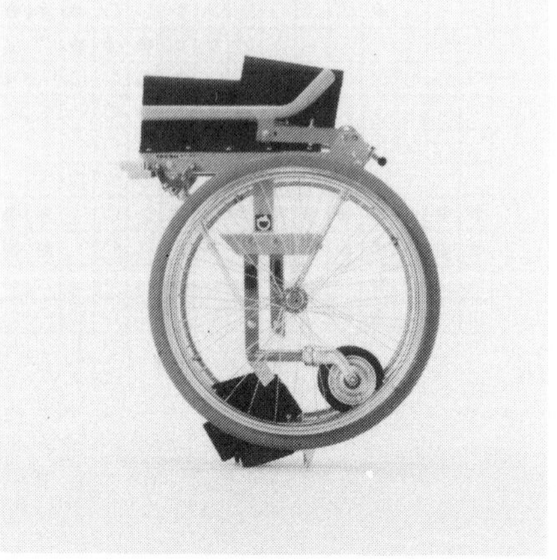

Figure 23.7 Compact wheelchair with 20-inch wheels

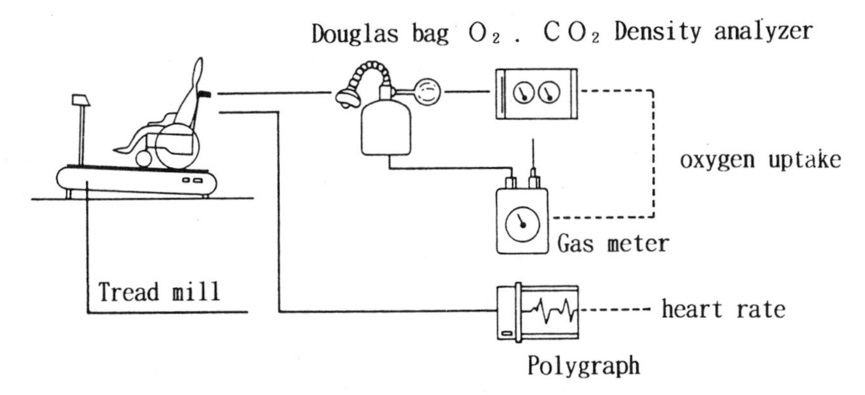

Figure 23.8 Experimental apparatus

wheelchairs are similar at 10 m/min, but not at 60 m/min. The oxygen uptake and heart rate are greater with the 20-inch than 24-inch wheelchair diameters.

The correlation between the 20-inch and 24-inch wheelchairs is shown in Figure 23.10. The oxygen uptake with the 20-inch wheelchair is plotted along the *x*-axis, and that of the 24-inch wheelchair is plotted along the *y*-axis. The coefficient of correlation at all speeds is high. The regression analysis indicates that the workload with the 20-inch wheelchair begins to exceed that of the 24-inch wheelchair, with oxygen uptakes of about 400 ml/min.

Heart rates with the 20-inch and 24-inch diameter wheelchairs are strongly correlated (see Figure 23.11). The workload with the 20-inch wheelchair begins

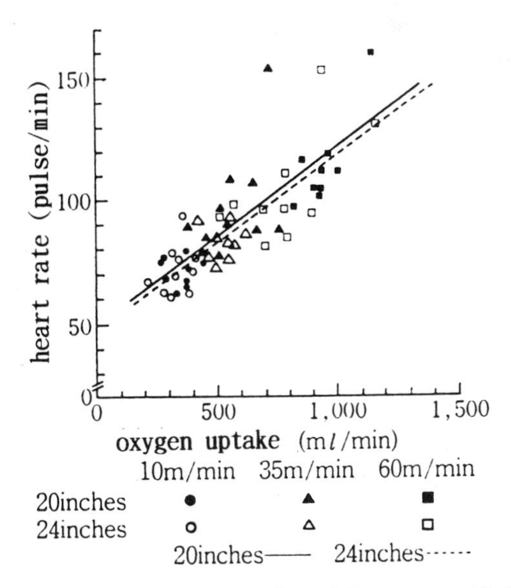

Figure 23.9 Correlation between oxygen uptake and heart rate with 20-inch and 24-inch wheelchairs

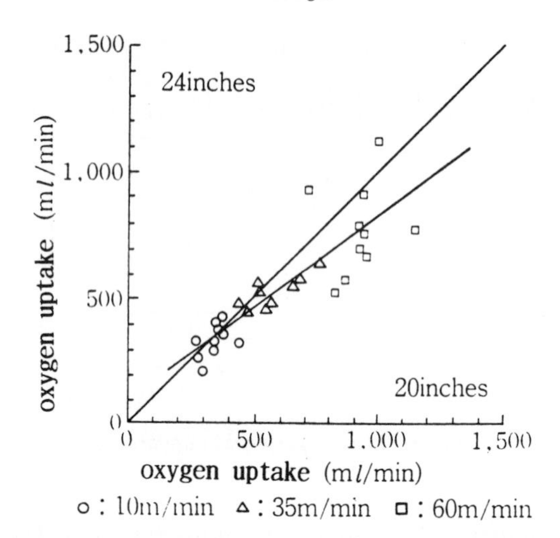

Figure 23.10 Correlation between 20-inch and 24-inch wheelchairs at different speeds with respect to oxygen uptake

to exceed that of the 24-inch wheelchair at heart rates of about 75 beats per minute.

The factor analysis of the oxygen uptake and heart rate shows that wheelchair diameter exerts a significant effect on oxygen uptake, but not heart rate. As with oxygen uptake, heart rates are significantly greater with the 20-inch wheelchair than with the 24-inch wheelchair at the speed of 60 m/min. Consequently, wheel diameter affects workload when the wheelchair is operated at high speed. The physiological load with the 20-inch wheelchair at a speed of 15 to 20 m/min is however, less pronounced.

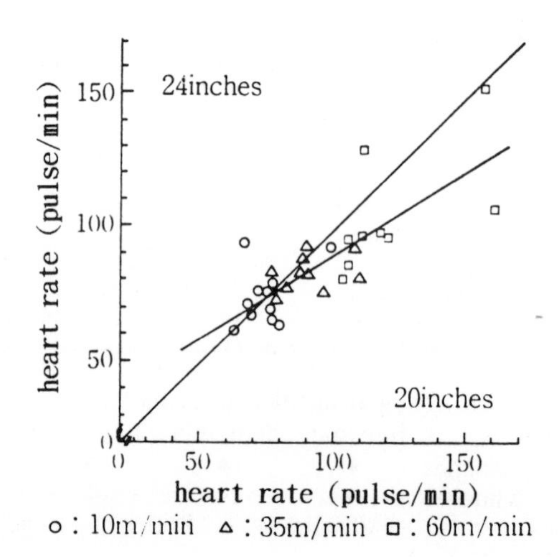

Figure 23.11 Correlation between 20-inch and 24-inch wheelchairs at different speeds with respect to heart rate

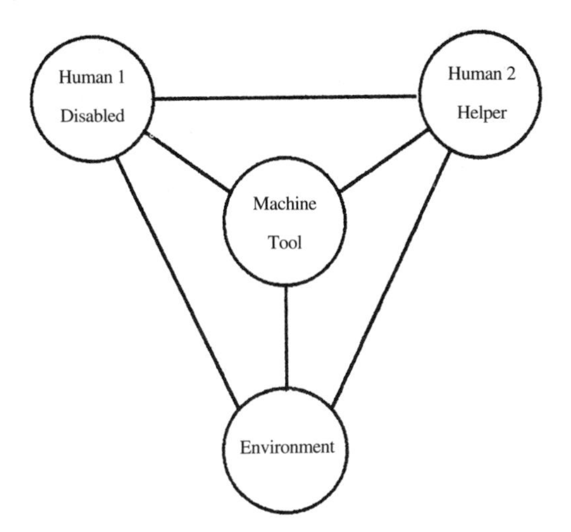

Figure 23.12 3M · E System approach

Based on the findings described above, the wheel diameter of the compact wheelchair was set at 20 inches, and the wheelchair was designed to fold in height, width and length.

Conclusions

The physically disabled differ in the type, degree and site of disability, and vary in age, sex, skill, and other factors. Disability characteristics may be classified by using a 6 by 6 matrix to facilitate the design of products and buildings for the physically disabled.

Four factors must be considered in the design of daily appliances and buildings for the physically handicapped; these are the physically handicapped person, helper, equipment, and environment. This combination is designated 3M · E (man 1-man 2-machine-environment), as shown in Figure 23.12.

References

Imai, G., (1981), *Spinal Injury Handbook*, (Giho-do Co. Ltd, Tokyo) (in Japanese).

Japan Architect Meeting, Standard of Building Design for the Handicapped User, Ministry of Construction, Department of Housing Japan (in Japanese).

JIS T 9201 and JIS T 9203, (1971), *Wheelchair*, (Japanese Standard Association, Tokyo) (in Japanese).

Koga, T., (1977), Study on the design of the portable wheelchair for the handicapped driver with ergonomics, *Sohgoh Rehabilitation*, **15**(5), 53–62 (Igaku Shoin, Tokyo) (in Japanese).

Koga, T., (1986), Study on the space planning of housing for the severely handicapped – on the classification for disabled persons, *Journal of Architectural Institute of Japan*, **362**, 94–101 (in Japanese).

Koike, F., (1974), *Medical Rehabilitation*, series 15 (Ishiyaku Publishing Co Ltd., Tokyo) (in Japanese).

Ministry of Health and Welfare, Japan, (1985), *Disabled Persons Welfare Law*, (in Japanese).

Mishima, H., (1975), *Rehabilitation for Hemiplegia*, (Igaku-shoin, Tokyo) (in Japanese).

Noro, K., (ed), Koga, T., (1990a), *Illustrated Ergonomics*, 85 (Japanese Standard Association, Tokyo) (in Japanese).

Noro, K. (ed), Koga, T., (1990b), *Illustrated Ergonomics*, 88–89 (Japanese Standard Association, Tokyo) (in Japanese).

Sasaki, T., Ishida, H., (1974), *Medical Rehabilitation*, series, 17 (Ishiyaku Publishing Co. Ltd., Tokyo) (in Japanese).

Steinbrocker, O., et al., (1949), Theraputic criteria in rheumatoid arthritis, JAMA, 140, 659, 1949.

Ueda, S., (1968), Rehabilitation Medicine Illustrated, (Takeda Medicine Industry Co. Ltd, Tokyo) (In Japanese).

24

Seating for pregnant workers based on subjective symptoms and motion analysis

Narumi Hirao and Mami Kajiyama

Purpose of study

The Equal Employment Opportunity Law was established in Japan in 1985, the last year of the United Nations Decade for Women. The law has drastically changed working conditions for female employees, such as the removal of restrictions on late night work of female government officials and increases in women workers having children. The recent explosion in office automation equipment has greatly altered the office environment of women workers (Women's Bureau, Ministry of Labour, 1986; Women Workers' Division, Womens Bureau, Ministry of Labour, 1986).

The number of females working has increased as a result of greater participation in higher education, changing values, increased leisure time, and other factors (Women's Bureau, Ministry of Labour, 1986). As a result, more women continue to work during and after pregnancy (*see* Figure 24.1, Ministry of Labour, 1986). Consequently, the importance of accommodating these workers has magnified.

The objective of this study is to evaluate the implications of pregnancy on seat requirements. This was accomplished via (*a*) an investigation into the social trends of pregnant female workers, (*b*) a subjective symptom survey of pregnant women at work, and (*c*) video analysis of these workers rising from and sitting on chairs.

Trends in pregnant workers in Japan

Recent trends

The Labour Force Survey of the Statistics Bureau, Management and Coordination Agency, indicated that the female worker population in 1985 was

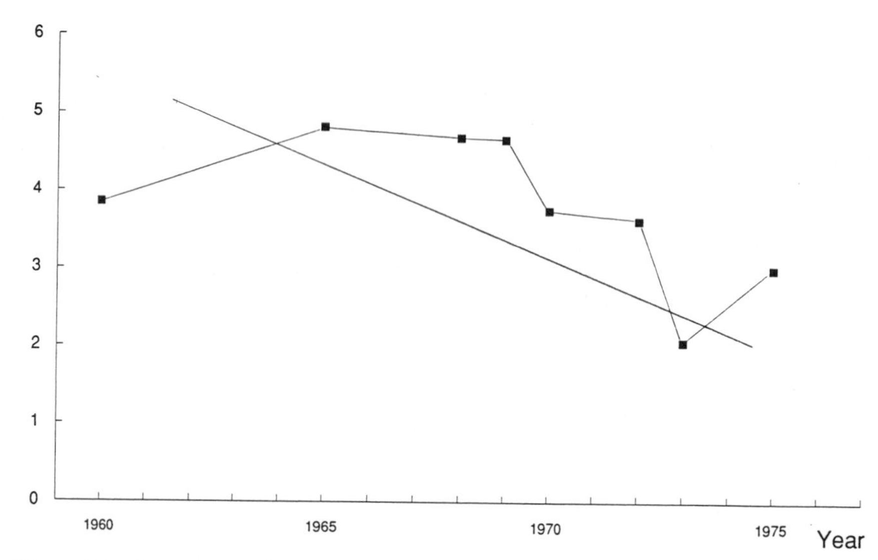

Figure 24.1 Percentage of women who did not continue work due to pregnancy or childbirth (pregnant and nursing women = 100 per cent)

23 670 000; this represents an increase of 200 000 or 0.95 per cent over the previous year. Since the increase in the male worker population was less (0.4 per cent), female workers account for 39.7 per cent of the total labour force in Japan. The percentage of female workers has been increasing since 1976 (Women's Bureau, Ministry of Labour, 1986). The Women's Welfare Division of the Women's Bureau, Ministry of Labour, reported that the number of pregnant female workers more than doubled, from 140 000 in 1958 to 340 000 in 1980 (Women Workers' Division, Women's Bureau, Ministry of Labour, 1986). According to a 1986 Female Worker Protection Survey, the number of female workers who retire as a result of pregnancy or childbirth has been decreasing yearly (Women's Bureau, Ministry of Labour, 1986). Consequently, many more female workers are pregnant than in previous years.

Women are working more because they are attaining higher levels of education, have more available time because they have fewer children and less house work, or need additional incomes to maintain living standards or prepare financially for old age (Women's Bureau, Ministry of Labour, 1986). Pregnant females today continue to work for both financial and psychological reasons. These rates are expected to increase further in the future.

Effect of employment on pregnant women

Sugawara *et al.* (1984) analyzed the results of the questionnaire responses of a survey conducted by the Ministry of Health and Welfare on 3689 working women and 8345 domestic women from September 1980 to October 1982. They found significant medical differences between the working women and domestic women in menstrual disorder, spontaneous abortion, premature birth, abnormal labour, low-birth-weight infancy births, and other disorders.

The causes of these disorders are not apparent, but employment was found to be a factor. Concerns regarding VDT work included the effect of radiation on pregnant women. The World Health Organization report maintained that levels of work-related radiation are not sufficient to affect the health of pregnant women adversely (Industrial Health Division, Labour Standards Bureau, Ministry of Labour, 1986). According to a questionnaire survey conducted by the National Institute of Industrial Health from 1984 to 1985, the spontaneous abortion rate of pregnant females was 9.9 per cent before working at a VDT, and 12.5 per cent afterwards; this difference was not statistically significant. McDonald *et al.* (1985) interviewed about 56 000 pregnant women and found no significant differences in the rate of spontaneous abortions and birth defects between VDT and non-VDT users. In contrast to these findings, the Microcomputer Survey Committee, General Council of Trade Unions of Japan (1985) reported that 36.4 per cent of VDT users had abnormal pregnancies. Causes cited included:

1. working for extended hours at a VDT;
2. stress and abdominal compression in constrained sitting;
3. overly stressful and intensive mental work;
4. excessive stress and inappropriate personal relations;
5. a lack of job control on the part of pregnant women, including the protection of maternity;
6. effect of low-frequency noise on the foetus;
7. air quality and 'sick building syndrome'; and
8. ionizing and electromagnetic radiation, electric power, and static electricity.

Different researchers report different factors affecting pregnancy of women workers, but these factors are not considered responsible for rates of spontaneous abortion.

Many of these studies have researched the effect of employment on pregnant working women by comparing these women with housewives, rather than investigating how employment affects pregnant women. Consequently, it is not clear how the work process and work environments can be improved. Cherry (1987) related work demands to health complaints from pregnant women. He indicates that the health complaints of pregnant women are systematically correlated with postural demands associated with work. For example, leaning forward while standing for long hours, twisting the torso, and lifting weights are very strenuous for women in the third trimester.

Consequently, working postures represent important considerations for pregnant women in the second and third trimesters (Microcomputer Survey Committee, 1985).

Subjective symptom survey of pregnant workers

Workplace survey

Pregnant workers were surveyed for subjective symptoms. VDT workplaces that employed pregnant workers were selected for the survey;

A: Workplace engaged in design of electronic office equipment (300 employees)
B: Workplace engaged in software development (30 employees)

C: Workplace engaged in communication service (100 employees)
D: Workplace engaged in accounting (30 employees)
E: Workplace engaged in common clerical work (30 employees)

These workplaces were each modern and air conditioned. One pregnant woman was selected from each workplace and tested for a length of time. Since woman A is investigated three times and women D and E were investigated twice each, a total of nine women were involved in the survey.

The sample of nine subjects is smaller than employed in ordinary surveys. Although considerably more pregnant women work than in the past, few pregnant workers are typically evident at any given workplace.

Work fatigue

The results of the subjective symptom survey are given in Table 24.1. The terms first trimester, second trimester, and third trimester in the table signify 2–4, 5–7, and 8–10 months of pregnancy, respectively.

In the first trimester, pregnant women often report symptoms indicating mental instability, such as 'feel sleepy', 'feel ill', and 'feel nervous'.

In the second trimester, pregnant women often report symptoms indicating physical fatigue or musculo-skeletal stress, such as 'want to lie down' and 'get tired in the legs'.

In the third trimester, pregnant women report that they 'feel low back pain' and 'feel difficulty in bending forward'.

Identification of workplace problems

The subjective symptom survey indicated that the problems facing pregnant women at their workplace can be characterized by the trimester of pregnancy. These problems are associated with:

1. Labour management;
2. Equipment (including chairs); and,
3. The environment.

Discussion of work fatigue

Pregnant women often report that they feel sleepy in the first trimester; this symptom is caused by the pregnancy-induced increase in the secretion of progesterone. The 'feel ill' symptom is due to morning sickness; as with women who feel ill but are not pregnant, they often require rest. Many women are under stress and most susceptible to spontaneous abortion in the first trimester. Workers in the first trimester may require special provisions such as beds, replacing them with other workers when they feel ill, and moderation of stress at work. Consequently, labour management measures, including job allocation and welfare, are critical for female workers in the first trimester.

Pregnant women are relatively stable in the second trimester. Their body shape changes markedly and they express musculo-skeletal discomforts associated with static postures. Management must be sympathetic to these requirements.

Table 24.1 Subjective work fatigue symptom survey of pregnant women

Trimester		Symptom	Frequency
First	1	Feel sleepy	6
	2	Feel ill	6
	3	Feel nervous	5
	4	Irritated	4
	5	Yawn	4
	6	Want to lie down	4
	7	Feel the head muddled	4
	8	Lack patience	3
	9	Feel difficulty in doing delicate work	3
	10	Feel low back pain	2
	11	Feel eye strain	2
Second	1	Want to lie down	6
	2	Feel tired in the legs	4
	3	Yawn	4
	4	Feel low back pain	4
	5	Feel tired over the whole body	4
	6	Feel sleepy	4
	7	Feel clumsy in motion	2
	8	Feel headache	2
	9	Feel difficulty in keeping neat	2
	10	Feel eye strain	2
	11	Feel ill	2
	12	Feel nervous	2
	13	Feel dizzy	
Third	1	Feel low back pain	4
	2	Feel difficulty in bending forward	2
	3	Feel tired in the legs	2

Nine pregnant working women rated their subjective symptoms of work fatigue on a five-point scale. Frequency was calculated by frequency = number of complaints × weight (1 for frequent complaint and 2 for less frequent complaint).

Women in the third trimester gain an average of about 10 kg in weight (Yamana *et al.*, 1981), and are subjected to considerably greater musculoskeletal load. The body weight increases differ from obesity, and are primarily in the abdomen. The protrusion of the abdomen leads to such complaints as lower back pain, and difficulty in bending forward. Postural loads of pregnant women in the second trimester must be reduced by improving workplace equipment and environment.

A preliminary interview survey found that many of the pregnant women expressed extreme difficulty on rising from, and sitting in their chairs.

Video analysis of rising and sitting

The rising and sitting motions of the nine pregnant women were video-taped with a standard video camera. The positions of the acromion (shoulder), iliac crest (hip) and patella (knee cap) in each frame were entered into a computer at 1/30-s intervals through a Sony MAW-8050 motion analyzer, and analyzed.

Rising and sitting

Rising and sitting motion durations of two pregnant women were compared with those of two women who were not pregnant. The time required to rise from and sit on a chair without armrests is shown in Figures 24.2 and 24.3, respectively. In each case, the mean and standard deviation of time were larger for the women who were not pregnant.

Rising from the chair

The angular velocity ω (Hasegawa *et al.*, 1985) when rising from chairs with and without armrests were compared. The angular velocity ω is the rate of change over time in the angle formed by the shoulder, hip and knee muscles (acromion, iliac crest and patella) shown in Figure 24.4. The results (shown in Table 24.2) indicate that all values are negative. This means that the angle decreases with time (Δt). The angular velocity ω is slightly smaller for the woman in the eighth month of pregnancy.

Pregnant women take longer to rise from the chair which shows that pregnant women hunch forward more when sitting in a chair without, than with armrests. The angular velocity ω is somewhat smaller for the woman in the

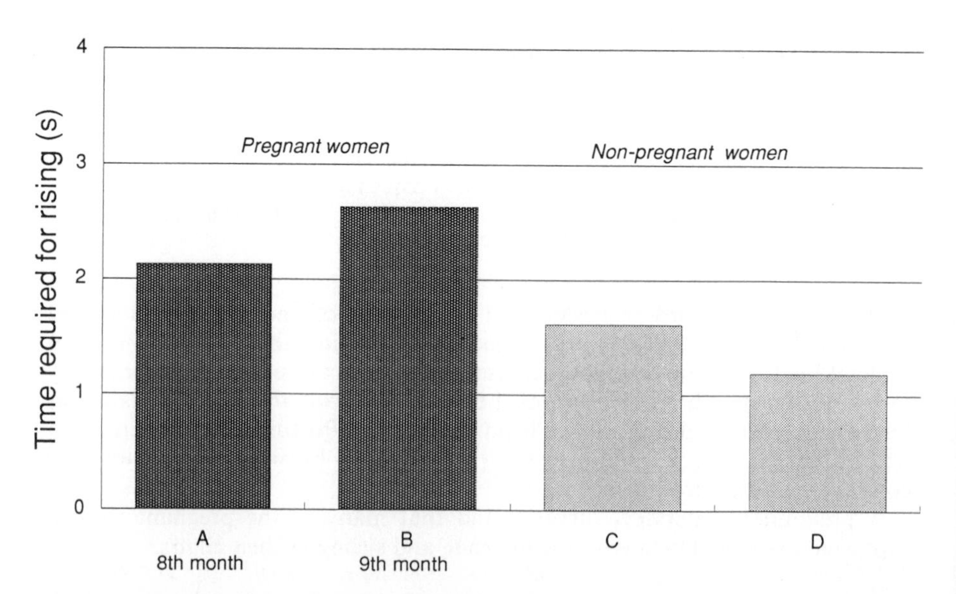

Figure 24.2 Comparison of pregnant women and non-pregnant women in time required to rise from chairs without armrests. The time required is the interval of time when the subject's upper body is vertical in the seated posture to the time when the subject's upper body is vertical in the standing posture. An average of 15 measurements was taken for each subject. The height of each bar is the mean value

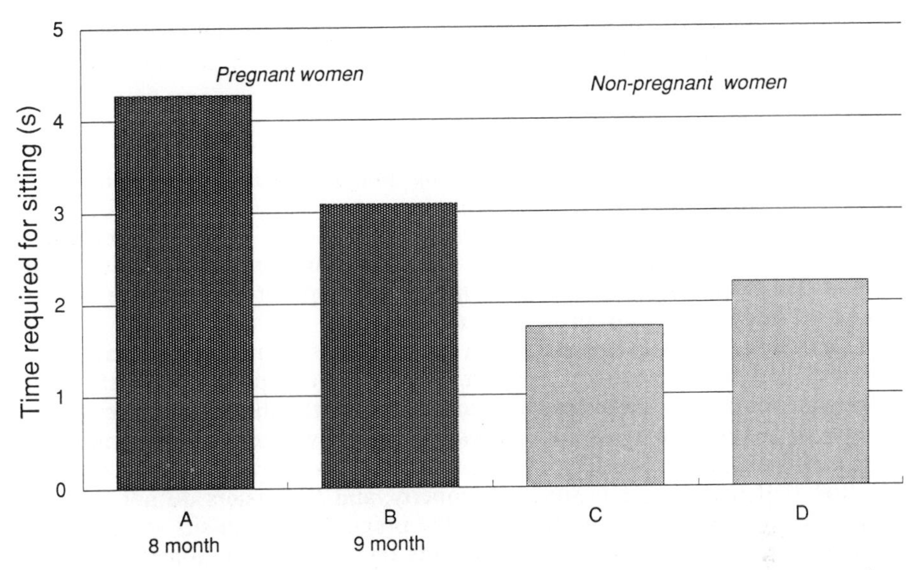

Figure 24.3 Comparison of pregnant women and non-pregnant women in time required to sit on chairs without armrests. The time required is the interval from the time when the subject's upper body is vertical in the standing posture to the time when the subject's upper body is vertical in the seated posture. An average of 15 measurements were taken for each subject. The height of each bar is the mean value

Figure 24.4 Angular velocity ω when the subject separates from a chair. The line drawing on the left shows the subject in the seated posture, and the line drawing on the right shows the subject bending the upper body forward to rise from the chair. The angular velocity ω is calculated by the equation $\omega = (\theta' - \theta)/\Delta t$

Table 24.2 Comparison of chairs with and without armrests in maximum average angular velocity ω (GRD/s)

Subject	Chair without armrests	Chair with armrests
Woman in 8th month of pregnancy	−95.69	−48.20
Woman in 9th month of pregnancy	−77.63	−47.95

ninth than the eighth month of pregnancy, probably as a result of their greater abdominal protrusion.

Study from photographs

The video records of rising from and sitting on a chair were examined in order to study the qualitative aspects of motions.

Figure 24.5 is a video print of the woman in the ninth month of pregnancy rising from a chair without armrests, while holding the seat pan with both hands. She apparently needed this support to keep her balance due to her increased body weight and protruding abdomen.

Figure 24.6 is a video print of the same female rising from a chair with armrests, which she used for support. These photographs indicate that the pregnant woman needed to bend forward less when the chair had armrests.

The pregnant woman visually reviews her chair (with or without armrests), before she sits on it. Since the chair has casters, this visual check is needed to confirm that the chair is positioned properly, and the casters do not slide so easily the chair becomes unsafe. When the pregnant woman sits on the chair with armrests, she visually reviews and sits on the chair by holding the seat surface with both hands, as when she rises from the chair. When armrests are available, she uses these for support. It can be inferred from Figures 24.5 and 24.6 that forward bending is reduced when armrests are provided.

Study of working posture

A video print of a woman in the ninth month of pregnancy is shown writing (Figure 24.7), and performing VDT work (Figure 24.8).

Figure 24.5 Video print of a pregnant woman in the ninth month of pregnancy rising from a chair without armrests

Figure 24.6 Video print of pregnant woman in the ninth month of pregnancy rising from a chair with armrests

Figures 24.7 and 24.8 depict specific task-related postures. One demand of VDT work is that workers need to maintain postures to support their visual requirements. A visual distance of about 30 cm to the desk surface and about 40 cm to the VDT screen surface are recommended for writing and VDT work, respectively (Japan Industrial Safety and Health Association, 1986). Visual distances of pregnant females were 39 cm and 53 cm for writing and VDT work, respectively (*see* Figures 24.7 and 24.8). The visual distance was 39

Figure 24.7 Video print of pregnant woman in the ninth month of pregnancy writing

Figure 24.8 Video print of pregnant woman in the ninth month of pregnancy performing VDT work

cm when the pregnant woman reclined and was 23 cm when she leaned forward while writing. If a more reclined posture is recommended as an alternative to the forward posture which compresses the abdomen of the pregnant woman (Hirao *et al.*, 1987; Noro, 1987), her visual distance becomes excessive for writing. In contrast, such postures meet the recommended guidelines for viewing distance when working at a computer. Pregnant women should have the VDT display height adjusted to accommodate them.

VDT work provides a more suitable posture for pregnant women than writing.

A prototype chair for pregnant women

The following recommendations for chairs for pregnant women are based on previous findings.

1. The chair should facilitate rising and sitting down.
2. The chair should have such armrests to reduce forward bending.

Evaluation of chairs by pregnant women

Commercially available chairs were evaluated by pregnant women (Kajiyama *et al.*, 1987). Three types of chair available on the market were selected and evaluated. Two were office automation chairs with seat heights which adjusted with one-touch, and the other was a conventional office chair which adjusted manually. The characteristics of the chairs are depicted in Table 24.3, with dimensions in Table 24.4. The subjects were five women in the sixth to ninth month of pregnancy. Their body dimensions were measured before they evaluated the chairs. They used an evaluation sheet to rate the chairs. The evaluation results of the three chairs are shown in Figures 24.9 to 24.11.

Table 24.3 Characteristics of chairs evaluated

Chair	Armrests	Material of seat	Seat height adjustment	Seat angle adjustment	Backrest height adjustment	Backrest angle adjustment	Rocking	No. of base arms	Casters
A	Yes	Cloth	Yes (One touch)	No	Yes (One touch)	Yes (One touch)	Yes	5	Double
B	No	Cloth	Yes (One touch)	Yes (One touch)	Yes (Screw)	No	No	5	Double
C	No	Vinyl	Yes (One touch)	No	No	No	No	4	Rubber

Table 24.4 Dimensions of evaluated chairs

Chair	Seat size	Backrest size (mm)
A	440 mm wide × 440 mm deep	420 mm wide × 330 mm high
B	450 mm wide × 420 mm deep	390 mm wide × 280 mm high
C	390 mm wide × 380 mm deep	320 mm wide × 240 mm high

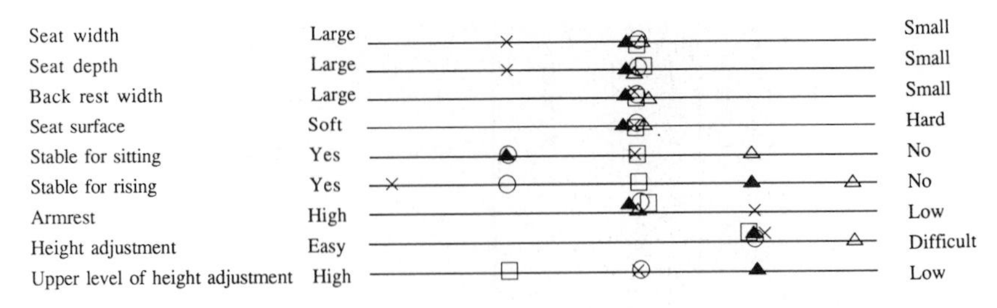

Subjects 1 × 2 □ 3 ○ 4 △ 5 ▲

Figure 24.9 Evaluation of chair A

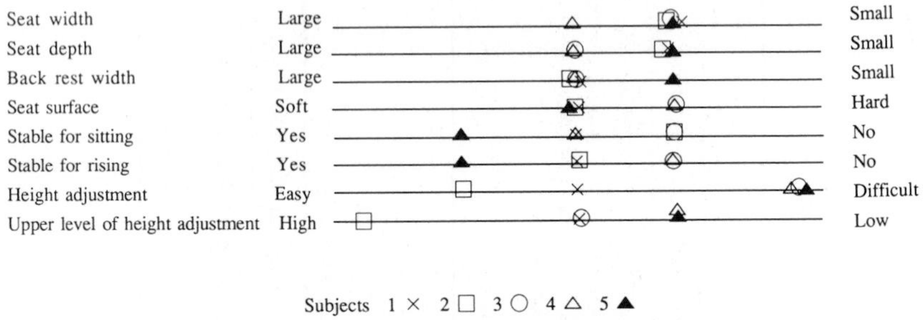

Seat width Large Small
Seat depth Large Small
Back rest width Large Small
Seat surface Soft Hard
Stable for sitting Yes No
Stable for rising Yes No
Height adjustment Easy Difficult
Upper level of height adjustment High Low

Subjects 1 × 2 □ 3 ○ 4 △ 5 ▲

Figure 24.10 Evaluation of chair B

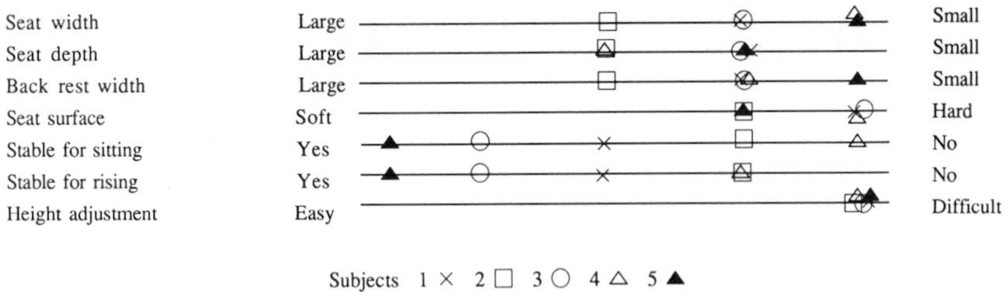

Seat width Large Small
Seat depth Large Small
Back rest width Large Small
Seat surface Soft Hard
Stable for sitting Yes No
Stable for rising Yes No
Height adjustment Easy Difficult

Subjects 1 × 2 □ 3 ○ 4 △ 5 ▲

Figure 24.11 Evaluation of chair C

Discussion of evaluation results

The seat width, seat depth, and backrest width of conventional office chair C were judged as small. Seat height adjustment was rated as difficult for some chairs. All subjects emphasized the need for armrests. The subjects differed in their opinion regarding chair stability. When asked again about chair stability, the subjects were found to evaluate chair stability from the following two viewpoints:

1. Sliding of casters on floor surface
2. Creaking of chair when they sit.

When the subjects evaluated the chairs in accordance with these criteria, they preferred rubber casters with minimum sliding and a stable structure.

Figure 24.12 Prototype chair developed for pregnant women

Chair requirements of pregnant women

Size:
Given the fact that chair A is not small compared to those now widely used, chairs of larger size are required.

Stability:
Pregnant women are worried whether they can safely sit on their chair, both because their body weight is increased, and because they carry the foetus. Consequently, they are concerned about chair stability and safety, that it should not suddenly move backward, turn upside down, or break. This highlights the importance of using stable materials and number of base arms.

Table 24.5 Attributes of subjects

Subject	1	2	3	4	5
Age (years)	28	26	25	26	26
Pregnancy (months)	6	6	7	8	9
Stature (cm)	153.0	159.6	163.7	157.4	157.4
Weight (kg)	50.0	48.0	60.5	68.5	70.5
Transverse diameter at hip (cm)	30.5	31.0	34.8	95.0 (abdominal circumference)	98.5 (abdominal circumference)

Armrests:
The protrusion of the abdomen becomes conspicuous for women in the second and third trimesters. Pregnant women must also avoid leaning forward execessively to prevent adverse effect on the foetus. The motion analysis of the subjects indicates that they must bend forward as they put their hands on the edge of the seat surface before rising from the chair without armrests. However, when the chair has armrests, they can rise without extreme forward bending. Armrests are now traditionally provided to higher status employee positions, but will be shortly required for pregnant women.

Seat height adjustment:
The seat height adjustments of evaluated chairs are all located below the seat, as is common for housing the adjustment mechanism. The subjects complained of difficulty in adjusting the seat height because they had to lean down to manipulate the seat height adjustment below the seat. The seat height adjustment must be placed so that it is easier to adjust. Since pregnant women tend to adjust the seat height over a wider range than women who are not pregnant, the seat height adjustment range must be also expanded for them.

Figure 24.12 shows a chair developed for pregnant women, based on all of the above research. Besides the features of conventional office chairs for female workers, the new chair has the following features:

1. seat height designed for easy adjustment by pregnant women;
2. armrests; and
3. large seat width.

Conclusions

Increasingly, pregnant women want to continue to work as long as they remain healthy. The authors identified ergonomic problems for pregnant women at work and produced a prototype chair for them. The prototype chair is now in use and undergoing a follow-up, including research on how it is actually used.

References

Cherry, N., (1987), Physical demands of work and health complaints among women working late in pregnancy, *Ergonomics*, **30**, 689–701.

Hasegawa, Y., Takahashi, I. and Salman, H., (1985), Robot pre-determined time standards system, *Proceedings of System Science Research Institute*, Waseda University, **16**, 17–34, (in Japanese).

Hirao, N., Hirasawa, N., Noro, K. and Aizawa, Y., (1987), VDT operation by pregnant workers at workplace, *Japanese Journal of Traumatology and Occupational Medicine*, **35** (Supplement), 133, (in Japanese).

Industrial Health Division, Labour Standards Bureau, Ministry of Labour, (1986), VDT and Industrial Health (Tokyo), 147–8, (in Japanese).

Japan Industrial Safety and Health Association, (1986), Occupational Health Guidelines for VDT Work (Japan Industrial Safety and Health Association, Tokyo), 6, (in Japanese).

Kajiyama, M., Nishiguchi, H., Hirasawa, N., Hirao, N., Shiraishi, M. and Noro, K., (1987), Evaluation of chairs by pregnant women, *Proceedings of 1987 Annual Meeting of Japanese Ergonomics Research Society* (Japanese Ergonomics Research Society, Tokyo), 59–62, (in Japanese).

Koshi, S., Yamamoto, S., Saito, S., Matsui, K. and Ishii, T., (1986), *Survey and Study of Health Condition and Visual Load of VDT Operators* (National Institute of Industrial Health, Kawasaki), 83–4, (in Japanese).

McDonald, A. D., Cherry, N. M., Delorme, C. and McDonald, C., (1985), Survey of health hazards of work with VDTs in pregnancy, in K. Noro (ed.), *Occupational Health and Safety in Automation and Robotics* (Taylor & Francis, London), 269–79.

Microcomputer Survey Committee, General Council of Trade Unions of Japan, (1985), Final Report of Survey of VDT Work and Health – Guidelines for Regulation of VDT Work (Japan Workers Safety Centre, Tokyo), 133–57, (in Japanese).

Ministry of Labour, (1986), *Survey of Current Situation of Maternity Protection*, (in Japanese).

Noro, K., (1987), Characteristics of VDT work posture, *Proceedings of the 7th Symposium of Human Posture* (Japan Institute of Human Posture Research, Tokyo), 5–16, (in Japanese).

Sugawara, T., Hayashi, H. and Ichinohe, K., (1984), Effect of employment on pregnancy and childbirth, *Perinatal Medicine*, **14**, 5.

Women Workers' Division, Women's Bureau, Ministry of Labour, (1986), *Textbook for Maternity Protection of Female Workers* (Japan Industrial Safety and Health Association, Tokyo), 9–43, (in Japanese).

Women's Bureau, Ministry of Labour, (1986), *Actual Situation of Female Workers* (Printing Bureau, Ministry of Finance, Tokyo), 1–8, 20, 32–52, (in Japanese).

Yamana, N., Okabe, K., Nakano, C., Zenitani, Y. and Saita, T., (1981), The body form of pregnant women in monthly transitions, *Japanese Journal of Ergonomics*, **20**, 171–8, (in Japanese).

PART X

Design applications

25

Seating and access to work

E. N. Corlett and H. Gregg

Introduction

Sitting has been a subject of scientific interest for many decades, and fascinated designers for centuries. It is amazing that anything is left to say about sitting, but the new designs of seats demonstrate that the subject is by no means dead. They also demonstrate that there is little or no agreement concerning how a seat should be designed, or even what the sitter should experience. Further, no universally accepted operational definition of comfort exists (Lueder, 1983). Considering that it is so widespread, we still have much to learn about sitting.

At this point in a paper, after the preceding statement, the wise author would quietly withdraw. However, it is necessary to revisit old problems and reassess the situation if we are to progress. We will review some of the common arguments about seat design and their validity.

There have been some recent reviews of sitting (Lueder, 1983; Eklund, 1986; Corlett, 1989) so some major points only will be mentioned here. Akerblom (1954) was amongst the first to point to the importance of lumbar support, whilst Keegan (1953) illustrated how the movement of the thigh to a position at right angles to the trunk caused a rearward rotation of the pelvis, leading to a flattening of the lumbar curve.

Work by Nachemson and Elfström (1970), and Andersson *et al.* (1975) demonstrated that this configuration leads to an increase in pressure within the lumbar discs, presumably because they are under increased load from ligaments and from wedging by adjacent vertebrae.

This research also demonstrated that increasing the angle between the seat and back rest, reduced pressure on the discs. This increased angle also increases lumbar lordosis (Akerblom, 1954).

However, the back rest functions as more than a means of improving lordosis. It also reduces loads on the spine by transmitting some of the weight of the trunk, head and arms through the chair structure, rather than through the lumbar spine (Corlett and Eklund, 1984). This is an important function, particularly for those enduring extended periods of sitting. Brief opportunities to use a back rest during a work cycle also afford the possibility of intermittent muscle relaxation; these may modify an otherwise static (postural) load to a more dynamic load, and thereby reduce the muscular strain considerably (Corlett and Manenica, 1980).

Variations on a theme

These and other studies have led designers to draw their own conclusions. Today, only misinformed advertisers presume that an ergonomic seat should support a single idealised posture. It is widely recognised that although some postures may be better than others (here we must specify our criteria for judgement) – good seats must allow adoption of as many postures as possible without preventing sitters from achieving their purposes. Work seats must promote postural changes yet facilitate task demands.

With the heightened recognition that in general change is beneficial and stasis is detrimental to health, it becomes evident that many designs are inadequate.

For desk work, most men and half of the women are accommodated by adjusting their conventional seat. Shorter people require a footrest; otherwise the horizontal seat surface would leave their legs unsupported.

In assembly, sewing or checkout work, operators must frequently sit on the front edge of the seat so that they can use their feet to operate controls or stabilize when reaching. This is because work-place height requires a chair height too high for them; they may also need to reach forward.

Frequently in such situations, the combination of a horizontal work surface with seat and reach constraints renders the back rest unusable.

Although it is generally believed that a conventional seat can accommodate many people who work at a desk, Mandal (1975) has pointed out that forward (anterior) postures at a horizontal surface introduce considerable loads on the spine. He recommends that the desk surface be raised with a small slope, and with a forward slope to the seat pan. This latter change reduces pelvic rotation, thus aids retention of the lumbar curve; the sloping desk top reduces the need to lean as far forward to work.

Mandal's presentations have had a major influence on some designs of seats which permit a forward sloping thigh. Critics of Mandal point out that it is difficult to remain stable on a forward sloping seat, as muscular activity in the legs must counteract the forward component of the seat reaction force, which tends to push the sitter off the seat. This component is reinforced if the sitter uses a back rest. Mandal argues that a back rest is not essential, as the lumbar curve is promoted by the sitting posture arising from his sloping forward seat. However, the functions of the back rest include more than promotion of lumbar lordosis.

To remain seated on a forward-sloping seat pan, it is necessary to counteract the component of force acting downwards, and parallel to the seat surface. There are several ways in which this may be done, and many may be combined.

A common method is to increase the friction, most usually by upholstery. The combination of padding and fabric cover can virtually counteract shallow angles of tilt. The force is still there, however, and can drag the sitter's clothing upwards, to their discomfort. If the sitter is wearing synthetic materials which have low friction characteristics, this effect is noticeable much earlier during the sitting period; indeed, it may make any considerable period of comfortable sitting impossible.

Additional resistance to sliding forwards off the seat can be provided by the sitter's feet. On a conventional height seat, about 10 per cent of the body weight is carried on the feet. For a forward sloping seat, to counteract any tendency to slide forward, synergistic muscle activity is needed to transmit force to the ground. In a pilot experiment with six subjects (3 male, 3 female) Eklund, Houghton and Corlett (1982) tested different angles and heights for a flat sloping seat pan. They used a force plate to measure the load on the feet. The test involved only short periods (a few minutes) of sitting. They noted that

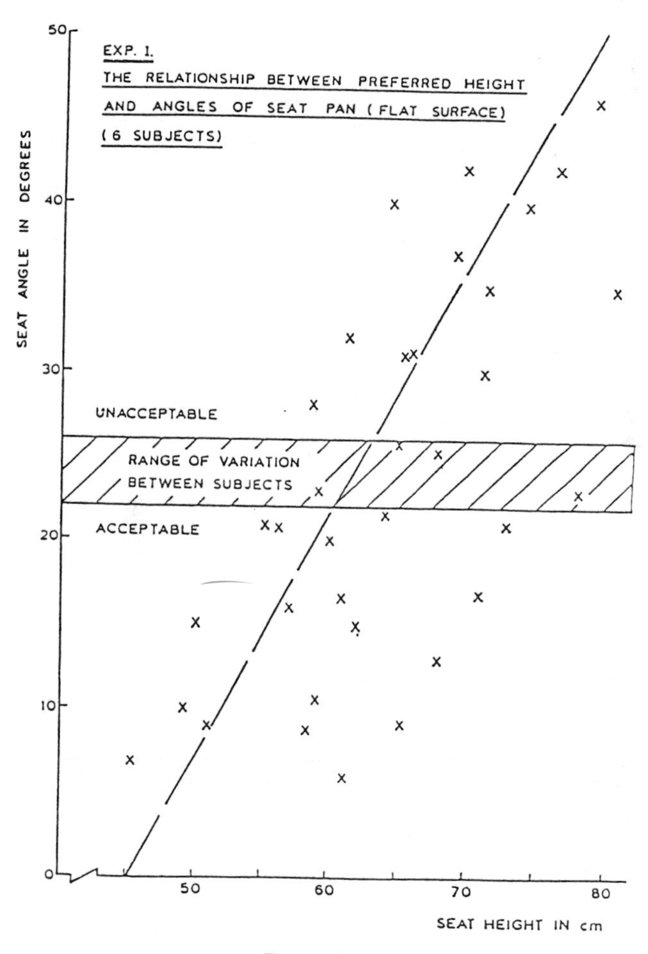

Figure 25.1

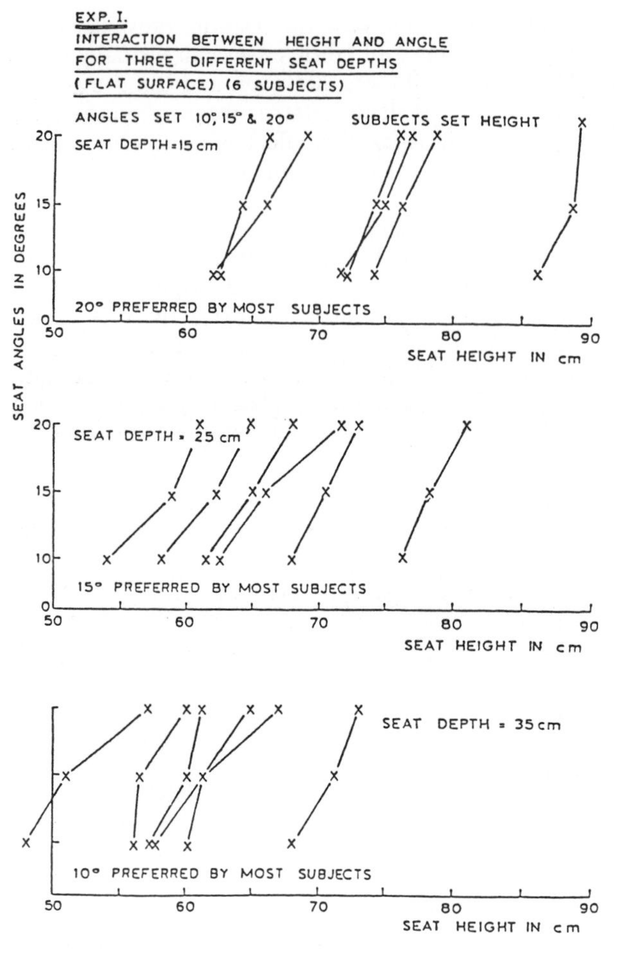

Figure 25.2

the seat was not acceptable when loads on the feet exceeded 25–30 per cent of body weight. They further noted a relationship between seat angle and seat height (see Figure 25.1).

It will be seen that 20° is the maximum angle acceptable, although a range of heights was possible.

Figure 25.2 depicts the spread of seat heights that subjects selected for any given angle. This was caused by an interaction between leg length and seat depth; taller subjects found it possible to use a higher seat. As seat depth increased, the usable height decreased, indicating that the front of the higher seat cut into the thigh. The possibility of using a narrow seat at some 20 cm higher than the deeper seat will be noted; several high work seats have been built to this dimension, resulting in a 'perch' for their users.

Another way to counteract the effect of the forward slope is by using the 'kneeling' seat. Here the forward thrust is counteracted by support from the sitter's knees (or shins) on suitably placed pads. Here again, one problem is overcome at the cost of possibly introducing others. Continued pressure on the knees is not desirable. As changes in posture are desirable, seating arrangements which require continuous muscular activity, sustained pressure, constraints to certain limb positions, or which provide difficult access or egress are not optimal.

Industrial seats

If the demands of the task and work-place constrain the use of the work seat, every effort must be made to promote postural change. Many workers must choose between perching on the front of the seat, or standing, particularly where the work surface is high and/or restricts knee placement.

Where reaching, lifting or some force production is needed, the chair should assist. This may be accomplished by permitting foot placements to extend reach, allowing use of the back rest to counteract relevant forces, or by affording opportunities to stand at will without requiring an awkward posture to continue to work.

Whilst some of these requirements may be achieved by using a high seat with a footrest, this should be considered a last resort. The footrest is acceptable only where other changes cannot allow the sitters to have their feet firmly on the floor.

A new seat design

Few chairs currently available in the market address both the needs of the user and demands of the task. A horizontal surface for the seat promotes free use of the back rest, stabilizes posture, and reduces leg muscle activity. To sit at different heights requires under-thigh clearance so that the thighs may slope downwards and support the feet on the floor. If the front-to-back dimensions of the seat were reduced, body support and the feeling of postural stability would be hindered. The worker sitting on the front edge of a raised seat narrows the seat depth to permit a downward sloping thigh.

An alternative means of addressing job and postural considerations is to curve the seat surface, front to back, and incorporate appropriate upholstery (Figure 25.3). This seat rotates forward as it is raised, providing clearance for the increasing slope of the thighs, whilst retaining a horizontal surface below the main weight bearing areas of the ischial tuberosities. Some contact under the thighs is maintained, giving security to the sitter as well as reducing point loadings on the seat.

Eklund *et al.* (1982) demonstrated that when starting from a horizontal seat position with right angles at the trunk and knee ('right angled sitting') the change in preferred thigh angle with increase in seat height was almost constant as seat height increased, regardless of leg length (Figure 25.2). This relative constancy allows the forward tilt of the seat to be linked with change in seat height. To accomplish this requires some initial height setting to account for the user's anthropometry; this adjustment is done in the 'right angled sitting' position at a low level of the seat.

Figure 25.3. Although working at a high table, this tall woman can keep her feet firmly on the floor and use the backrest. Note the curved seat profile

However, automatic angle change is not necessary; conventional lever-operated tilt mechanisms are adequate. For the shorter person using a conventional height bench, or a keyboard on a desk, the seat height and angle can be set to give a comfortable elbow height and arm position which, with a conventional chair, would lift the sitter's feet off the ground and require a footrest. Because the seat of this new chair is rotated forwards, the sitter's feet will be firmly on the ground, whilst the body support will remain at a comfortable working height (Figure 25.4). This forward rotation thus replaces much of the vertical adjustment without sacrificing comfort or stability.

For the user who must occasionally stand at work or needs wide reaches, the chair will permit the person to sit at about $\frac{3}{4}$ of standing height (Figure 25.5). Thus when the user stands (Figure 25.6), the relationship between the trunk and work-place is little changed, and the task can be pursued either seated or standing without discomfort.

The concept underlying this seat has been incorporated into seats in various situations. A recent trial (Gregg and Corlett, 1990) evaluated it, using a range of criteria, for sewing tasks in a clothing factory. The testing procedure was

Figure 25.4. A short keyboard operator is well placed in relation to the keyboard, at a conventional height desk, but the forward slope of the thigh is evident, whilst her feet are on the floor

described by Corlett (1989), but included the comparative investigation of seat comfort using magnitude estimation methods. This allowed the magnitude of the difference in comfort between conventional sewing seats and the new seat to be assessed.

The principal requirements for such a design in the sewing task are given in Table 25.1 (Gregg and Corlett, 1990). Conventional scaling techniques were employed in the evaluation protocol, as was a structured questionnaire, magnitude estimation methods and informal interviews with users. In summary, the new seat performed significantly better over all parameters of measurement, than did conventional designs and was well liked by its users. Table 25.2 depicts the results for one subject group for the magnitude estimation assessment.

An inherent difficulty in measuring 'comfort' is that the term means different things to different people. In a sense, however, this problem with semantics can be overcome if one employs a technique which does not purport to define comfort, but relies instead upon each subject's own construct of what it means. Such an approach can be problematic with inter-subject comparisons.

Figure 25.5. The higher seat and sloping thigh give greater clearance whilst enabling access to the task. Note the links between the seat to give automatic rocking of the seat as height is increased

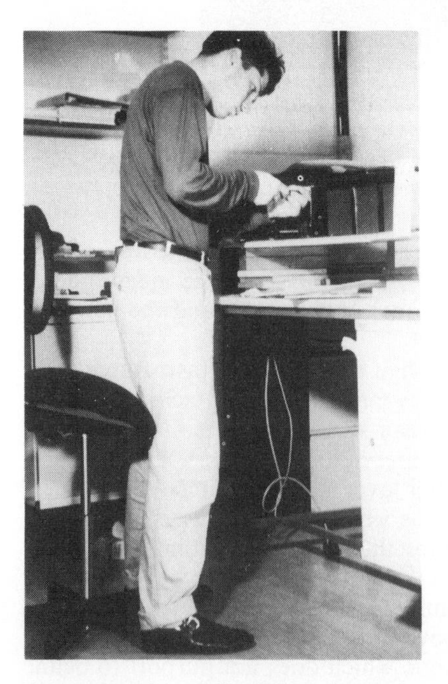

Figure 25.6(a) and (b). When the worker stands ready access is available without undue postural strain. The hand operated and automatic tilt versions of the seat are illustrated in this figure

Table 25.1 Indications for the use of the Nottingham Seat in industrial sewing workstations

1. Decreased spinal loading due to maintained lumbar lordosis.
2. Reduced loading on shoulder musculature through correct sitting height (reduced abduction of arms).
3. Reduced discomfort/increased comfort in the neck, shoulder and lumbar regions, seat contact area and feet.
4. Improved haemodynamics (circulation), including decreased blood pooling at the ankles through minimisation of postural constraint and low thigh compression values.
5. Superior subjective ratings.
6. Better suited to task.
7. Improved ease of ingress – egress.

However, this method is useful as an adjunct to the more conventional approach of 'reductions in discomfort', because the experimental design for the sewing investigation compared new and traditional seats.

Each subject used both chairs; consequently, whatever operational criteria a subject used in evaluating comfort, would likely be employed in the same way for both seats.

Laboratory investigations have also used objective measurements. Eklund (1986) investigated spinal shrinkage (a derived measure of spinal loading), reaction forces, postures and subjective ratings for a conventional seat, and a

Table 25.2 Magnitude estimation results group geometric means (n = 10). The group geometric mean represents the mean of the logs (base 10) of individual results. Thus, for instance, in comfort measurement, the original difference between the two measures (1.09 conventional seat, and 1.39 new seat) may be found by taking the antilogs, i.e. 12.3 and 24.5. A higher number indicates an estimated improvement in the factor by the subject

Seat feature	Conventional seat	Change	New seat
Comfort	1.09	+0.3	1.39
Stability	1.37	+0.15	1.52
Weight on feet	1.11	+0.4	1.51
Weight on seat	0.97	+0.24	1.21
Backrest position	0.80	+0.83	1.63
Backrest angle	0.76	+0.87	1.63
Overall seat change	0.84	+0.6	1.44
Seat angle	0.88	+0.57	1.45
Seat length	1.06	+0.38	1.44
Seat width	1.19	+0.19	1.38
Seat padding	0.92	+0.28	1.26
Pressure distribution under thighs	0.79	+0.58	1.37
Ease of rising	1.07	+0.43	1.50
Suitability to job	0.89	+0.60	1.49
Ease of adjustment	0.53	+1.14	1.67

*Table 25.3. A comparison of a conventional chair with a sit-stand seat for an assembly task with restricted knee room. The measurement results from shrinkage and reaction forces are given as means, followed by the standard deviation. The ratings are given as the total sum of the discomfort score for the eight subjects. Significant differences are marked with *. (Eklund, 1986)*

	Conventional chair	Sit–stand seat	Standard deviation of individual differences
Shrinkage (mm)	2.41 (1.20)	0.93 (0.71)	0.80*
Backrest force at rest (N)	109 (36)	61 (25)	20*
Backrest force at work (N)	8 (7)	—	—
Force upon feet at work (N)	131 (23)	136 (39)	28
Position of backrest force above seat at rest (mm)	210 (12)	209 (15)	9
Increase in discomfort score for the worst body part	381	262	*
Sum of overall discomfort score	357	279	
Number of discomfort statements	33	26	
Performance (screws)	83	81	
Preference	1	7	*

sit-stand design for an assembly task with restricted knee room. The sit-stand configuration provided superior results (Table 25.3).

Field trials in other work-places have been undertaken, and will form the basis of future papers. European and foreign patent applications have been filed for this seat, and it is currently manufactured under licence in the UK.

References

Akerblom, B., (1954), *Chairs and sitting In Human Factors in Equipment Design.* Lewis, London.

Andersson, B. J. G., Örtengren, R., Nachemson, A. L., Elfström, G. and Broman, H., (1975), The sitting posture, an electromyographic and discometric study, *Orthopaedic Clinics of N. America*, **6**(1), 105–20.

Corlett, E. N., (1989), Aspects of the evaluation of industrial seating. The Ergonomics Society Lecture. University of Reading, 3–7th April. Taylor & Francis, London.

Corlett, E. N. and Eklund, J. A. E., (1984), How does a back rest work? *Applied Ergonomics*, **15**(2), 111–4.

Corlett, E. N. and Manenica, I., (1980), The effects and measurement of working postures, *Applied Ergonomics*, **11**, 7–16.

Eklund, J. A. E., (1986), Industrial seating and spinal loading. PhD thesis, Department of Production Engineering and Production Management, University of Nottingham.

Eklund, J. A. E., Houghton, C. S. and Corlett, E. N., (1982), Industrial seating. Report on some pilot studies. Internal report. Department of Production Engineering and Production Management, University of Nottingham.

Gregg, H. D. and Corlett, E. N., (1990), The evaluation of a new workseat for industry and commerce International report, Department of Production Engineering and production management, University of Nottingham.

Keegan, J. J., (1953), Alterations of the lumbar curve related to posture and seating, *Journal of Bone and Joint Surgery. A*, **35**, 589–603.
Lueder, R. K., (1983), Seat comfort. A review of the construct in the office environment, *Human Factors*, **25**(6), 701–11.
Mandal, A. C., (1975), Workchair with tilting seat, *Lancet*, **1**, 642–3.
Nachemson, A. and Elfström, G., (1970), Intravital dynamic pressure measurements in lumbar discs, *Scandinavian Journal of Rehabilitation Medicine*, Suppl. 1.

26

An ergonomic study of dynamic seating

Y. Suzuki, T. Sugano and T. Kato

Concept of dynamic seating

The postwar evolution of office chairs in Japan took place over four generations, as depicted in Figure 26.1.

1. *First generation.* Steel chairs replaced those made of wood in the 1960s. Mechanisms remained simple, with simple rocking actions adopted for seat and back rest movement. Air conditioning became prevalent in the office.
2. *Second generation.* Revisions of a Japanese Industrial Standard (JIS) for office desks and chairs reflected advances in ergonomics research, and greatly improved dimensions and performance. The chair mechanisms became somewhat more complex, and incorporated seat pan and back rest height adjustment. Super high-rise office buildings were introduced.
3. *Third generation.* The introduction of office automation was accompanied by new kinds of seating. Forward tilting seats, and designs with front support (FS) rock or sliding capabilities appeared in the first half of the 1980s. Chairs with synchronized seat pan and back rests became available in the latter part of this decade. Although these designs were intended to accommodate dynamic patterns of movement, adjustments were still operated manually.
4 *Fourth generation.* By the late 1980s, chairs that provided continuous and dynamic support were developed, as exemplified by Herman Miller's Equa and Steelcase's Sensor chairs. Some versions of these chairs allow the seat and back rest to lock in many positions without intervention by the user. In Japan, the fourth generation dynamic seating are expected to become prevalent in offices' in the 1990s.

The Dynafit chair represents an example of this new generation of office seating (Figure 26.2). A description of the research and development of the Dynafit chair follows.

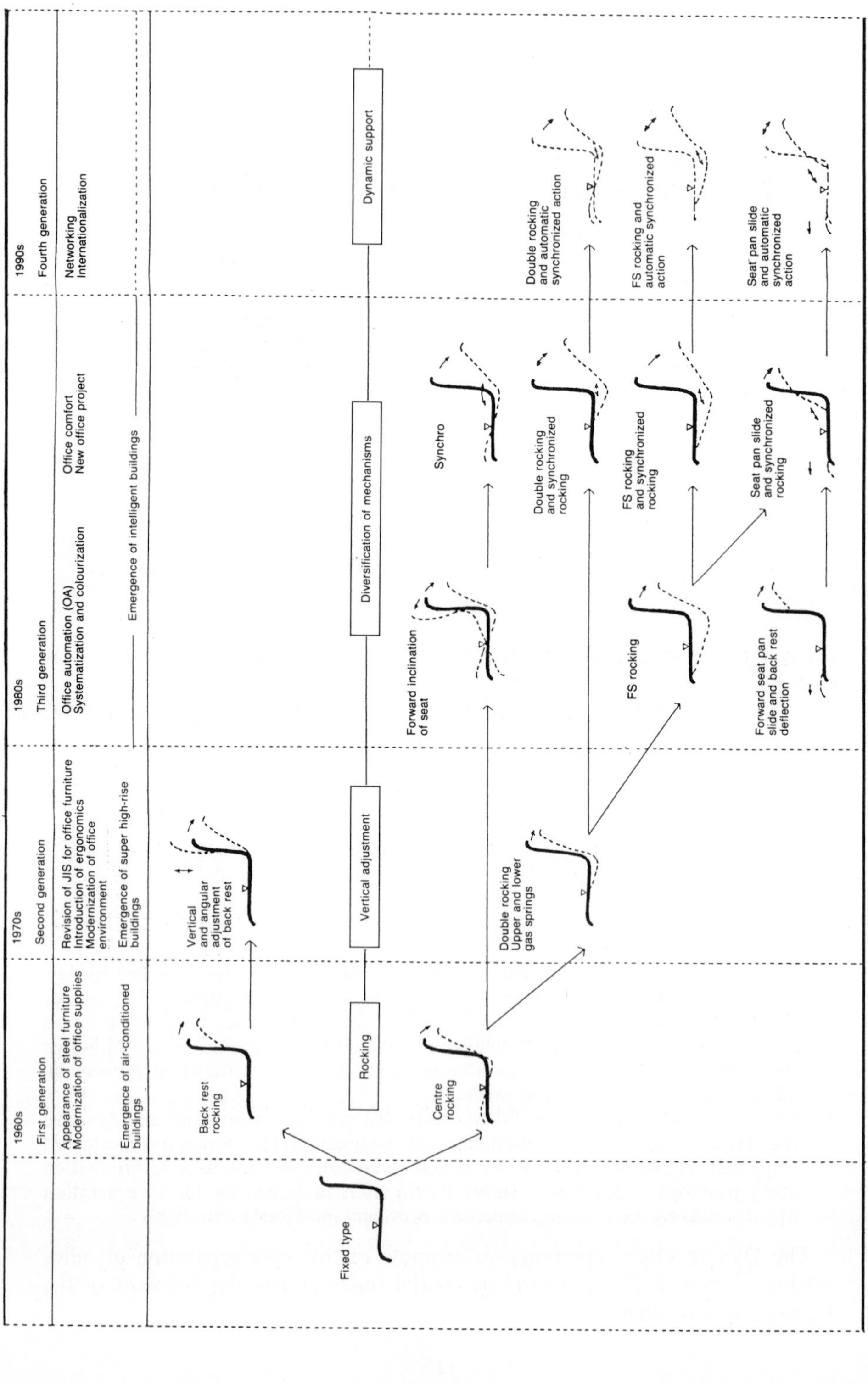

Figure 26.1 History of office chair development in Japan

Figure 26.2 Dynafit chairs

Characteristic concepts and mechanisms of the Dynafit chair

The Dynafit chair is characterized by a sliding seat pan and synchronized seat back movement. Its design is based on:

1. An emphasis on horizontal, rather than vertical, movement to accommodate the VDT user's movements (Figure 26.3). For example, the seat pan slides forward when the user leans back, and the back rest deflects horizontally at the upper torso when leaning back. In contrast, previous ergonomic seating emphasized the vertical adjustments of seat pan and back rest.
2. Springs and joints that move the seat and back rest in unison to support user postures, and to promote movement. Differences in body weights are accommodated (Figure 26.4).
3. Supporting lumbar lordosis at the L2/L3 vertebrae and distributing of back rest pressures at the lumbar region (Figure 26.5).
4. Contour and cushioning of the seat pan and back rest to provide support, yet feel soft (Figure 26.6).
5. Avoidance of projections in order to promote safety, and simplify the operation of controls. The five-arm base is moulded from an engineering plastic to save weight (Figure 26.7).

Characteristics of the Dynafit chair mechanisms include (Figure 26.8):

1. Link mechanism. This mechanism connects the back rest and seat pan with a single shaft, and allows the seat to 'float'. Forces applied to the back rest are transmitted to the seat, which move it forward.
2. Balance mechanism. The seat moves forward as the locus of support of the user changes. This mechanism plays a role in balancing the sitter's body weight. When the back rest tilts rearward, the user's centre of gravity moves back. This change in centre of gravity is countered by slight movements of the linkage between the seat pan and back rest.

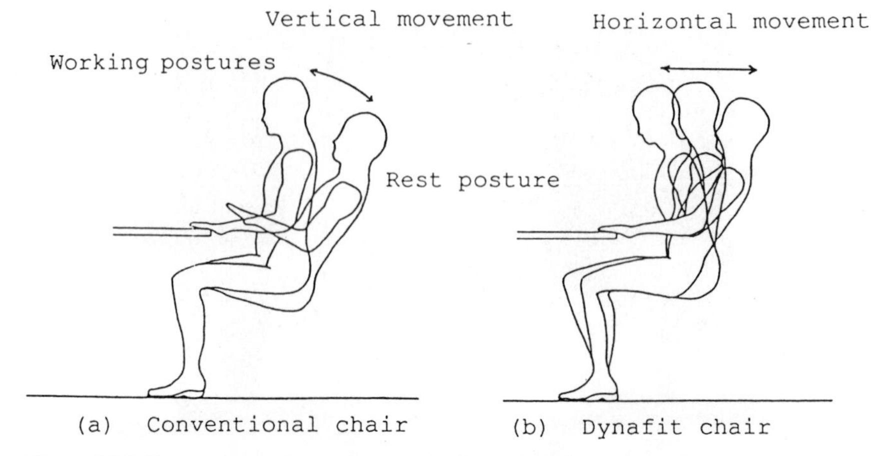

Figure 26.3 Comparison of a conventional chair (a) and the Dynafit chair (b). With the conventional chair, changes in posture move the user away from his work. The sitter can assume a number of task-related postures with the Dynafit chair

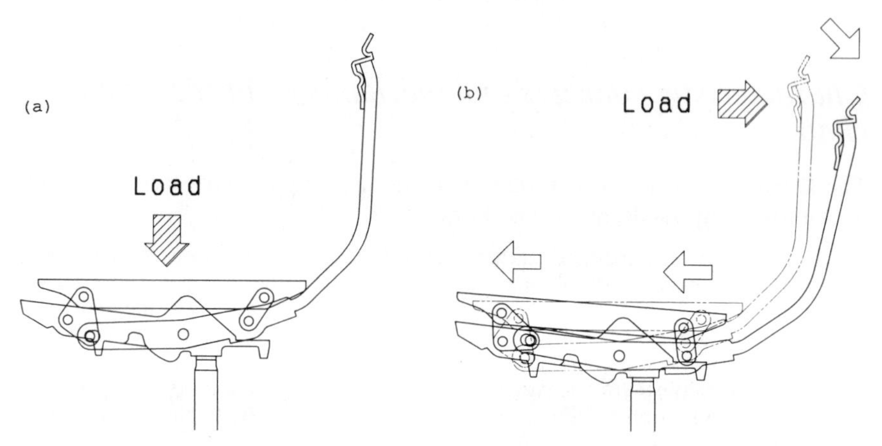

Figure 26.4 Link mechanism of Dynafit chair for (a) vertical load and (b) horizontal load. The seat and back move in unison

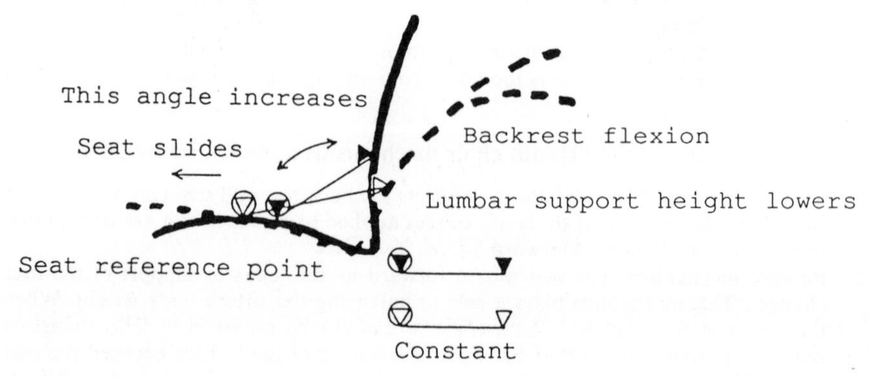

Figure 26.5 Distance between the seat reference point (SRP) and lumbar support height remains constant as the back reclines

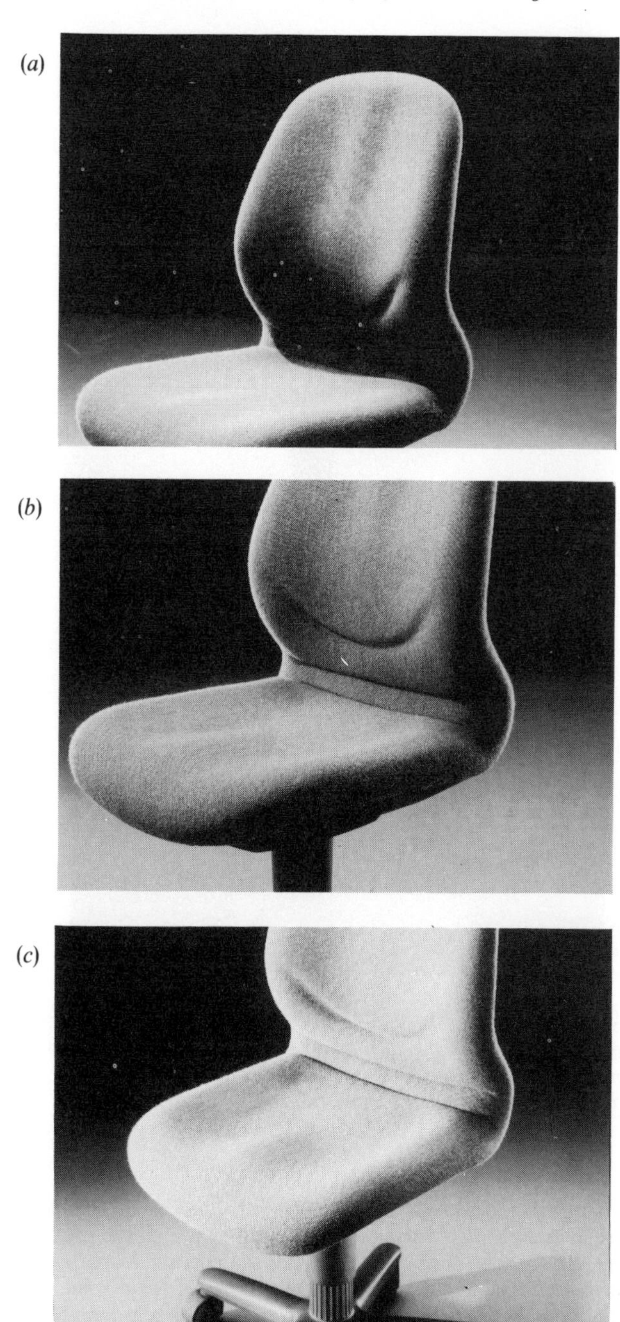

Figure 26.6 Geometrical design of Dynafit chair. (*a*) *Back shape: the upper back is slightly convex at the centre to increase postural support;* (*b*) *Lumbar support; the interface between the back rest and seat pan curves to securely support the lumbar region;* (*c*) *Seat shape; the seat pan is contoured and depressed to stabilize thighs and improve comfort. The seat pan 'waterfalls' to facilitate blood flow to thighs and provide freedom of leg movement*

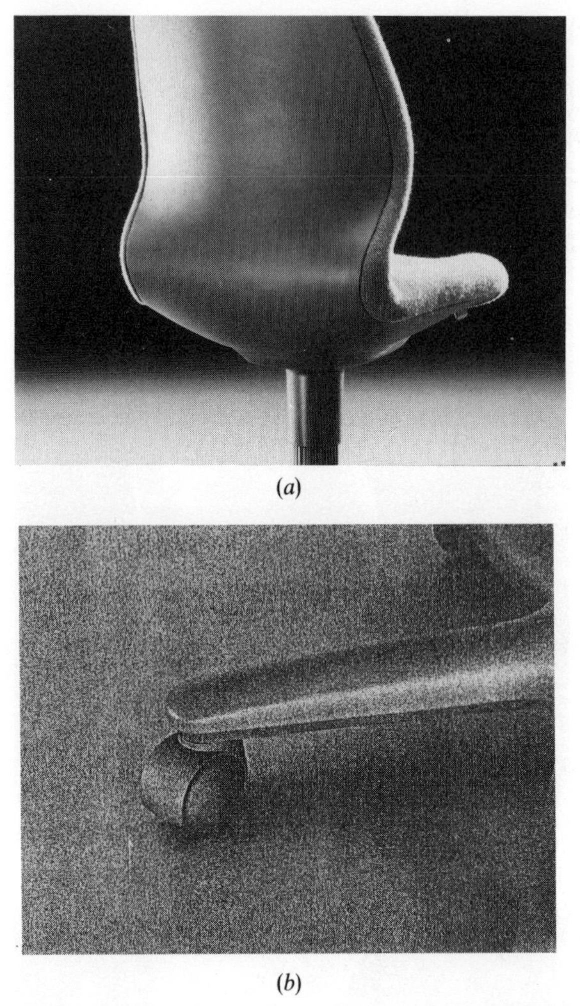

(*a*)

(*b*)

Figure 26.7 Back shell and base. (*a*) *Outer shell: the back shell design is simple and rounded. It is moulded from plastic to prevent damage to surrounding furniture from contact;* (*b*) *Base: the base is moulded from standard engineering plastic*

3. Springs. Two springs operate in conjunction with the link mechanism to keep the user's posture in equilibrium. Spring tension is adjustable to the user's body weight.
4. Adjustable height lumbar support. When the seated user leans back, their pelvis rotates, and lowers the height of the lumbar curve of their spine. An adjustable height-lumbar support was incorporated into the seat to compensate for this lowering of the user's lumbar curve. The inside of the back rest expands and contracts, like bellows. When the back rest reclines, the lumbar support increases in depth, and lowers.
5. Flexible back rest shell. The back rest shell is moulded from resin. Deflection of this shell effectively increases the seat-to-back angle and promotes small movements of the user.

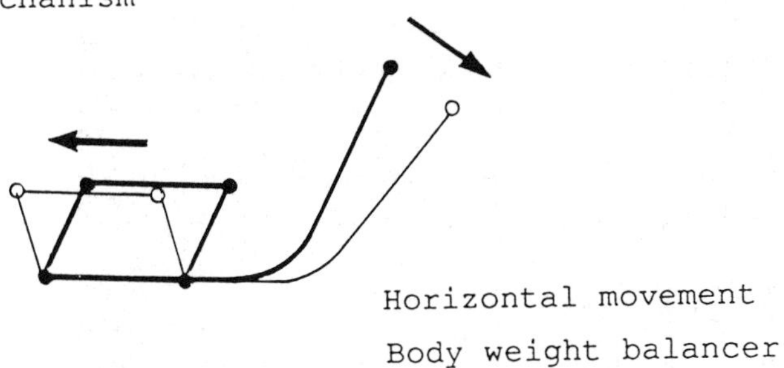

Balance mechanism

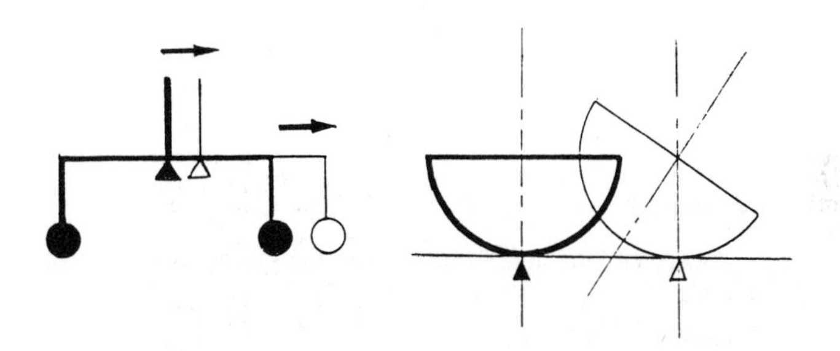

Seat-back balance

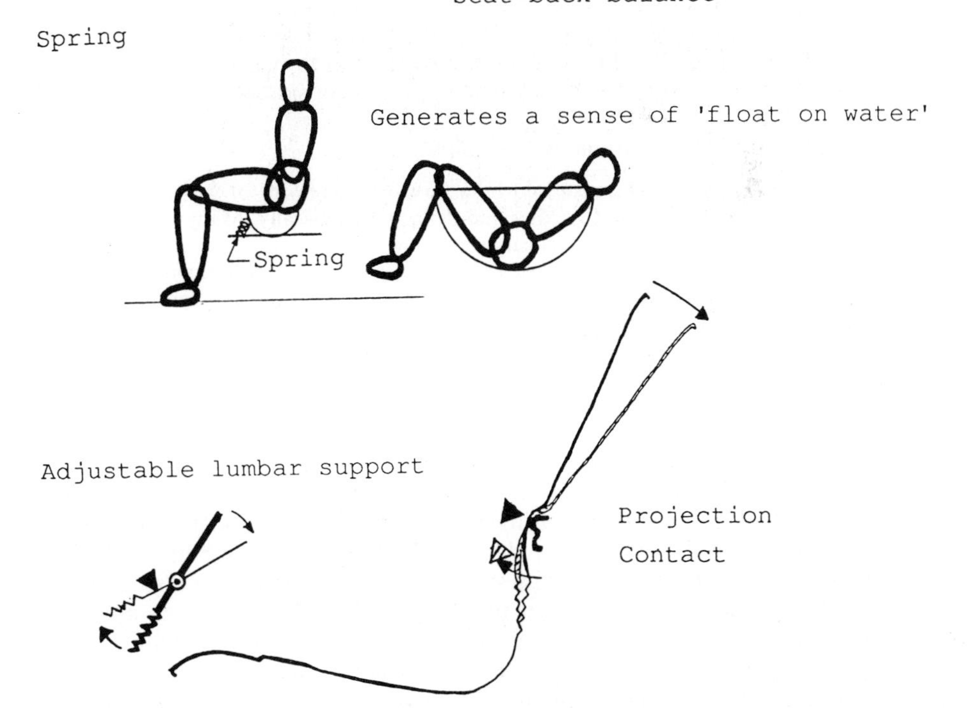

Figure 26.8 Mechanisms of the Dynafit chair

Experimental research

The early developmental stages of the Dynafit chair involved the use of mock-ups and prototypes to correct functional deficiencies. The subsequent design was again evaluated and refined.

Dynamic changes associated with the interlocking action of the seat pan and back rest were also analysed.

Measurement of synchro-ratio

The rate of change of the angles between the seat pan and back rest during recline is called the synchro-ratio. The synchro-ratios of the upright and unsat-upon chair and the reclined posture were measured for the Dynafit chair (*see* Figure 26.9).

The seat pan angle changed by 5°, the back rest angle changed by 15°, and the seat-to-back rest angle changed by 10° as the seat moved from upright to recline. The synchro-ratio was thus 1 : 3.

Patterns of postural change

Postures and movements of one user seated on the Dynafit chair were video-taped and analysed by computer (*see* Figure 26.10). These changes are also modelled in Figure 26.11. The Dynafit chair was found to stabilize leg positioning and promote torso movements more smoothly than with other chairs tested.

The sitter's postural changes were also measured in relation to the desk (Figure 26.12). The Dynafit chair minimizes vertical movements and promotes horizontal movements. Such movements were considered suited for VDT work.

The results may be summarized as follows.

1. The Dynafit chair is characterized by a forward movement of the seat pan, some 'float' and a back rest that reclines and deflects horizontally. Many other dynamic seating are characterized by vertical, or rocking movements.
2. The seated individual tested showed stable leg positions, and horizontal deflection of the upper torso with recline. As a result, the vertical head motion is small enough to maintain working postures. The other chairs increase vertical head motions, which is a disadvantage for continuous work.
3. The Dynafit chair provides an effective synchro-ratio and a large seat-to-back rest angle. This helps support users' movements, and opens the torso angle.

Stopping capability

An experiment was conducted using a three-dimensional model of the human body to evaluate the stability of the Dynafit chair, and its ability to lock in place. The operating force and its smoothness of motion were also assessed. This phase indicated that the seat provided good ability to lock in any position, appropriateness of seat tension, and adapted to the user's weight.

The body weight of a three-dimensional human model was set at three levels (40, 63 and 80 kgf), and measured with a deflection meter and load cell. Seat tension was determined by noting the load which initiated back rest movement.

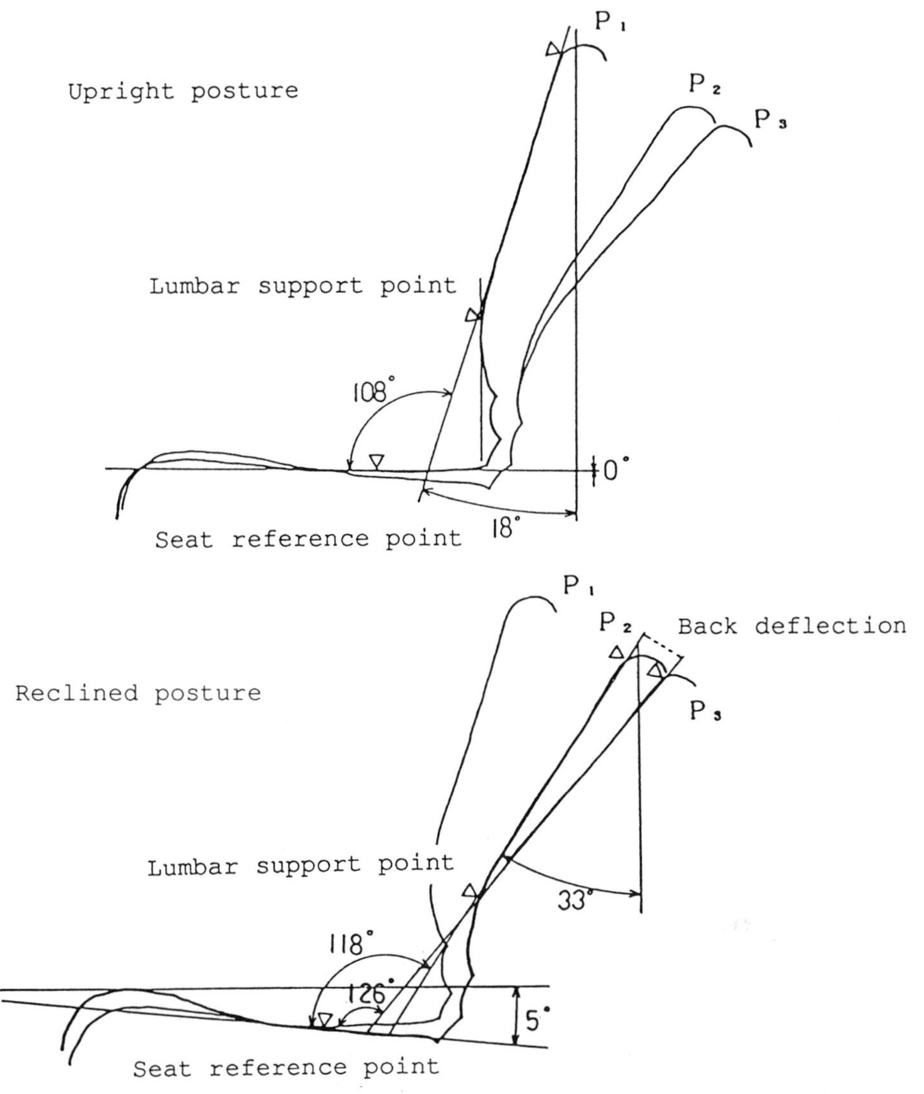

Figure 26.9 Angular change of cushion contour. The Dynafit chair is characterized by a large change in seat-to-back rest angle

Figure 26.13 compares body weight required to initiate the back rest motion with the three chairs (Dynafit, A, and B). Conclusions were:

1. When seat tension forces are set for body weights from 40 to 80 kgf-such users can stop the back rest at any position.
2. The force required to move the back rest from upright to recline is about 1.4 to 10.16 kgf, despite differences in body weight.
3. Seat tensions of 0.7 to 6.3 kgf are required to return the back rest from recline to upright.
4. Subjective preferences of 55 subjects were measured regarding tension adjustments, and compared with their body weights. The survey found that the spring tension adjustment could accommodate these users by varying the number of turns.

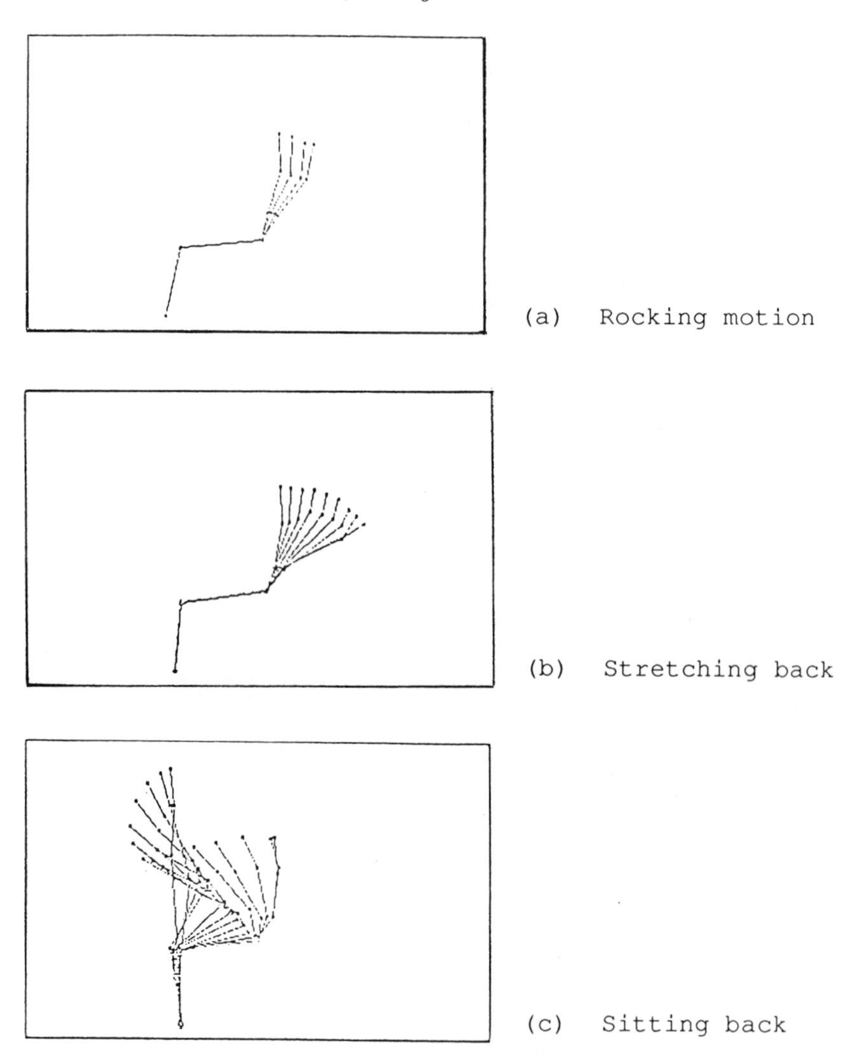

Figure 26.10 Motion analysis by computer

Muscular activity

EMGs of 5 muscle groups were conducted to evaluate the functional performance of the Dynafit chair. These were the straight and external oblique abdominal muscles, the broadest muscle of the back (latissimus dorsi), the mesogluteus, and the straight muscle of the thighs (rectus). The electrical potentials from these muscles were led by a polygraph at a time constant of 0.01 to 0.03 s and amplitude of 1 kHz, and integrated over the measurement time. Some results are shown in Figure 26.14.

EMG tracings indicated that muscle activity levels were minor because of the high levels of support afforded by the back rest, even when it is temporarily still. The Dynafit chair is associated with smaller muscular loads of the

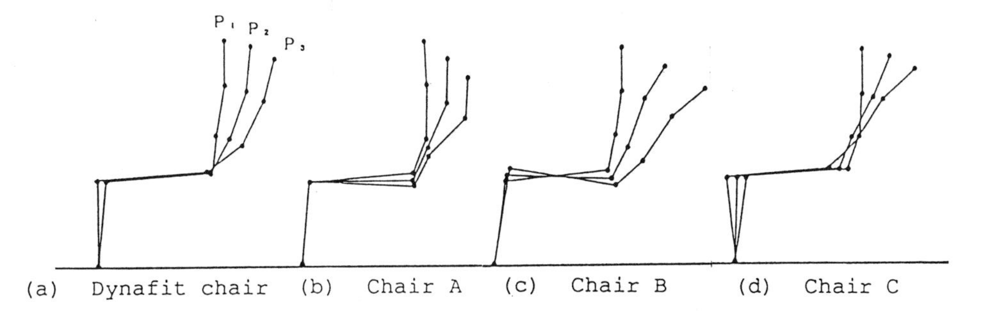

(a) Dynafit chair

- Sitter's upper body mainly moves back and forth.
- This chair accommodates good working postures
- Vertical head movement is small

(b) Chair A

- Considerable up and down movements of torso
- This chair is not suited for work when reclined
- Vertical head movement is large

(c) Chair B

- Sitter's upper body moves to large angle
- This chair is suited for relaxation but not work
- Vertical head movement is large

(d) Chair C

- Sitter's body made extensive back and forth movements
- This chair is not suited for reclined work postures
- Sitter maintains unnatural torso postures

Figure 26.11 Motion analysis

sitter during movement than with other chairs investigated. As a result, the Dynafit backrest can move smoothly with little activation force.

Sensory evaluation of seating comfort

The seating comfort of the Dynafit chair was evaluated by the Semantic Differential (SD) method. Six types of chair, including the Dynafit chair, were evaluated for a short time (1–2 min) and for a long time (30 min). Twenty-five

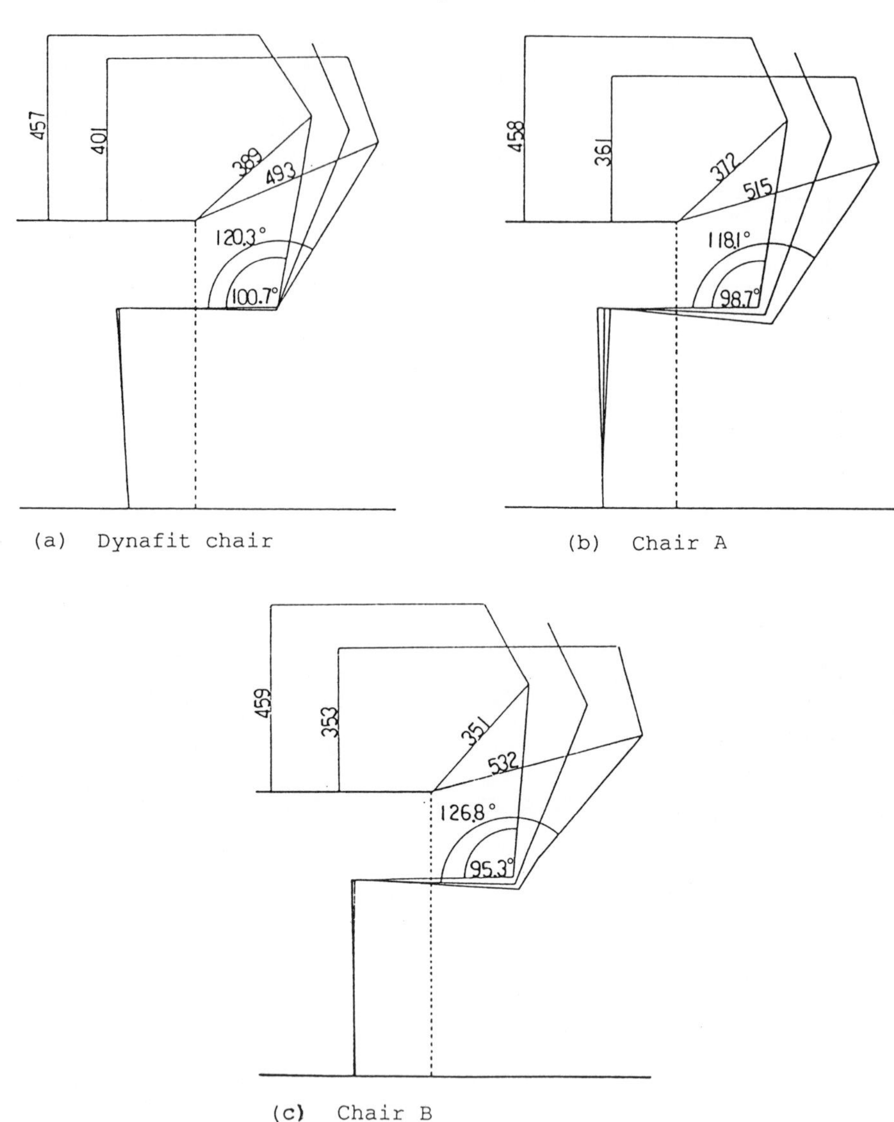

Figure 26.12 Deflection of chair in combination with desk. Vertical change of the Dynafit chair is reduced with VDT work

subjects participated in each experiment. The following 10 factors were evaluated on a seven-point scale.

1. Cushioning (seat and back rest): hard–soft, deep–shallow, good–poor
2. Degree of support (seat pan and back rest): loose–tight, compatible–incompatible, good–poor
3. Smoothness: heavy–light, smooth–rough, good–poor
4. Postural equilibrium: good–poor
5. Ability to relax: good–poor
6. Support of task: good–poor

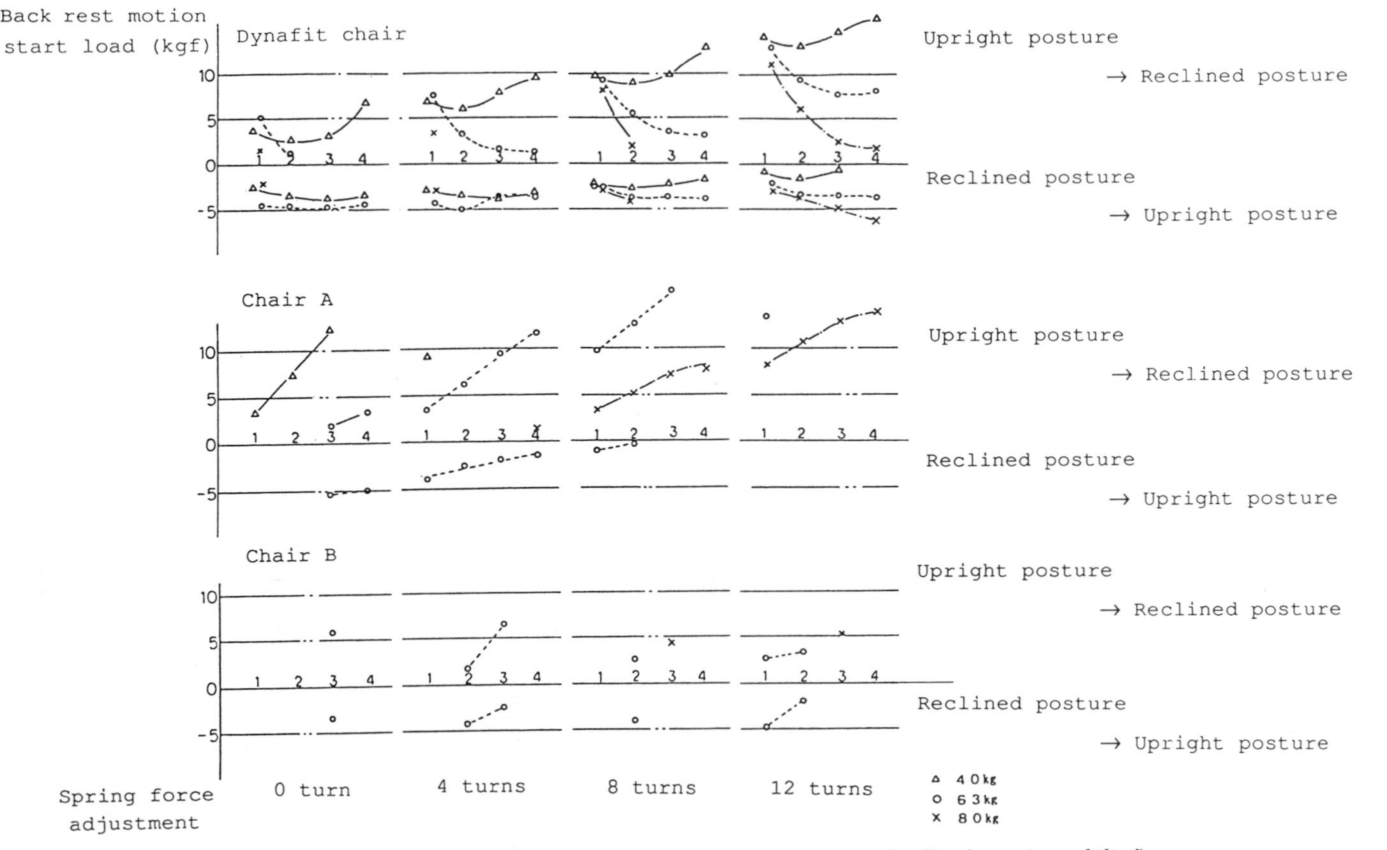

Figure 26.13 Load which initiates back rest movement (as a result of back rest 'stopability')

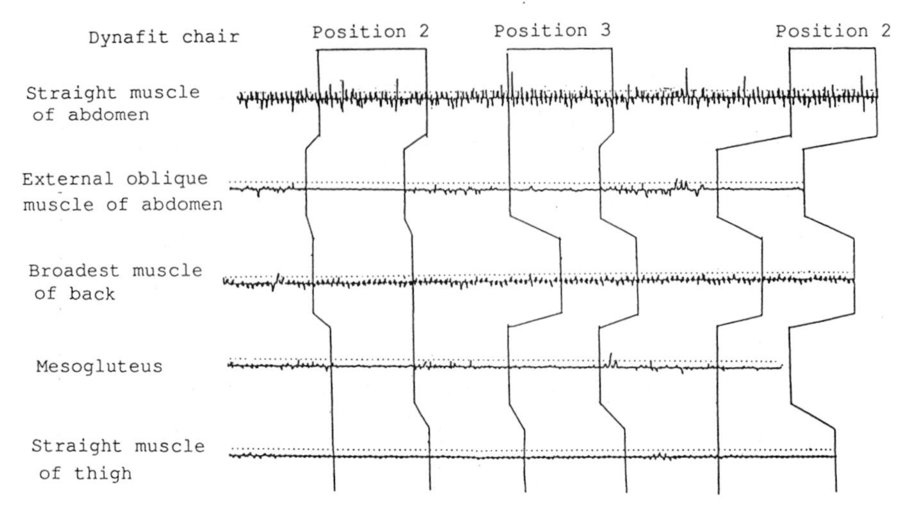

Figure 26.14 Evaluation of Dynafit chair by electromyography

7. Ability to stretch/lean back: back stretchable–not stretchable, good–poor
8. Fitness: upper back fit good–poor, lumbar fit good–poor, hip fit good–poor
9. Body pressure distribution: upper back compression strong-weak, lumbar compression strong–weak, overall compression strong–weak
10. Overall evaluation: good–poor

The mean sensory evaluation scores of the Dynafit chair for the two durations are shown in Figure 26.15. On many scales, the mean evaluation is higher for 30 min than 1–2 min trials. The overall score is also higher for the longer than for the shorter trials. The Dynafit chair was rated good after the longer trial. This may be mainly explained by the cushioning or work factor.

The overall chair evaluation scores for the two trial durations are summarized in Figure 26.16. The Dynafit chair receives positive evaluation scores for the 30 min trials.

Evaluation scores varied greatly with short-term trials of 1–2 min. The decreased variability of the long-term evaluation scores may have resulted from habituation. The cushioning and degree of support of the back rest were strongly associated with the overall evaluation of the chairs. The supportiveness, smoothness and rest/ability to relax properties are considered to relate to the short-term overall evaluation of the chairs, and ratings of postural and task support are considered related to the long-term overall evaluation of the chairs.

Body pressure distribution and final stable posture

The seated pressure distribution and postural stability were measured to evaluate the degree of support with the dynamic back rest.

Seated pressure distributions

The seated pressure distributions were measured when the user sat upright and reclined, with a simple two-liquid mixture method (Figure 26.17).

Cushioning: Seat pan

●Long time
○Short time

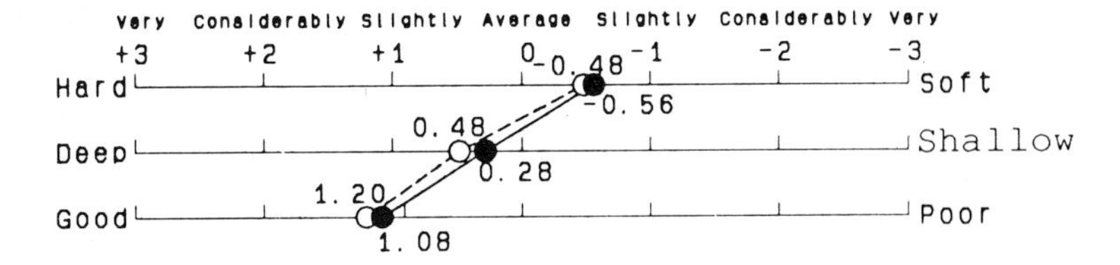

Cushioning : Backrest

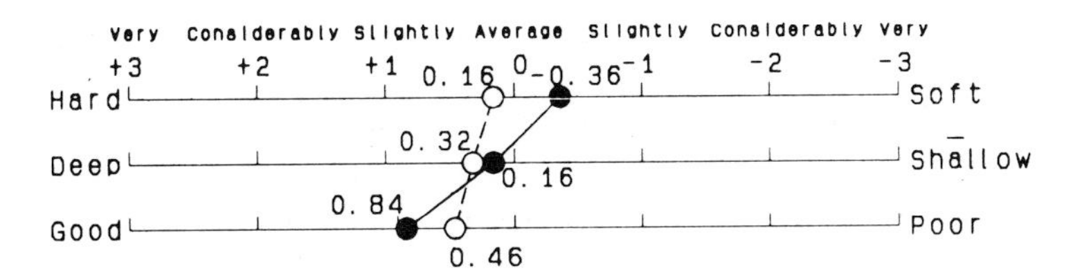

Supportiveness: Seat pan

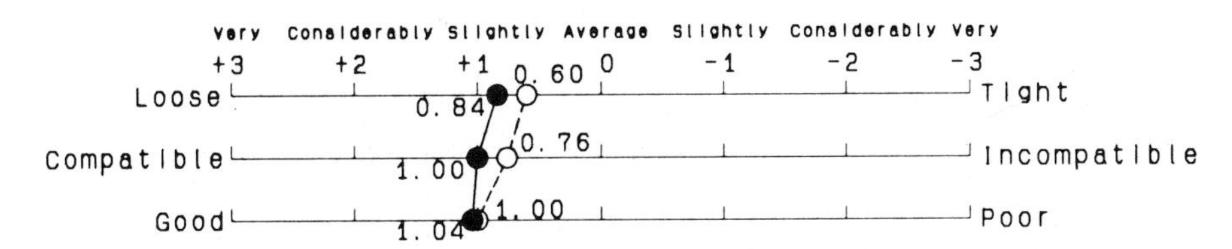

Supportiveness: Backrest

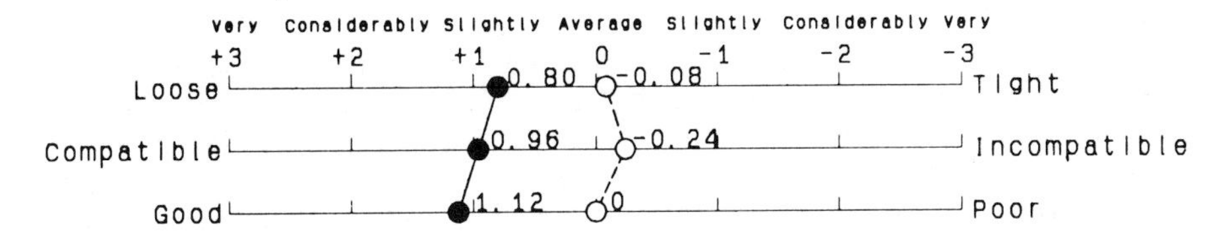

Smoothness

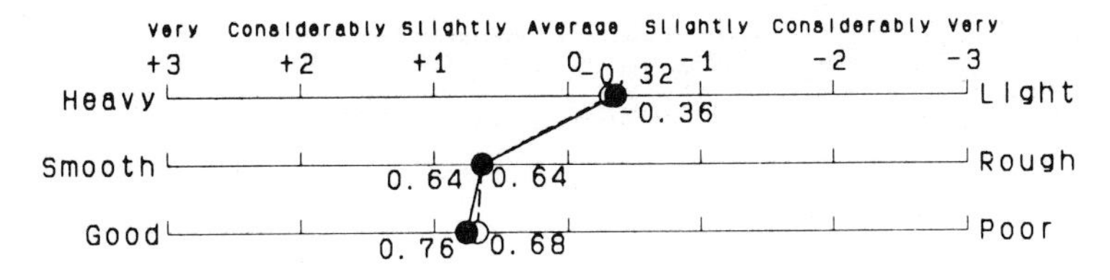

Figure 26.15 Seat comfort evaluation of Dynafit chair

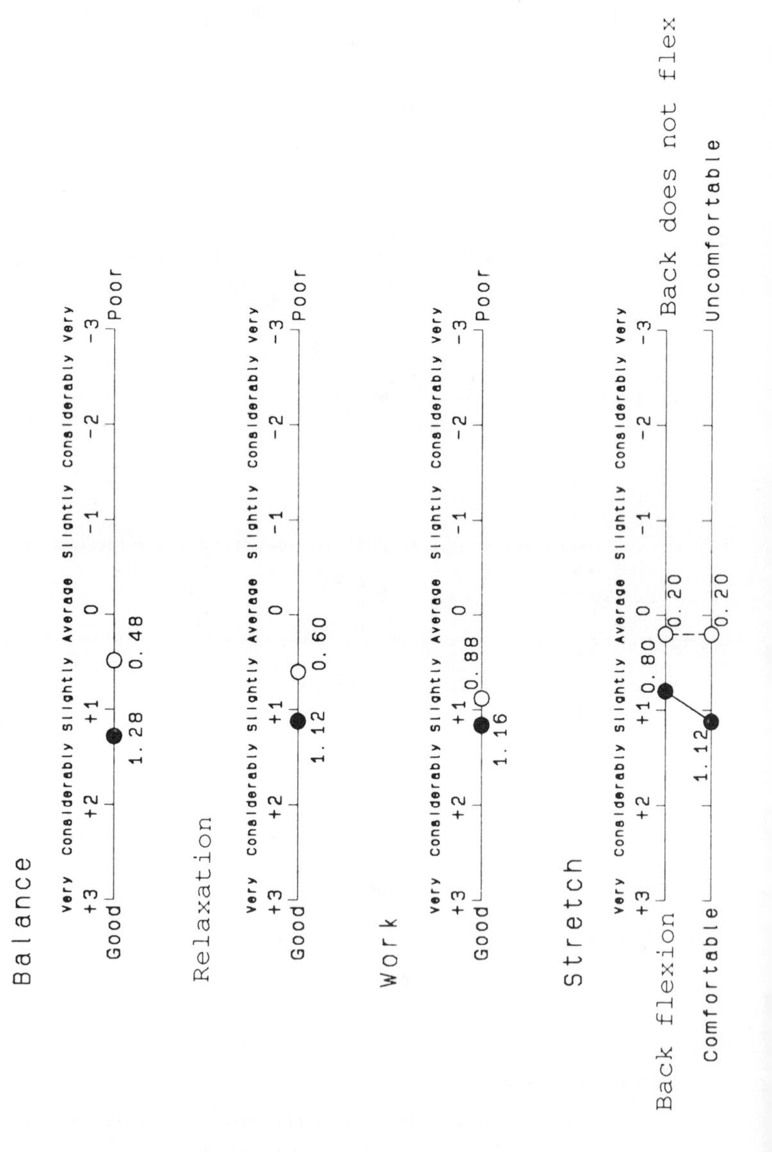
Balance
Very Considerably Slightly Average Slightly Considerably Very
+3 +2 +1 0 -1 -2 -3
Good
Poor
1.28
0.48
Relaxation
Very Considerably Slightly Average Slightly Considerably Very
+3 +2 +1 0 -1 -2 -3
Good
Poor
1.12
0.60
Work
Very Considerably Slightly Average Slightly Considerably Very
+3 +2 +1 0 -1 -2 -3
Good
Poor
1.16
0.88
Stretch
Very Considerably Slightly Average Slightly Considerably Very
+3 +2 +1 0 -1 -2 -3
Back flexion
Back does not flex
0.80
0.20
Comfortable
Uncomfortable
1.12
0.20

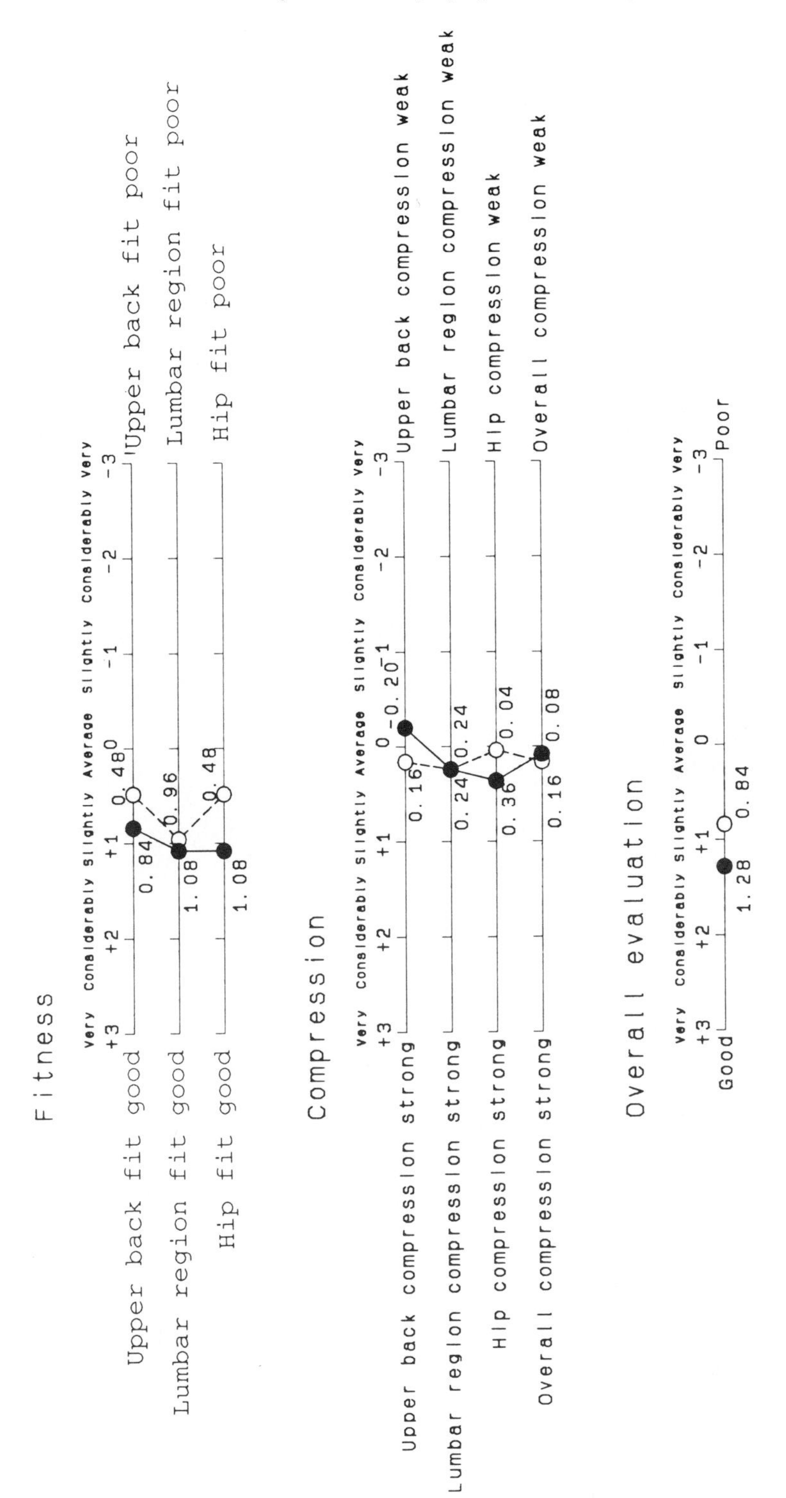
Fitness
Very Considerably Slightly Average Slightly Considerably Very
+3 +2 +1 0 -1 -2 -3
Upper back fit good
Upper back fit poor
Lumbar region fit good
Lumbar region fit poor
Hip fit good
Hip fit poor
0.84
0.48
1.08
0.96
1.08
0.48
Compression
Very Considerably Slightly Average Slightly Considerably Very
+3 +2 +1 0 -1 -2 -3
Upper back compression strong
Upper back compression weak
Lumbar region compression strong
Lumbar region compression weak
Hip compression strong
Hip compression weak
Overall compression strong
Overall compression weak
-0.20
0.16
0.24
0.24
0.36
0.04
0.16
0.08
Overall evaluation
Very Considerably Slightly Average Slightly Considerably Very
+3 +2 +1 0 -1 -2 -3
Good
Poor
1.28
0.84

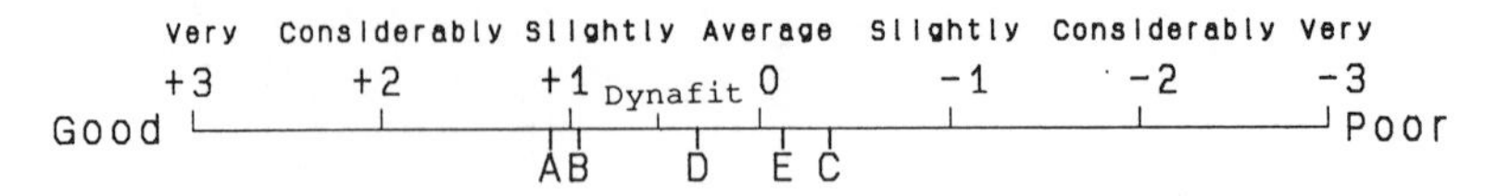

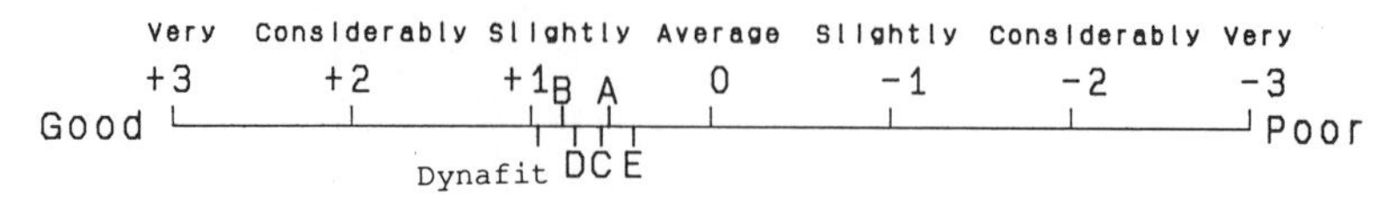

Figure 26.16 Comparison of overall scores of the sensory evaluation. Chairs A to E are made by other manufacturers

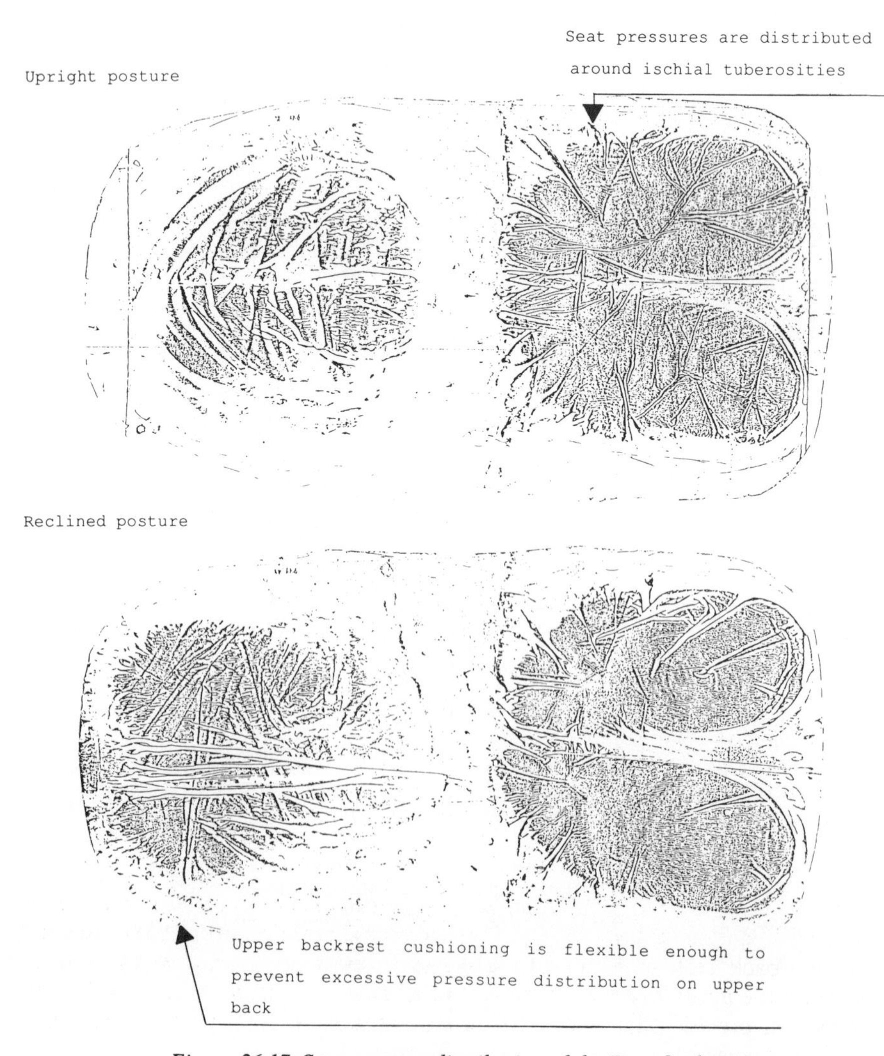

Figure 26.17 Seat pressure distribution of the Dynafit chair

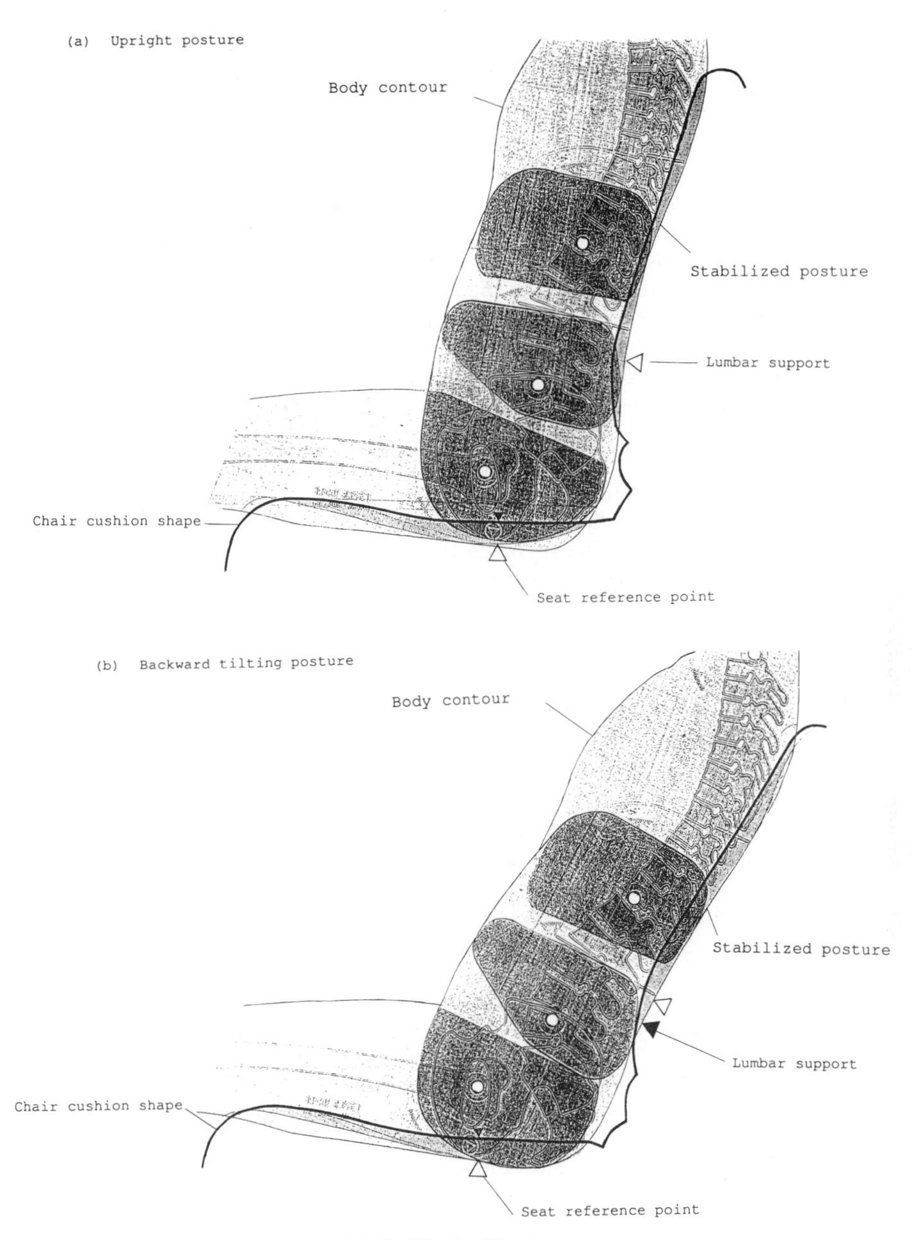

Figure 26.18 Check of lumbar support

Seat pressure distributions do not differ appreciably between the upright and reclined positions, and are approximately optimum in each posture. Seated back rest pressures are distributed more uniformly when reclined, because the upper back rest flexes with the user. This suggests that the Dynafit chair can provide adequate pressure distribution despite dynamic changes in back rest position.

Postural stability

Postures were found stable when the user sinks into, and compresses, the seat cushion. A gypsum mould was used to measure the user's postural stability when seated upright and reclined.

The upright posture is characterized by a seat cushion compression of 25 mm at the seat reference point (SRP), and a back rest cushion compression of 16 mm at the centre of the lumbar support (vertically). In this case, the users adjusted the seat pan angles to 8°, and the seat pan–back rest angles to 98°.

The reclined posture was characterized by a maximum seat cushion compression of 30 mm at the seat reference point, and back rest cushion compression of 16 mm at the centre of the lumbar support (vertically). When the postures were stabilized users adjusted the seat pan angle to 15°, and the seat pan–back rest angle to a mean of 116°. The average change in the torso and back rest angles was 7° and 25°, respectively. The synchro-ratio is about 1 : 3.6, which approximates that of the three-dimensional model.

The two-dimensional model was subsequently set to duplicate the seat cushion profile of user's final stable posture, to evaluate the degree of back rest lumbar support (Figure 26.18). When the user is reclined, the back rest lumbar support deflects downward relative to the user's back, and the locus of support moves to support the user's lumbar (L2 and L3) vertebrae. This beneficial effect is associated with the compensatory motion of the lumbar support.

Sensory evaluation of fabrics

Sensory evaluation was used to guide the selection of the Dynafit upholstery. This process is complicated by the fact that seat fabrics different to different body parts. The semantic differential and paired conparison methods were employed to assess factile feel. Steps were taken to keep aesthetics from influencing the ratings. Characteristics of seat fabrics used in the experiment are depicted in Figure 26.19.

Eighteen adjective pairs were selected to describe the five factors of heat, moisture, smoothness, hardness, and preference, as Semantic Differential scales.

1. Heat: Warm–cold, glowing–chilling
2. Moisture: Sweaty–parched, wet–dry
3. Smoothness: Smooth–gritty, rough–fine, polished–uneven, slippery–scratchy, pliable–stiff
4. Hardness: Flexible–inflexible, soft–hard, swollen–crushed

The results of the sensory evaluation (*see* Figure 26.20) may be summarized as follows:

1. Unlike the woven fabrics, the knit fabrics are felt as warm, flexible, and soft. Smoothness is affected by yarn number. These subjective responses must be compared with the objective physical properties of upholstery fabrics.
2. The knit fabric of the Dynafit chair was rated more highly than the upholstery fabrics of the other chairs.

Evaluation of dynamic seating

A dynamic seating with the interconnected mechanisms and springs was developed by using such ergonomics techniques as the analysis of dynamic user

Chair	Fabric	Type	Material	Yarn number and count
Dynafit		Knit	Acrylic 75% Ester 15% Wool 10%	No. 7 and 13 gage
A		Woven	Nylon 60% Wool 40%	No. 10, 2 and 3 double-folded yarns, and 21 X 16
B		Knit	Polyester 92% Nylon 8%	17 gage
C		Woven	Acrylic 100%	No. 3.5 , and No. 6, double-folded yarns, 11 X 11
G		Woven	Acrylic 100%	No. 11, and 20 X 21

Figure 26.19 Upholstery fabric samples

motions, evaluation of 'stopability', electromyography, subjective comfort ratings, measurement of seated pressure distribution and postural stability, and sensory evaluation of chair upholstery. These techniques helped to qualitatively and quantitatively evaluate the functional characteristics of the dynamic seating.

It is proposed that an ergonomics checklist be developed and based on these studies to serve as a tool for evaluating dynamic seating. This research agenda adds dynamic factors to existing techniques used with conventional chairs. Dynamic seating has dynamic, support, and basic functional requirements. General approaches for evaluating such requirements are listed in Table 26.1.

Checklists were also used to evaluate the Dynafit chair. The results may be summarized as follows.

Dynamic function:
The horizontal movement of the back rest of the Dynafit chair is synchronized with the sliding and rocking of the seat pan to stabilize leg/thigh positions as the user changes posture. As a result of vertical back rest movement the sitter is afforded considerable freedom of upper body movement, and receives continuous back support. When the sitter sits on the Dynafit chair, the back rest automatically locks in a preset position, and promotes postural stability. The maximum initial force required to move the back rest is 10 kgf, which is low enough to operate without the sitter's awareness. This engenders a sense that the chair moves automatically with the user. When the seat tension adjustment is set to the weight of the sitter, the Dynafit chair can move with the user.

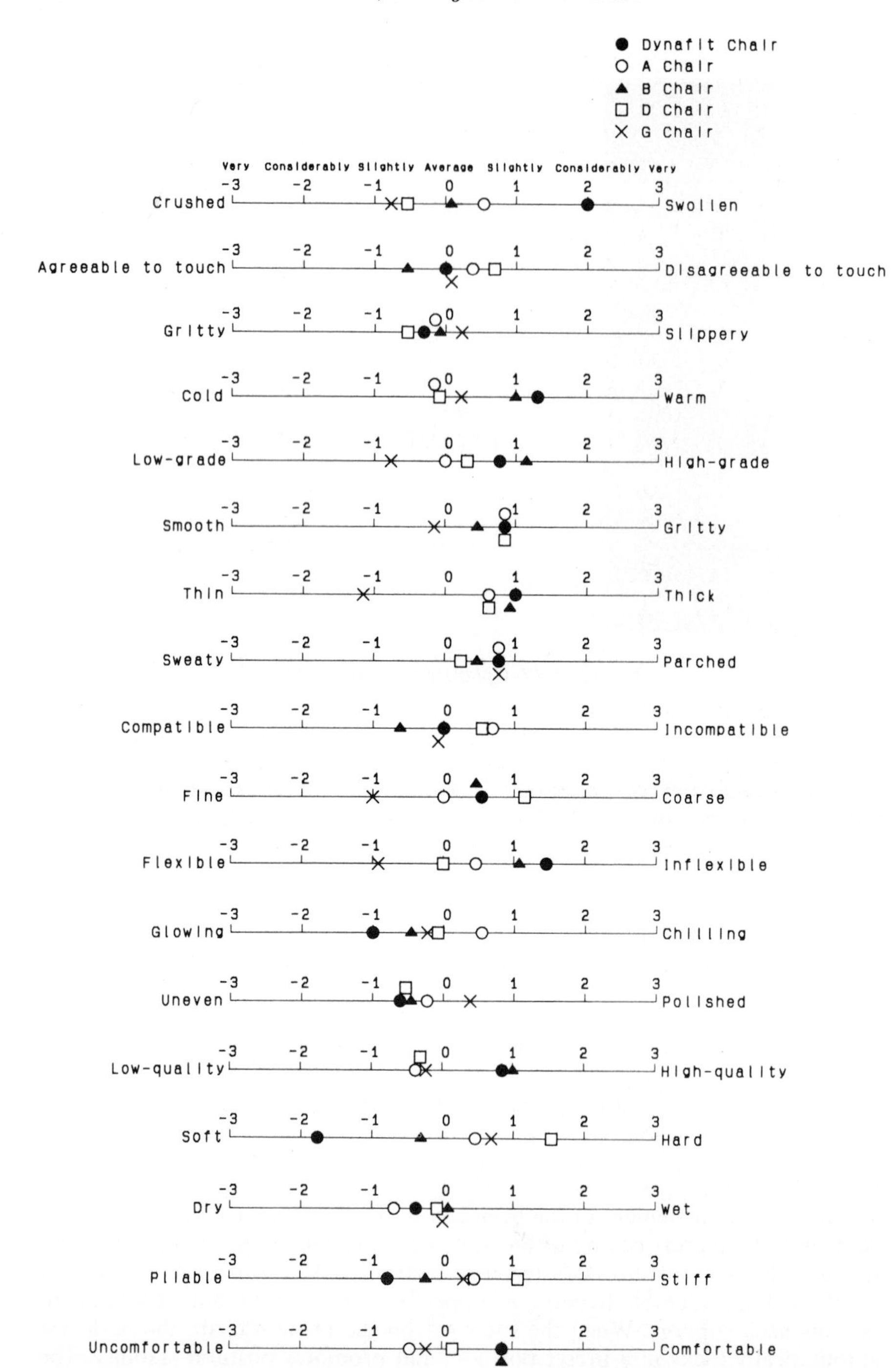

Figure 26.20 Sensory feel evaluation of upholstery fabric samples by Semantic Differential

Table 26.1 Ergonomics checklist for chairs

Function	*Check item*	*Yes/No Weight*
Dynamic function	A-1. Back moves in unison with seat	1. Interlocked seat-back movement
	A-2. Chair moves smoothly	2. Smoothness of motion
	A-3. Chair moves appropriately	3. Appropriateness of movement
	A-4. Back rest can stop in any desired position	4. 'Stoppability'
	A-5. Back rest can stop and move without sitter operation	5. No sitter operation
	A-6. Back rest is stable in stop position	6. Postural stability
	A-7. Back rest motion start force is small	7. Back rest motion start force
	A-8. Chair can accommodate sitters of different body weight	8. Body weight difference accommodation
	A-9. Miscellaneous	
Support function	B-1. Back size and angle are appropriate	9. Dimensional adaptability
	B-2. Back cushion is properly contoured	10. Cushion shape
	B-3. Back properly cushions sitter	11. Cushioning property
	B-4. Back properly fits sitter	12. Fitting property
	B-5. Back properly holds sitter	13. Holding property
	B-6. Body pressure distribution and compression are adequate	14. Body pressure distribution
	B-7. Lumbar support point responds to back movement	15. Lumbar support
	B-8. Chair is flexible enough for horizontal swaying and micro-motion	16. Flexibility
	B-9. Miscellaneous	
Basic function	C-1. Chair helps sitter to maintain correct posture	17. Posture
	C-2. Chair imposes no adverse muscular load on sitter	18. Muscular load
	C-3. Chair is adjustable to suit body of sitter	19. Adjustability
	C-4. Chair feels good to sitter	20. Feel
	C-5. Chair is stable (does not turn over)	21. Stability
	C-6. Chair facilitates seating, standing and moving of sitter	22. Mobility
	C-7. Chair is free from projections	23. Safety
	C-8. Chair is easy to operate	24. Operability
	C-9. Miscellaneous	

Table 26.2 Evaluation of Dynafit chair by ergonomics checklist

Function	*Check item*	*Evaluation*	*Remarks*
Dynamic function	A-1. Interlocked seat-back movement	Seat sliding and synchro rocking	Increased openness of thigh-torso angle
	A-2. Movement smoothness	Smooth horizontal movement	High mobility
	A-3. Movement appropriateness	Leg and thigh stability, and horizontal movement	Small head movement as suited for VDT work
	A-4. Stopability	Back rest stops in any desired position	
	A-5. No sitter operation	If body weight of sitter is pre-set, back rest automatically stops in suitable position	
	A-6. Postural stability	Load difference of 6–13 kgf	Posture stable in normal position
	A-7. Back rest starting force	Body weight Force 40 kgf: 3–6 kgf for tilting 2–4 kgf for returning 63 kgf: 2–8 kgf for tilting 4–5 kgf for returning 80 kgf: 1–10 kgf for tilting 3–6 kgf for returning	Back rest starts to move under force of 10 kgf or less
	A–8. Body weight difference accommodation	Lever is turned to suit body weight 40–50 kgf: 0–2 turns 50–60 kgf: 4–6 turns 60–80 kgf: 8–10 turns	Less force is required to return back rest

Support function	B-1. Dimension adaptability	Approximately good	
	B-2. Cushion contour	Transverse section of seat: Convexity at centre Longitudinal section of seat: Swell at front edge Horizontal section of back rest: Convexity at centre Shape of back rest: Saddle	Good 'fit'
	B-3. Cushioning property	Seat: Good (slightly soft) Back rest: Good	Good long-term evaluation
	B-4. Fitting property	Upper back: Good Lumbar region: Good Hip: Good	Good long-term evaluation
	B-5. Holding property	Good	Good-long term evaluation
	B-6. Body pressure distribution and compression	Body pressure distribution; Dynamically good Compression Upper back: Slightly weak Lumbar region: Ordinary Hip: Ordinary Overall: Ordinary	No special body pressures are observed in front seat edge and upper back rest in backward tilting posture
	B-7. Lumbar support point	Lumbar region between L2 and L4 vertebrae is continuously supported in upright and backward tilting postures	Lumbar support point moves in backward tilting posture
	B-8. Flexibility	Backrest deflects horizontally and from backward tilting posture	

Basic function	C-1. Posture	Eyes and acromions do not greatly move away from front edge of desk	This condition is suited for VDT work Back rest bends to help sitter adopt rest posture
	C-2. Muscular load	Muscular load is low when sitter moves	
	C-3. Adjustability	Seat height adjustable over range of 375 to 465 mm Lumbar support point adjusts according to load and cushioning property	
	C-4. Feel		
	C-5. Stability	Fine arm base assures Stability	
	C-6. Mobility	No special interference with movement of sitter. Rotation is also good	
	C-7. Safety	Free from projections and rounded on whole	
	C-8. Operability	Easy to operate	

Support function:
The Dynafit chair accommodates office work in the upright posture. It increases the seat–back rest angle, and imparts a sense of freedom of movement. The seat cushioning is contoured, and was rated highly in the sensory evaluation. The Dynafit chair rated highly on the fitting, compression and support after extended sitting. Seat pressures are distributed during dynamic changes in movement. To compensate for the displacement of the lumbar support relative to the user when leaning back, the Dynafit incorporated a movable lumbar support mechanism to continuously support the lumbar region between the L2 and L3 vertebrae. The back rest was moulded as a flexible plastic shell to accommodate horizontal movements.

Basic function:
The Dynafit chair accommodates postures and minimizes muscular loads. The seat pan height, tension, back rest angle, and lumbar height adjust. The chair has a stable five-arm base, is free from any potentially unsafe projections, and is easy to operate.

The evaluation of the Dynafit chair is shown in Table 26.2.

27

The functional requirements of forward-tilting office chairs

Mitsuaki Shiraishi and Yoshiyuki Ueno

Introduction

Humans are distinguished from four-legged animals by the shape of their spine and larger pelvic bones. Unlike animals, we also sit on chairs for many hours a day. This posture is perceived as comfortable because (among other factors) leg muscle activity is reduced. However, when sitting, stresses on the spine increase and the spine deviates from its natural S-shaped configuration (Jiro Kohara *et al.*, 1975; Mandal, 1974, p. 33). Yet, most office chairs have seats that tilt backward, which is frequently inappropriate for performing work. Some reasons include:

1. a long-held assumption that chairs are intended for rest; and
2. frequently, no clear distinction is made between chairs for work and chairs for rest.

When the Japan Industrial Standard (JIS) was introduced, back rests were recommended to reduce stress on the spine.

Aim

Alternative seating approaches suggested to reduce stresses include (Mandal, 1974, p. 30):

1. forward seat pan tilt to increase the thighs and rotate the pelvis forward; and
2. back rest contouring, particularly lumbar supports, to extend the lumbar curvature.

The objective of this study was to evaluate forward-sloping seats for the office and their relation to inclined work surfaces by analysing the kinds of posture assumed during office work including the frequency of hunching forward (hereafter, the forward-leaning posture).

Method

Experiments were conducted to examine seat pan angles and associated back rest positions. Inclined work surfaces and their relation to office chairs were also investigated.

Seat angles

Fixed seat condition

Subjects were seated on chairs with seat angles which remained fixed when they changed their postures (Figure 27.1). Changes in subjective comfort, posture, seat pressure distributions and muscular activity were monitored. Subjective ratings consisted of comfort ratings associated with variations in seat angles.

Electro-myograms were used to measure the activity of the abdomen, back and shoulder muscles (rectus abdominis, erector spinae, trapezius, and the rectus femoris). Seat angles in the experiment ranged from 4° back to 6° forward, with two pitches. 17 males and 15 females participated in the study.

Chair with variable seat angle

Users sat on chairs with seat angles which varied with their postures (Figure 27.2). For this condition, the time spent sitting in different seat angles was measured as subjects performed office work. Subjects rated their subjective comfort and sense of postural stabilities associated with different seat angles on a five point scale.

Experiments were conducted to analyse seat design factors, such as seat tension and seat pivot point. Three seat tensions were used: 1 kg/mm, 2.5 kg/mm, 7 kg/mm. The seat pivot point could adjust from 12 cm forward to 14 cm behind the seat reference point (SRP).

A total of 30 subjects were employed; these were 18 males and 12 females.

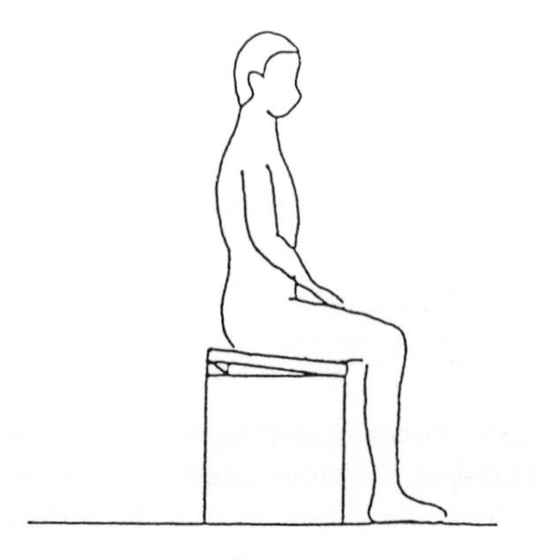

Figure 27.1

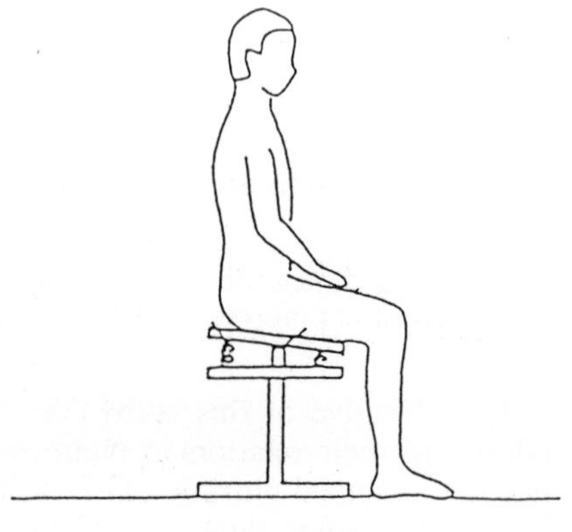

Figure 27.2

Back rest

Lumbar height

The experimental seat device is depicted in Figure 27.3. The seat pan angles were set at zero, 2° and 4° forward tilt and 2° and 4° backward tilt. Lumbar heights were 15 cm, 17 cm, 19 cm, 21 cm and 23 cm from the SRP. These were varied randomly.

Subjects were assigned an English-Japanese translation task for 10 minutes and asked to rate the chair's performance on a five point scale from 'high–low' and 'good–bad'. A total of 25 subjects participated in the experiments.

Relation between seat and back rest (lumbar point)

It is often assumed that seat back rests should support the spine while leaning forward during office work. In order to test this hypothesis, we examined whether the back rest should move in synchrony with the seat to support the person leaning forward, or move independently from the seat pan to remain close to their backs as they lean forward.

Two chairs were used as experimental devices; one with the seat attached to the back rest (device A) and another with the seat separated from the back rest (device B). The subjects performed 10 minutes of English–Japanese translation while seated in the experimental chairs, then scored the chairs on 'good–bad' five point scale. Twenty subjects participated in the experiments.

Fore–aft position of the lumbar support

The position of the lumbar support during forward-tilting was examined. The experiments were conducted under the conditions described below. These examined subjective ratings, muscular activity, posture change, and body pressure distribution when the relative height and fore-aft position of the lumbar

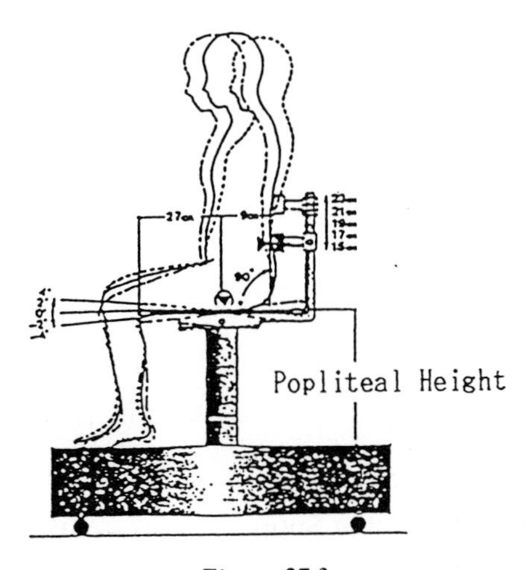

Figure 27.3

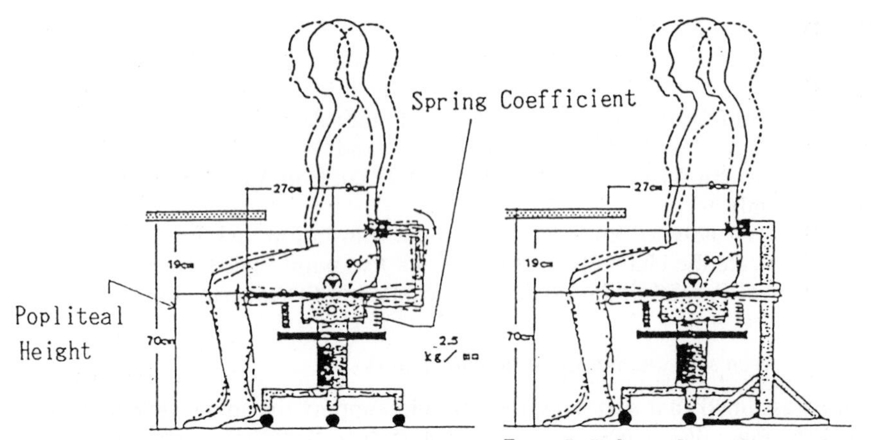

Figure 27.4

support were varied according to each seat angle. Subjective evaluations consisted of ratings of the positions in front of and behind the lumbar support on a good–bad five point scale.

Seat angle: 0°, −2°, −4°, −6°,
(minus signifies forward seat inclinations)

Back rest height: 15, 16, 17, 18, 19 cm
Positions in front of and behind back rest lumbar point: −3.5 ∼ +2.5 (0.5 cm pitch)

Thirty subjects were involved in the subjective evaluations and in other experiments.

Forward-tilting chair and desk

Most Japanese offices are big rooms laid out so that workers sit facing each other, and use desks of identical size and height, regardless of the user's height or sex.

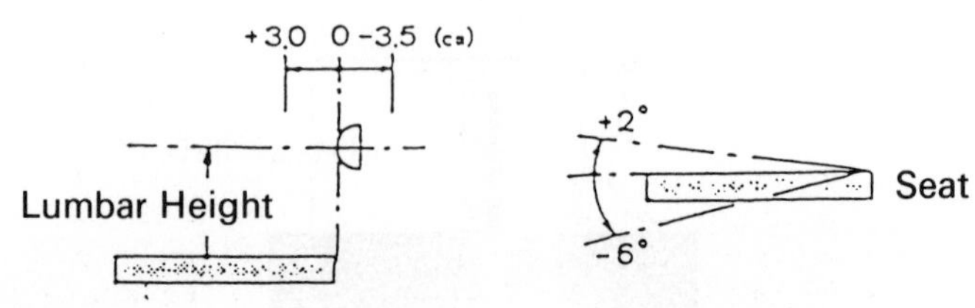

Figure 27.5 Shows the lumbar height and fore–aft positions

However, because office desks must be evaluated in conjunction with the chair, the relationship between the gradient of the desk surface and that of the seat is of critical importance. Desks with tilting surfaces (Masamitsu Ooshima *et al.*, 1971; Mandal, 1974 p. 28; Ichiro Yano, 1979) have been favoured; these now appear to be more useful than previously believed, particularly when used with forward-tilting chairs. Experiments were conducted to evaluate desk surface inclinations when used in conjunction with forward tilting chairs.

Experimental chairs and desks were manufactured. These chairs had seat pans and back rests which moved in synchrony, up to 4° forward and 5° back. The desks had adjustable surfaces which inclined from zero degrees (horizontal) up to 20°. The tests were carried out as follows:

Subjective evaluations: the seats of the chairs were set at 4° (forward inclination), 0° (horizontal) and 3° (backward). The horizontal and backward-tilting seat pans were compared with ones that tilted forward. The desk surface inclination was varied with each seat angle under three task conditions: no work, writing and reading. In the case of no work, the subjects were asked to rate the chairs and desks in terms of (touching) comfort, and writing or reading comfort. The degree of tilt of the desk surface was also rated in comfort. Ten subjects participated.

Changes in position, degree of muscle activity: We investigated the forward arm pressure distribution, change in position (during a 60 minute writing period, recorded by an 8 mm camera), muscle activity (during a 60 minute writing period involving four positions (abdomen (the rectus abdominis), back (erector spinae), shoulder (trapezius) and arm muscles (bicase brachii)) with three participants). Object Stability: A problem associated with slanted surfaces is that objects roll off. The extent of object movement slipping associated with different desk inclinations (*see* Table 27.1) was recorded.

Results

Findings of the study

Findings of the office study are shown in Table 27.2. Most subjects sat so that their upper bodies leaned forward while performing conventional office tasks, with lower limbs in a forward position. This forward-oriented posture was adopted in about 70 per cent of the office operations.

Table 27.1 Supplies used for experiments

1. pencil (circular) horizontal	11. report paper
2. pencil (hexagon) horizontal	12. pocket calculator
3. pencil(circular) vertical	13. scissors
4. pencil (hexagon) vertical	14. stapler
5. ball-point pencil	15. cutting knife
6. mechanical pencil	16. hole puncher
7. rulers (bamboo and plastic, 30 cm)	17. inking pad
8. triangle rulers	18. book (paperback)
9. ink bottle	19. data book
10. eraser	

Table 27.2 Ratio of different postures observed in office

	upper body	Bottom	Legs
office operations	1.6 Bending backward; Up-right 25; Bending forward 73.4	Forward 3.4; 17.2; Center 79.4	Backward 7.9; 9.9; Forward 38.8; Upright 43.4
conversation and rest	15.6; 47.8; 36.6	2.2; Backward 23.3; 74.6	10.1; 16.5; 36.3; 37.1
typing and writing	14.5; 46.5; 3.9	32.7; 67.3	1.6; 98.4
others	7.8; 44.5; 47.7	1.7; 12.5; 85.8	4.9; 20.5; 42.4; 32.4

In our other survey, the forward-leaning posture was frequently observed in women performing input duties in VDT operations.

Findings of the tests

Fixed seat angle chair

Subjective ratings: chairs with their seats tilting 2–3° forward were rated most positively. The greater the forward inclination, the more unstable the subjects rated their posture.

Postural variation: few postural changes were recorded when the seat was inclined forward about 2° (*see* Figure 27.6).

Seat pressure, distribution: seated pressures were evenly distributed when the seat inclined forward about 2°. Stress was observed in the front of the femur

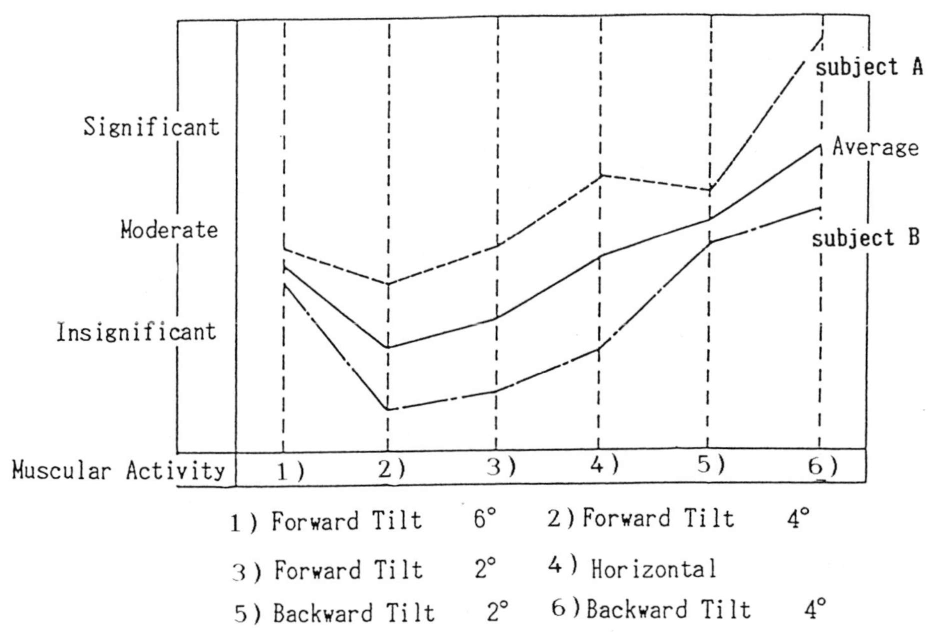

Figure 27.6

region of the person bending forward on seats reclined 4° backward. High pressure levels were recorded at the centre of the femur region when the seat slanted 6° forward. These increased stresses probably resulted as the user tried to avoid slipping forward.

Muscular activity: Figure 27.7 depicts EMG records of the two subjects. Muscular activity levels were lowest when the seat pan was tilted forward between 2° and 4°.

Chair with variable seat pan angle

Most subjects preferred seat inclinations from the horizontal (zero degree) to 4° forward. Figure 27.8 depicts recorded seat angles when subjects using the experimental chairs performed reading and writing tasks. Many set their seat pan angles from 2° to 4° forward.

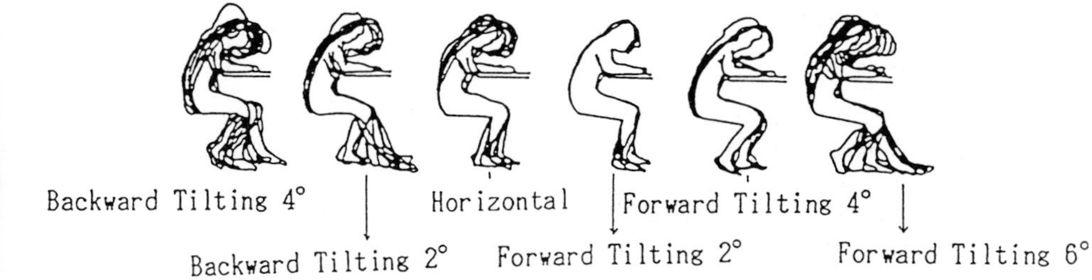

Figure 27.7

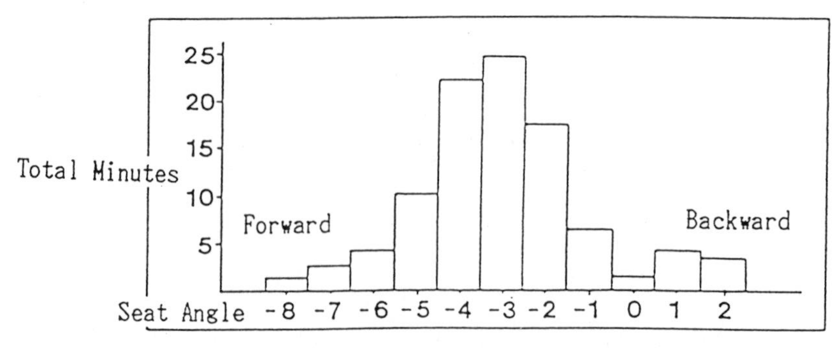

Figure 27.8

Subjective ratings indicated the most comfortable location for the seat angle pivot ranged from 2 cm in front of and behind the Seat Reference Point (SRP). 2.5 kg/mm was rated the most comfortable seat tension. Further evaluations on seat tension are now in process.

Back rest lumbar support

Lumbar height

As shown in Figure 27.9, no single 'good' back rest height was found. Back rest heights of 19 cm were preferred for forward tilting seats and 21 cm for seats that were horizontal or reclined. This 19 cm height was somewhat lower than the range of 20 to 25 cm set forth in the JIS (Japanese Industrial Standards).

Relation between seat pan and lumbar supports

When subjects sat in chairs with independent back rests, they frequently did not use the back supports (Figure 27.10). These findings are in agreement with

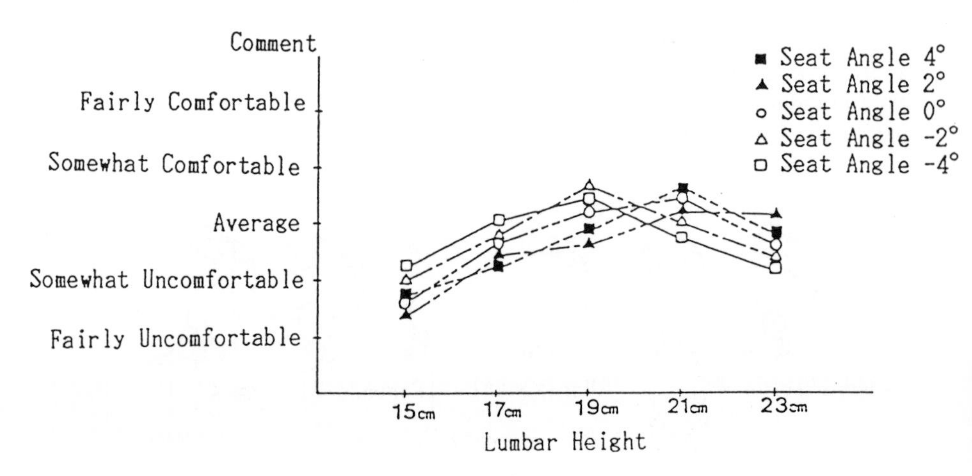

Figure 27.9

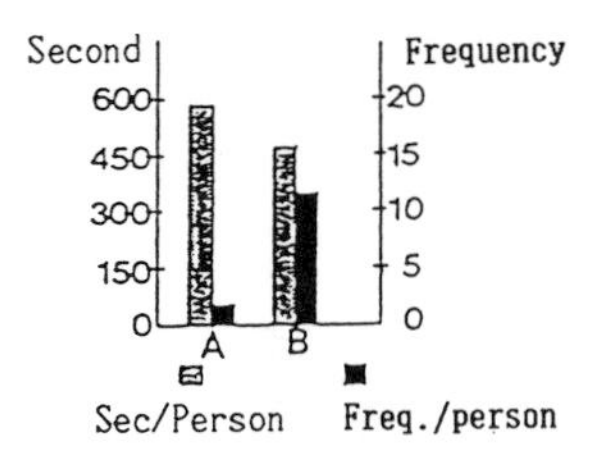

Figure 27.10

their subjective ratings; most of the subjects indicated that they preferred the all-in-one type chair since their backs were better supported (Figure 27.11).

Fore-aft position of lumbar support

Subjective ratings: these are depicted in Figure 27.12. The back rest moves forward when the seat pan inclines forward, and its height lowers somewhat relative to that of the seat.

Postural variation: postures changed the least when seat pans inclined forward 4°. Muscular activity was lowest when seat pans were horizontal or inclined forward 4° (Figure 27.13).

Seated pressure distribution: seated pressures were most evenly distributed when seats were tilted 2° or 4° forward.

Forward-tilting chair and desk

Subjective ratings: preferred work surface inclinations had no apparent relation to seat pan angles, and approximated 10° during rest and when writing.

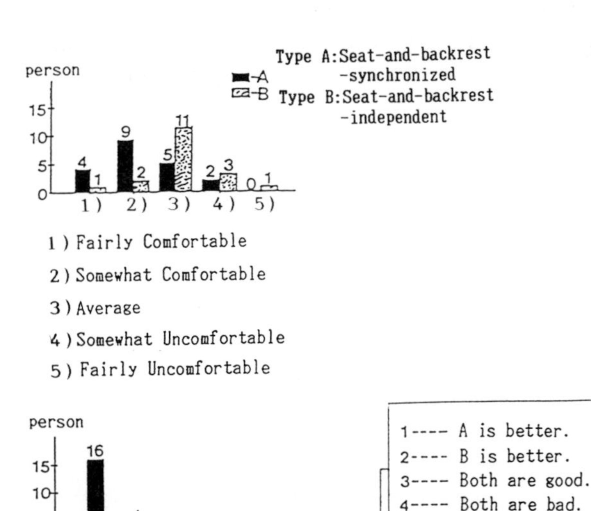

Figure 27.11

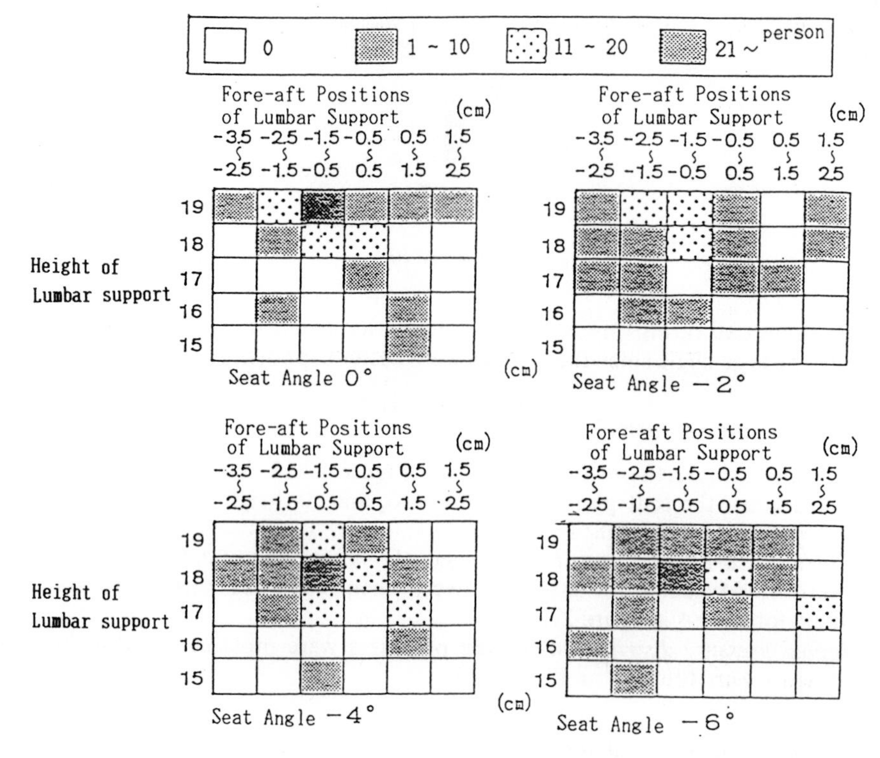

Figure 27.12

Users preferred more acute angles of the slanted work surface when reading (Figures 27.14, 27.15 and 27.16). Pressure distribution on front of arm rest: no difference was observed, but pressures were more evenly distributed when the work surface was inclined 7° than 0° or 15°. (Figure 27.17).

Muscular Region \ Seat Angle	0°	−2°	−4°	−6°
Trapezious	—	—	—	+
	----	----	----	↗↘
Erector Spinae	+	++	+	++
	→	↗		↗
Rectus Femoris	—	—	—	—
	----	----	----	----
Gastrocnemius	—	—	—	+
	----	----	----	----
Overall Rating	+	++	+	+++

Figure 27.13

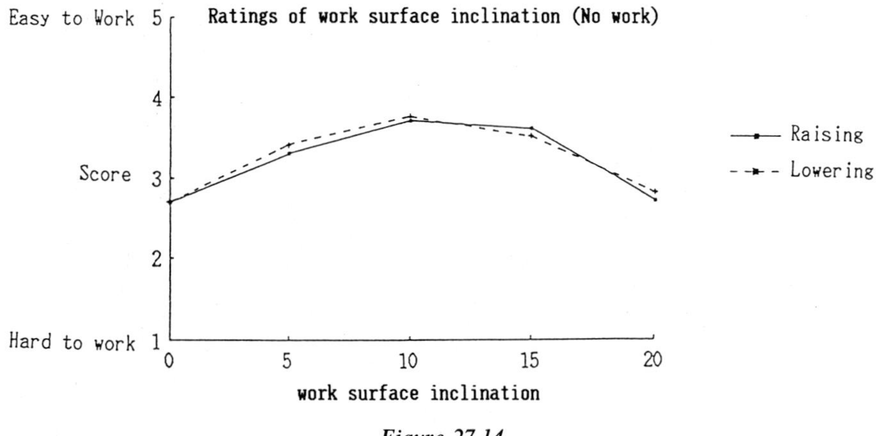

Figure 27.14

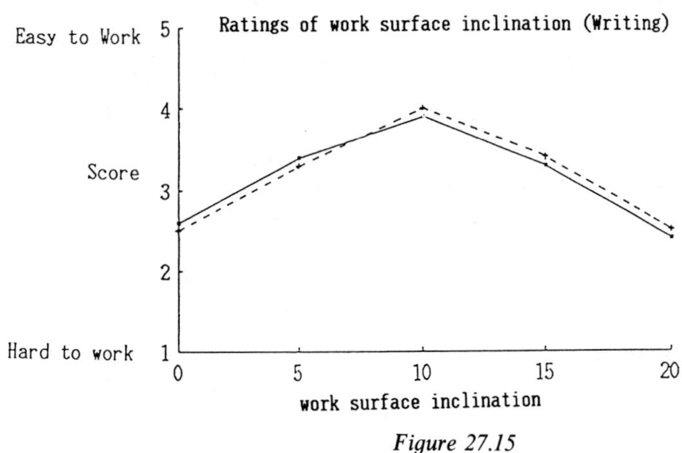

Figure 27.15

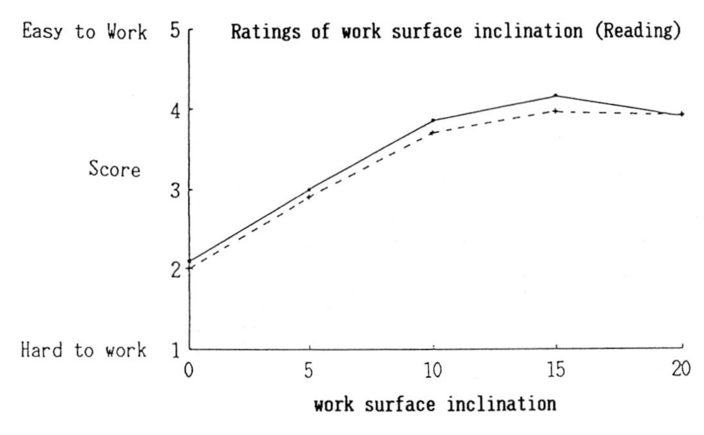

Figure 27.16

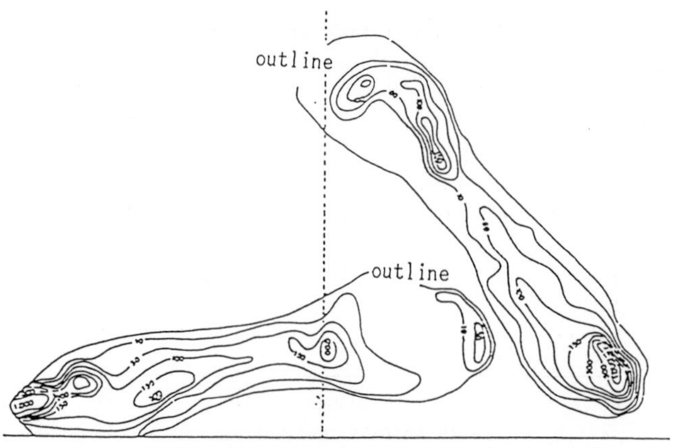

Inclination of work surface 0 (horizontal)

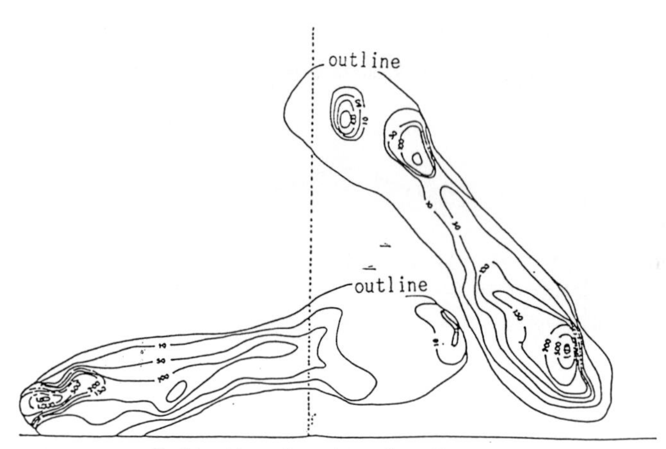

Inclination of work surface 7

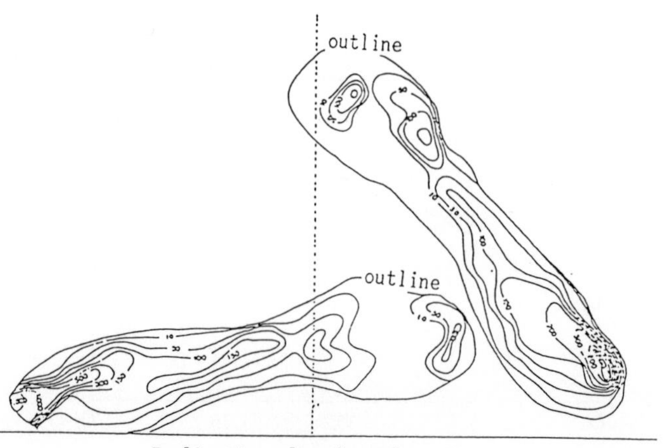

Inclination of work surface 15

Figure 27.17

Muscular activity levels: EMGs were recorded of the trapezius muscles. When the desk surface was inclined at 0° (horizontal) and 7°, muscular tension became noticeable after approximately 40 minutes of writing. When work surfaces were inclined at 15°, high levels of muscular tension were recorded as soon as writing was initiated.

Postural variation: the more acute the work surface gradient, the less the posture changed. This may, in part, be attributed to the fact that there was less available space left for the free movement of the user's upper body.

Stability of objects: most of the objects placed on a desk surface covered with vinyl cloth did not slide or roll down if the inclination was within 15°. However, with inclined work surfaces one must also consider whether coffee cups and other loose objects are stable, and the psychological effects of the tilt.

Summary

1. The optimum seat pan angle for traditional office work (reading and writing) is between 0° (horizontal) and 4° forward. However, these angles are specific to stable postures to sit on the seat.
2. The back rest angles should move in synchrony with the seat pan inclination.
3. The work surface angle should generally range from 5°–10°, irrespective of the seat pan angle.
4. Although most of today's chairs are able to tilt back, they should tilt forward as well.

This research was also used as a basis for the development of an office chair which tilts forward up to 4°.

References

Japan Standards Society, Steel Office Chairs, JIS-S1011-1969.

Jiro Kohara, Yoshichika Uchida, and Hidetaka Uno: Kenchiku shitsunai ningenkougaku, Kajima publishing, p112, 1975, Japan.

A. C. Mandal: The Seated Man, 1974, Denmark.

Masamitsu Ooshima, Yoshiomi Yamaguchi, Kunihiko Kimura, Tuneo Kubota: Studies of Chairs and Tables, Collected Symposium Papers for Posture, pp248–249, 1971, Japan

Ichiro Yano: Shisei to kenkou, Nihonkeizaishinbun publishing, p77, 1979, Japan.

PART XI

Industry perspectives

28

Evaluating office chairs with value analysis

Kozi Morooka and Hiroyuki Takeshita

Value analysis

Value analysis evaluates the benefits associated with the costs of imparting functions to a product. Wasteful components of the cost are identified and eliminated to optimize product usage. With value analysis, products are evaluated by comparing the benefits of each function with its cost. Value analysis identifies the function required of a product, establishes corresponding cost objectives, generates alternatives, and selects the one with the greatest value ratio. Wasteful functions are commonly eliminated, and the viability of market alternatives, methods of production, and price points are clarified.

This form of analysis was initially applied to materials, then gradually introduced to eliminate redundant functions during product design, simplify manufacturing methods, and avoid unnecessary costs associated with improper product engineering. Value design is advocated as a new concept to maximize the function of a product. As such, value, rather than cost, drives the initial stages.

Value analysis of seating

The costs of the experimental chair are shown in Figure 28.1. Responses to the subjective seating questionnaire (Table 28.1), its functions (Table 28.2), and cost analysis by function (Table 28.3) are presented. The functions of the chair are evaluated and ranked in the worksheet shown in Table 28.4. The function of each part is rated by comparing the associated cost with its benefit. The sum of the ratings of the parts that comprise the function of postural support (e.g., Parts 1, 3, 5, 13, 19, 22, 23, 29–31 and 33–48 as shown in Table 28.3) is obtained and divided by the total cost of the parts to determine the value ratio.

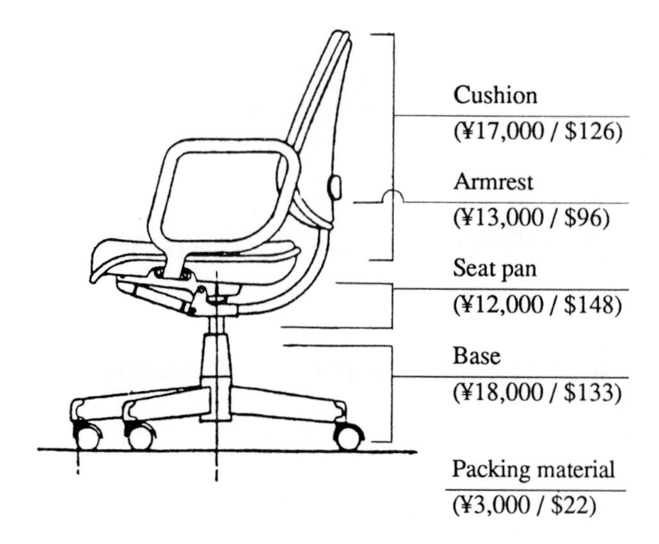

Figure 28.1 Seat feature costs

Table 28.1 Seat evaluation

Item	Score Good 4	3	2	1	0	−1	−2	−3	Poor −4
1. Seat comfort	*								
2. Seat pan height	*								
3. Seat pan size							*		
4. Seat cushion	*								
5. Backrest angle					*				
6. Backrest cushion			*						
7. Lumbar support				*					
8. Armrest size					*				
9. Armrest usage			*						
10. Armrest height	*								
11. Caster	*								
12. Ease of seat pan height adjustment	*								
13. Ease of backrest height adjustment	*								
14. Chair rotation	*								
15. Physical fatigue	*								
16. Operation			*						
17. Style			*						
18. Colour		*							
19. General	*								

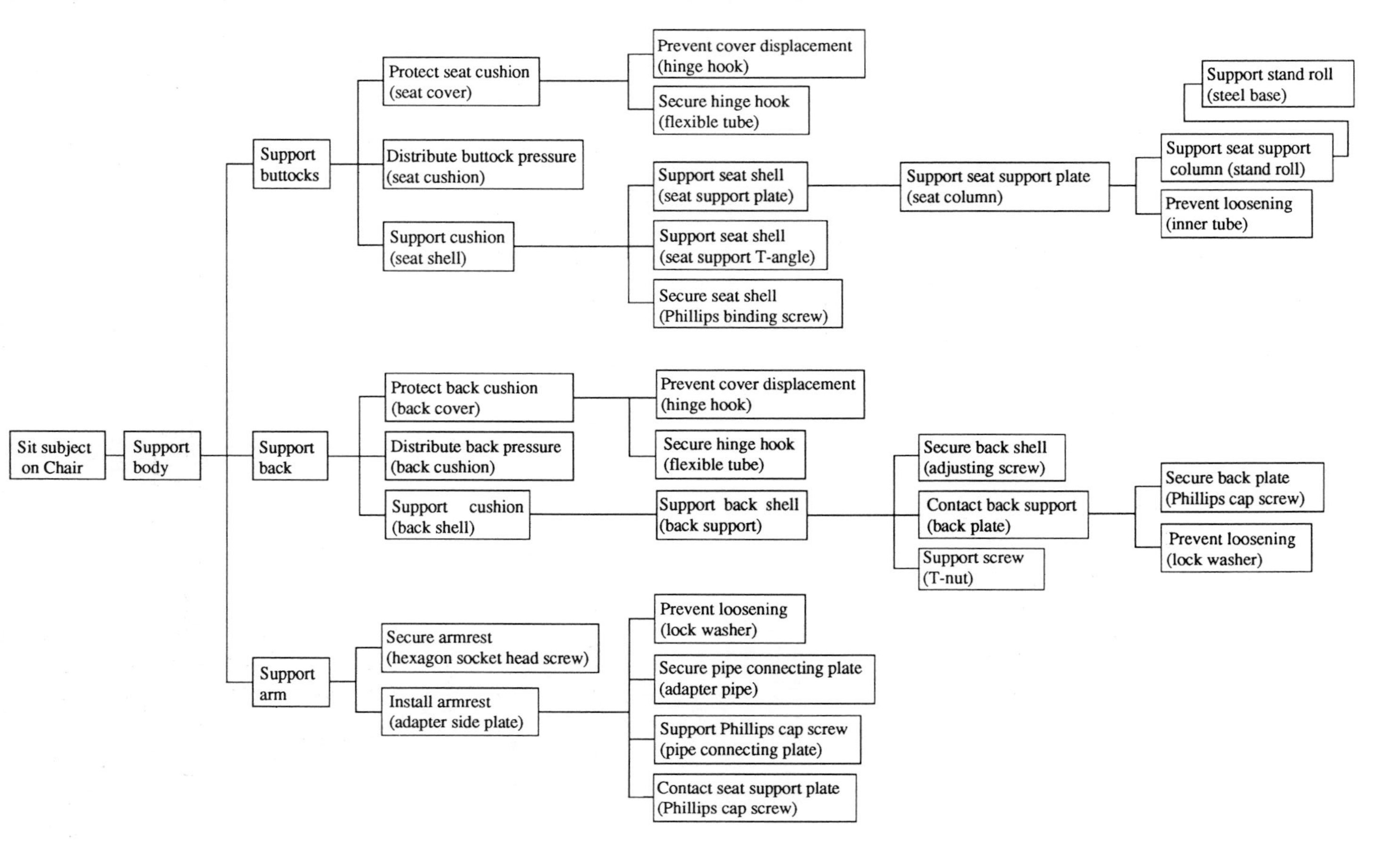

Figure 28.2 Hierarchy of chair functions

Table 28.2 Definitions of chair functions

No.	Component	Function
1	Stand roll	Support seat column
2	Caster clamp (5)	Install caster
3	Steel base arm (5)	Support stand roll
4	Cover ring cap (5)	Protect foot
5	Inner tube	Prevent loosening
6	Caster (5)	Facilitate movement
7	DX bearing (2)	Facilitate rotation
8	Spring bush	Lock compression spring
9	Spring guide (2)	Protect compression spring
10	Compression ring	Absorb shock
11	Jump ring	Prevent loosening
27	Type C snap ring	Secure gas spring (upper)
28	Gas spring	Push seat column
29	Phillips binding screw (4)	Secure seat shell
30	Lock washer (10)	Prevent loosening
31	Phillips cap screw (6)	Secure back plate in contact with seat support plate
32	Thrust bearing	Facilitate rotation
33	Hinge hook	Secure cover
34	Flexible fixer	Secure hinge hook
35	Back cushion	Distribute back pressure
36	Back shell	Support cushion
37	Back plate	Contact back support
38	T-nut	Support adjusting screw
39	Adjusting screw	Secure back shell
40	Seat pan cushion	Distribute buttock pressure
41	Seat shell	Support cushion
42	Armrest (2)	Support arm
43	Adapter pipe (2)	Secure pipe connecting plate
44	Pipe connection plate	Support Phillips binding head/cap screw
45	Adapter side plate (2)	Install armrest
46	Hexagon socket head cap screw (4)	Secure armrest
47	Backrest cover	Protect back cushion
48	Seat pan cover	Protect seat cushion

Functional evaluation of chairs with value analysis

Seven types of chair that are commonly available in Japan were used in the experiment to evaluate functions (*see* Figure 28.3). The adjustability range of each chair was measured with a digital planimeter*, which traced the movement of luminous tape attached to the side of each chair (*see* Figure 28.4). The adjustability range of the seat and backrest of the seven chairs is shown in Figure 28.5. Table 28.5 lists the subjects' reported fatigue symptoms while sitting at each chair and writing. Table 28.6 shows the subjects' incidence of

* Footnote: an instrument used to measure the plane of a figure by tracing its area.

Part	F1	F2	F3	F4	F5	F6	F7	F8	F9	F10	F11	F12	F13	F14	F15	F16	F17	F18	F19	Cost
										Function										
1 Stand roll	300			55															100	455
2 Caster clamp (5)		20					5													25
3 Steel base arm (5)	950			300															750	2,000
4 Cover ring cap (5)			50																	50
5 Inner tube	40			20															90	150
6 Caster (5)		1,500					5,000													6,500
7 DX bearing (2)		40						10						250						300
8 Spring bush			140								60									200
9 Spring guide (2)			35								25									60
10 Compression ring			15								65									80
11 Jump ring			7								3									10
12 Type C snap ring			7								3									10
13 Back support	300				500												200			1,000
14 Installation shaft B (2)		30								20								10		60
15 Back support shaft mount		201.1								600										801.1
16 Grip (2)		10							15	15						30		30		100
17 Hand lever		7							8							15				30
18 Trip releasing catch		3								7								20		30
19 Seat support plate	270			130																400
20 Release lever		7								8								15		30
21 Spring pin		0.1							0.1	0.1						0.3		0.3		0.9
22 Seat support plate	1,000			600																1,600
23 Seat column	250			50																300
24 Installation shaft A		3								7								15		25
25 Back gas regulator		100								400								2,500		3,000
26 Type E snap ring (7)		5							25	25								15		70
27 Type C snap ring		0.3							0.7											1
28 Gas spring		100							400							2,500				3,000
29 Phillips binding screw (4)	15			25																40
30 Lock washer (10)	10				8	4											8			30
31 Phillips cap screw (6)	5				20	10											25			60
32 Thrust bearing		40						60							200					300
33 Hinge hook	4			8	8															20
34 Flexible fixer	20			40	40															100
35 Back cushion	70				130								400							600
36 Back shell	684				3,000															3,684
37 Back plate	25				75												200			300
38 T-nut	5				15												30			50
39 Adjusting screw	50				100												200			350
40 Seat pan cushion	70			130								400								600
41 Seat shell	684			3,000																3,684
42 Armrest (2)	1,000					5,000														6,000
43 Adapter pipe (2)	100					500														600
44 Pipe connection plate	34					100														134
45 Adapter side plate (2)	50					350														400
46 Hexagon socket head cap screw (4)	10					30														40
47 Backrest cover	5				45															50
48 Seat cover	5			45																50
Total	5,956	2,066.5	254	4,403	3,941	5,994	5,005	70	448.8	1,082.1	156	400	400	250	200	2,545.3	663	2,605.3	940	37,380

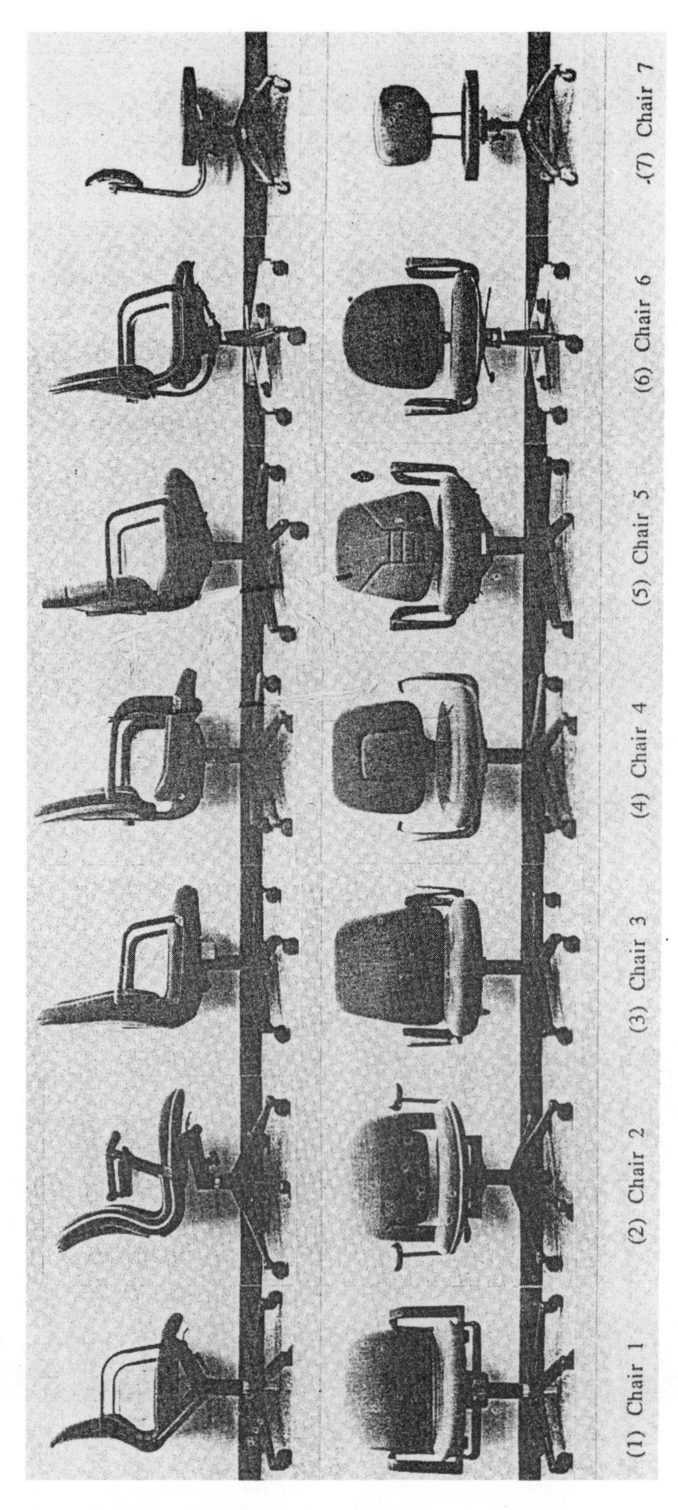

Figure 28.3 Seven types of chair used in value analysis

Table 28.4 Functional seat evaluation worksheet

Function	F	V = F/C	Ranking
F1: Support body	7273	1.22	9
F2: Allow free body motion	5309	2.57	13
F3: Protect body	1091	4.30	18
F4: Support buttocks	4363	0.99	8
F5: Support back	2546	0.65	7
F6: Support arm	364	0.06	1
F7: Move chair	2182	0.44	2
F8: Change direction	218	3.11	15
F9: Change height	1454	3.44	16
F10: Change angle	1454	1.34	11
F11: Absorb shock	1091	6.90	19
F12: Distribute buttock pressure	727	1.82	12
F13: Distribute back pressure	509	1.27	10
F14: Facilitate rotation*	109	0.44	2
F15: Facilitate rotation†	109	0.55	4
F16: Push seat column	1454	0.57	6
F17: Support back shell	2037	3.07	14
F18: Apply force	1454	0.56	5
F19: Support seat column	3636	3.87	17

* DX bearing, † Thrust bearing.

leaning back while writing. Table 28.7 shows the frequency of leaning back while seated. The results of Analysis of Variance on the number of times the subjects leaned back are depicted in Table 28.8.

Subject leg positions

Leg movements of the subjects were analysed. The most frequent leg position was associated with writing (*see* Table 28.9) with the leg oriented at the same angle, or symmetrical.

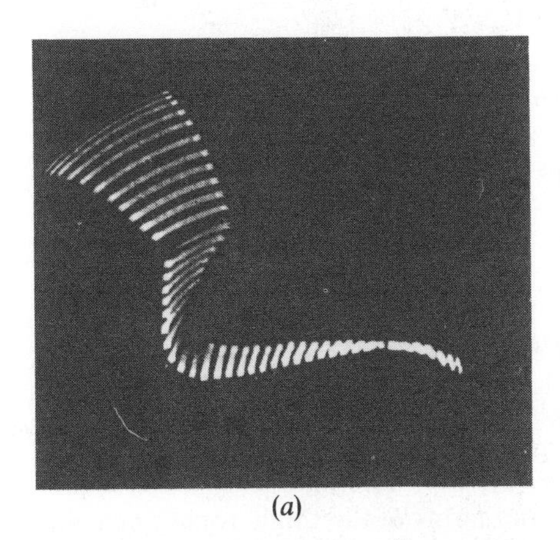

(*a*)

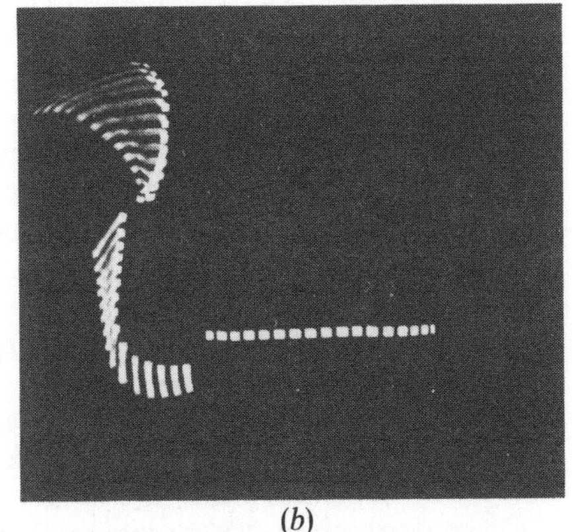

(*b*)

Figure 28.4 Tracing of luminous tapes attached to chair 1(a) and chair 2(b)

Table 28.5 Results of questionnaire survey of subjective fatigue symptoms of subjects seated on evaluated chairs

	Chair						
Question	1	2	3	4	5	6	7
1. Neck stiffness?			**	*	**	**	**
2. Neck tiredness?	*	*		*	**		
3. Neck pain?					**		
4. Shoulder stiffness?		**	*		*	***	
5. Shoulder fatigue?						*	
6. Shoulder pain?							
7. Back pain?				*			*
8. Hot in back?							
9. Itchy in back?							
10. Back fatigue?							
11. Lower back pain?			*				*
12. Lower back fatigue?							
13. Buttock pain?		*					*
14. Hot in buttocks?						*	**
15. Itchy in buttocks?							
16. Thigh pain?							
17. Thigh fatigue?		*					**
18. Itchy in thighs?							
19. Hot in thighs?							
20. Leg fatigue?							*
21. Leg numbness?							
22. Cold in legs?							
23. Arm pain?		*					*
24. Arm fatigue?		*					*
25. Hand pain?							*
26. Hand fatigue?	*	*					*
27. Abdominal compression?							

Asterisks denote the significant level between before task and after task.
***: 1%, **: 5%, *: 10%

The leg positions of seated subjects are shown in Figure 28.6. Data were transformed into transition matrices, shown in Table 28.10. The rows depict the right leg, and the columns the left leg position. The upper section of Table 28.10 for chair 1 shows that both legs assumed the condition of -4 three times, condition of -3 42 times, and condition of -4 and -2 three times, respectively.

Measurement of seated postures

Working postures are affected by the relationship between the axis of the torso and the direction of gravity, the thigh–torso angle and curvature of the spine. This study addresses the first two factors.

Office working postures were analyzed in terms of the thigh–torso angle and back inclination. The curvature of the spine is associated with the angle of the

Table 28.6 Time record when subject leaned back while seated on evaluated chair

(*a*) Chair 1

	Test duration (min)						
Subject	0	10	20	30	40	50	60
1				* *			
2			*		*	* *	
3		*		*		*	
4							
5						*	*
6		*				*	
7		*		*	*		
8					*		
9					* * *		

(*b*) Chair 2

	Test duration (min)						
Subject	0	10	20	30	40	50	60
1			*	*	*	*	
2				*	*	* * *	
3			* * *	*	* *	* * *	
4			*	*			
5			*	*		*	
6			*		*	* *	
7		*	*	*	*	*	
8				* *	* *	*	
9				* *	*		

(*c*) Chair 3

	Test duration (min)						
Subject	0	10	20	30	40	50	60
1		* *		*	*	*	*
2			*	*	* *	*	
3			* *			*	
4							
5			*		*		*
6						*	
7			*	* *			
8				*			
9			* *	*	* *	* *	

torso/back. The neck angle, which is affected by the torso/back angle, was also measured. The neck and torso/back (Figure 28.7) were extracted at 30-second intervals from videotapes taken of subjects working while seated on the chairs under evaluation. The statistics of the recorded data are depicted in Tables 28.10 and 28.11.

Chair	Area of adjustable range (cm²) A		Length of basic line (cm) B		Rate A / B	
	Seat	Backrest	Seat	Backrest	Seat	Backrest
Chair 1	5.02		7.60		0.66	
Chair 2	4.22		7.50		0.56	
Chair 3	0.98	5.22	3.40	3.90	0.29	1.34
	6.20		7.30		0.85	
Chair 4	0.89	5.26	3.80	3.70	0.23	1.42
	6.15		7.50		0.82	
Chair 5	1.33	4.11	2.50	1.90	0.53	2.16
	5.44		4.40		1.24	
Chair 6	NA	3.81	NA	4.10	NA	0.93
	3.81		4.10		0.93	
Chair 7	NA	1.34/0.93	NA	4.00/4.20	NA	0.34/0.22
	2.27		8.20		0.28	

NA: Not Adjustable.

Figure 28.5 Seat adjustment range

Table 28.7 Number of times subjects leaned back while seated on evaluated chairs

	Subject										
Chair	1	2	3	4	5	6	7	8	9	Total	Average
1	2	4	3	0	2	2	3	3	3	22	2.4
2	4	5	9	2	3	4	5	5	3	40	4.4
3	6	5	3	0	3	1	3	1	7	29	3.2
4	4	8	2	0	0	2	2	3	3	24	2.7
5	4	7	2	0	0	2	3	0	3	21	2.3
6	3	4	1	0	3	1	4	3	1	20	2.2
7	4	6	5	2	5	5	6	6	1	40	4.4
Total	27	39	25	4	16	17	26	21	21		
Average	3.9	5.6	3.6	0.6	2.3	2.4	3.7	3.0	3.0		

Table 28.8 Analysis of variance on number of times subjects leaned back while sitting on evaluated chairs

Factor	Sum of squares of deviation	Degrees of freedom	Unbiased estimator of population variance	Variance ratio	Value of F-distribution
Chair (A)	50.444	6	8.407	3.475	3.205 (0.01)**
Subject (B)	103.651	8	12.956	5.355	2.907 (0.01)**
Error	116.127	48	2.419		
Total	270.222	62			

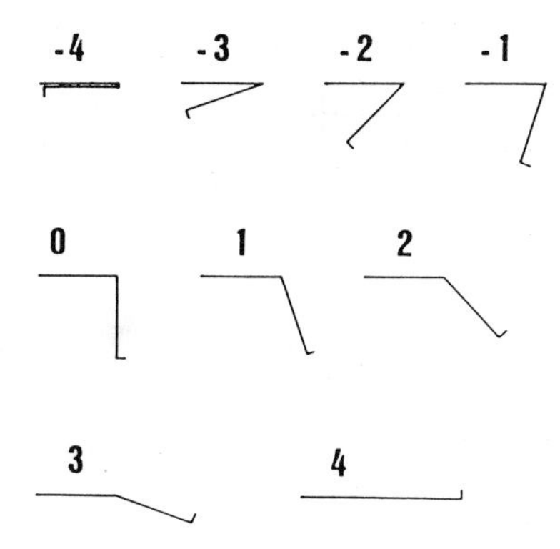

Figure 28.6 Classification of leg positions

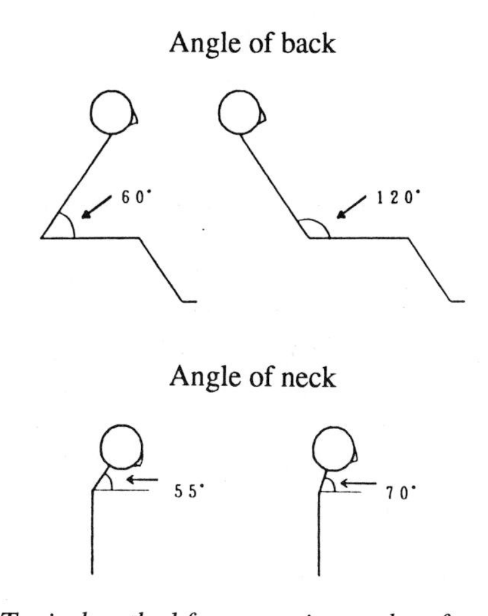

Figure 28.7 Typical method for measuring angles of working postures

Table 28.9 Frequency of identical leg positions

Chair	Frequency (%)
1	79.2
2	91.2
3	90.2
4	86.7
5	88.0
6	81.3
7	86.7

Table 28.10 Transition matrices of leg position for chair 1 (a) and chair 2 (b)

(*a*) Chair 1

	−4	−3	−2	−1	0	1	2	3	4
−4	3		3						
−3		42	4		6				
−2		6	145	1	1		1		
−1				72	6	9	1		
0			9	29	44	1			
1					18	46			3
2							35		
3			1					74	
4									

(*b*) Chair 2

	−4	−3	−2	−1	0	1	2	3	4
−4	12								
−3		4							
−2			194	5					
−1				69	13				
0				1	153	32			
1					1	41			
2						1	35		
3								38	
4									

Table 28.11 Survey of chair postures (in degrees)

Chair	Average angle of neck Standard deviation	Average angle of back Standard deviation
1	50.4	77.5
	14.44	10.85
2	48.6	77.5
	11.71	10.43
3	47.0	79.1
	14.12	9.39
4	50.2	77.7
	14.89	11.0
5	46.2	79.4
	12.32	10.45
6	46.9	77.7
	12.47	10.25
7	47.1	74.0
	14.56	9.47

Conclusions

Figure 28.8 shows the change in temperature distributions of the backrest and seat pan for 10 minutes after sitting down. A qualitative correlation was found between the ergonomic evaluation and temperature distribution of each chair.

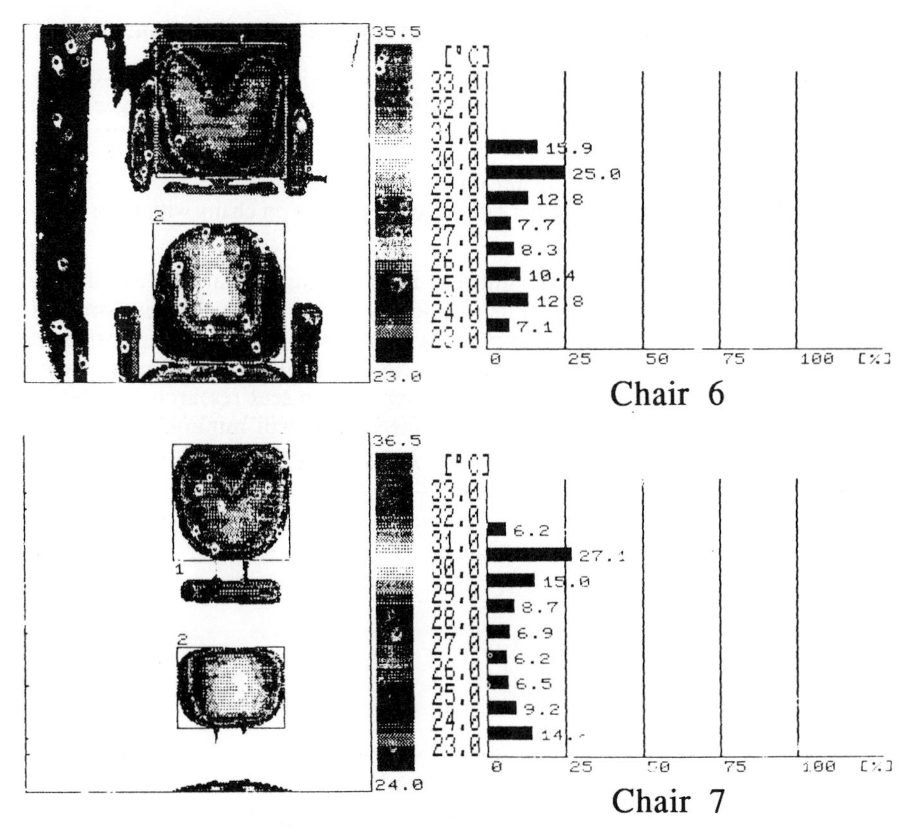

Figure 28.8 Temperature distribution: chair 6; chair 7

Table 28.12 Survey of subjects' posture (angles in degrees)

Chair	Average angle of neck Standard deviation	Average angle of back Standard deviation
1	55.7	70.6
	12.73	9.37
2	53.6	82.0
	9.08	9.37
3	38.7	84.8
	9.09	6.90
4	44.6	69.0
	8.78	4.72
5	29.7	81.0
	7.37	5.55
6	56.1	84.1
	9.75	9.00
7	45.1	80.1
	13.44	13.01
8	46.1	79.0
	9.54	8.37
9	62.9	67.5
	5.89	4.61

The following generalities can be made regarding the above experimental and survey results:

1. Upper body fatigue: the seated subjects experienced particularly high fatigue in the shoulders and neck.
2. The subjects expressed less subjective neck fatigue when they sat in chairs in which the seat pans and backrests were integrated.
3. The subjects reported less shoulder fatigue when they sat on chairs with large backrests.

These findings were used to weight and rank the functional seat evaluation worksheet used in the value analysis (Table 28.4); this led to the development of additional functions. For example, the functional field F6 Support arm ranks first on the functional evaluation worksheet (Table 28.4). This suggests that the armrest represents a primary consideration in seat research and development. The R&D effort will target an armrest that will minimize subjective fatigue symptoms and promote correct postures. Value analysis represents an ergonomic tool for evaluating office seating.

29

Towards systematic descriptions of chair performance

Christin Grant and Neil Goldberg

Introduction and premise

Corporations use various criteria in deciding what seats to acquire. These include cost, appearance, features, durability, and comfort. At present the most common method of determining whether a chair meets criteria is subjective: people try out chairs for short or long periods of time.

Not coincidentally, many seating designers base their work on equally subjective feedback. Typically, each design iteration of a chair is often evaluated by 'expert sitters' such as designers themselves, trying out working models of chairs, sensitively groping through a series of re-engineering efforts for something that feels just right. Eventually, panels of subjects are used as well.

Although subjective evaluation of seating is essential and irreplaceable, objective evaluation measures have much to contribute.

In this paper, we lay out two assertions. The first is that a vocabulary of chair descriptors should be developed that is quantitative, meaningful, and can be replicated. This vocabulary will be explored in the last section of this chapter, with reference to chair recline.

Second, it is necessary to reference this vocabulary to long-term health and the perception of comfort, in accordance with the physical, postural, and activity characteristics of users. This subject is not explored here but is the subject of further research.

Current context

The office seating standards and guidelines of the American National Standards Institute (ANSI), Deutsche Industrie Norm (DIN), the Canadian Standards Association (CSA), and the Japanese Industrial Standard (JIS) address only 14 dimensions of seating design. These include compressed seat height, seat depth, seat width, seat pan angle, angle between seat back and seat pan, backrest height, backrest contour, seat back width, clearance between armrests, height of armrests, casters, front lip contour, lumbar support, and footrests. They also prescribe adjustment ranges for some of these dimensions, including those for seat height, backrest height and backrest in-out position.

These are all fixed measures that describe a chair in its resting position. Variable measures that describe a chair's action or movement are usually absent from these important standards documents, which are sometimes the only human factors 'literature' that potential purchasers read.

Similarly, a review of 130 manufacturers' office chair brochures from the last three years revealed that descriptions of chairs are almost always limited to simple fixed attributes, such as dimensions, seat height range, the existence of backrest adjustability, and whether a chair tilts.

Variable descriptions, such as how far the chair tilts or the geometry of the tilt (such as knee tilt), were found in less than 5 per cent of the brochures, and less than 2 per cent mentioned more complex information such as the ratio between back and seat movement, or how much effort is required to recline. Yet to many users these are as important as seat height.

Contrast this state of affairs with, for example, the automotive industry. Knowledgeable customers are provided with a variety of measures of car performance to predict how a car will feel and whether it will suit their needs and preferences. Abstract concepts such as handling are quantified by braking distances, turning radii, and suspension characteristics. Car enthusiasts can often predict whether a new car is for them without driving it.

Consider also the sleep furniture industry, where mattresses are described to consumers in terms of relative hardness, independence of springs, and other details. While not yet standardized across manufacturers, sleep furniture descriptors may be considered more sophisticated than those in the office seating industry, which does not even note whether a given chair is soft, medium, firm, or hard.

It is entirely appropriate that the seating industry and the seating research community work toward similar levels of precision. After all, many people spend more time sitting in office chairs than driving or sleeping.

The lack of such information from many seating manufacturers means potential users, specifiers, and designers must sit in chairs in order to gauge their comfort. Under these circumstances, the feel of a particular chair is necessarily a subjective, somewhat mystical construct that is generally attributed to obvious factors (such as an impression of softness), rather than to subtler characteristics and unconscious responses to chair attributes.

For example, the impression of cushion softness is influenced by several dimensions. These include material density, rate of compression, recovery from compression, section thickness, contour, elasticity of the covering material, and even breathability. In addition, some of these dimensions, notably those having to do with rate of compression, are complex measures that are variable. They vary with degree of compression during the measurement process, temperature, time, and other changes that occur during normal use. They also interact with posture and body type.

In the following example, one subjective chair characteristic, ease of recline, is explored further. In the example, it is shown how objective measures of this characteristic can be used to compare different chairs and, with further research, to predict comfort.

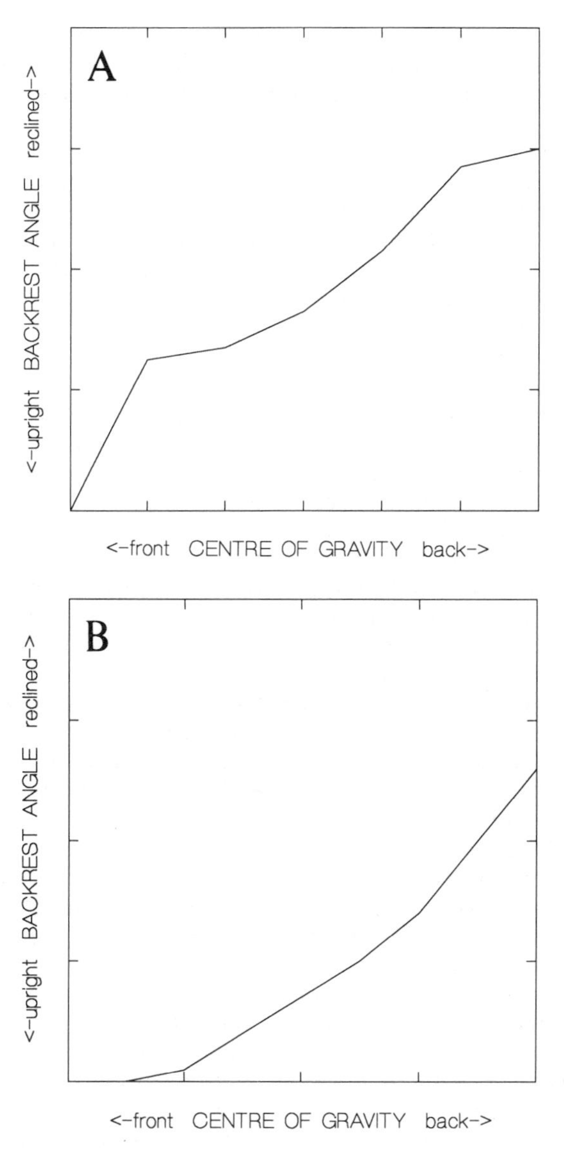

Figure 29.1 Relative backrest deflection graphs for two different chairs. Although both chairs have approximately the same recline range, chair A starts to recline readily, with little weight shift. Chair B does not reach a similar degree of recline until a much greater load has shifted from the front of the chair to the back. However, chair A becomes difficult to recline at the end of its range

Working example: how chairs tilt

Chairs vary widely in the ways they move. In the last two decades, much of the emphasis in new seating development has been directed towards finding new means of making chairs move in ways that respond to the anatomy, posture, and behaviour of individuals.

For example, some reclining chairs now rotate around a point directly behind the knees while others have a virtual rotation point near the ankles. Some chair backrests recline at different rates from the seat pans; others drop as they recline. Some seats slide forward as their backrests recline. There are also chairs whose seat pans both incline forward and recline back. Some chairs use mechanisms while others use flexible materials to achieve a recline.

Certainly, the measures described below are not the only worthwhile measures, nor do they perfectly measure their underlying constructs (such as recline stiffness or dwell). They have been of use, however, in chair design and evaluation, providing a helpful starting point.

Recline stiffness

A major difficulty in designing dynamic seating is allowing large and small people alike to move easily through the entire recline path.

Two chairs may have identical tilt ranges, but the amount of effort required to tilt them through their full range of recline may vary. Further, the pattern of that effort may change in different ways throughout the recline range. For example, a person may find a chair tilts too easily at the end of its recline range but requires too much force to tilt when upright. Another chair may give exactly the opposite experience. The two chairs may have identical average recline stiffness, but the patterns through the recline range are different.

This can be seen in the smoothed graphs in Figure 29.1. They show the angle, relative to the starting position, of the backrests of two chairs as weight is shifted from the front to the back of the chairs, similarly to the process used by a human being to make a chair recline. Both chairs had fixed backrests, and both were set at their softest tilt tension setting.

The chairs have similar tilt ranges (29° and 27°) for this load. However, chair A reclines a great deal with a slight shift in weight while chair B does not reach an equivalent recline until more than half of the weight shift is accomplished. Although they both end up at about the same point, their variable behaviour between maximum and minimum recline points is very different.

The procedure was performed using a modified H-point sitting machine, or dummy, developed by the Society for Automotive Engineers (SAE). This device allows weight to be shifted from front to back of the seated figure in a way that approximates the weight shift of a human body reclining in a chair. In this exercise, weights corresponding to a 50th percentile male were used. Ideally, a full test of recline stiffness would provide information regarding a broad range of users, over a series of tilt tension settings.

Which pattern is preferable, or even acceptable? Subjective opinions must be systematically obtained. Preliminary results indicate that chair A appeals to lighter people who prefer a reclining posture, while chair B is preferred by heavier people and people who like to sit upright. It can be hypothesized that a chair that is liked by both heavy and light people may have a recline graph that approaches a straight line.

Back/seat co-recline

A person's height and weight affect how a chair moves. A light person may be able to deflect a seat pan as much as a large person by adjusting the tilt

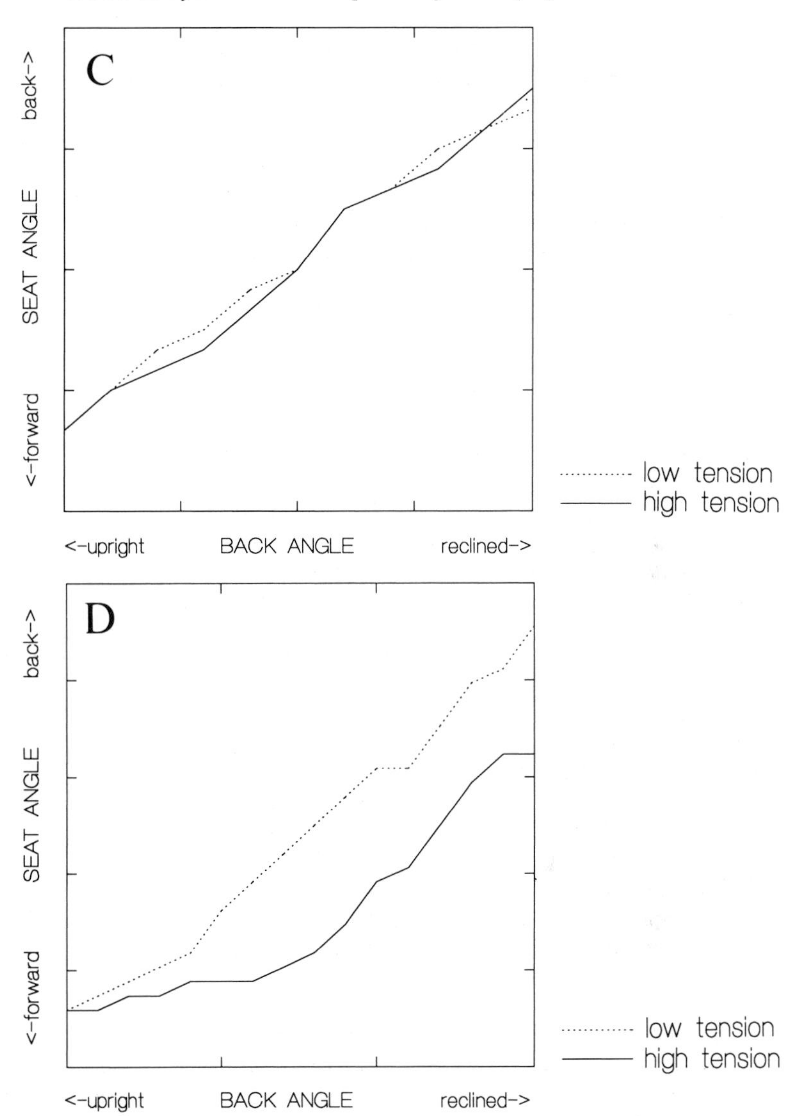

Figure 29.2 Relative angles of backrest and seat pan, as the backrest is pulled back from upright, at low and high tension settings. Chair C, with a backrest angle mechanism, shows little change when its tilt tension is changed from its softest to its stiffest point. The backrest of chair D, with a flexible shell rather than a backrest mechanism, behaves differently at low and high tension settings

tension, but the lighter person may not be able to deflect the backrest as much as the heavier person unless it, too, adjusts.

This can be seen in the smoothed graphs in Figure 29.2. They depict the angle of the seat pan and backrest of two chairs as their backrests were pulled back from the unloaded position to the fully reclined position. This measurement was performed with the tilt tension set at both its stiffest and loosest points.

Chair C's mechanism allows the backrest to move at about a 2 : 1 ratio compared to the seat, while chair D has a flexible shell rather than a backrest angle mechanism. The seat and backrest of C move consistently at both high and low tilt tension settings. Chair D behaves very similarly to C at low tension. At high tension, however, the backrest of D reclines more, relative to the seat. The flexibility of this chair's shell allows the backrest–seat pan angle to open, even when the seat pan is nearly immobilized.

Preliminary information indicates that some users prefer a chair whose seat can be optimally stabilized while retaining movement in the backrest, while other users want a backrest to move regardless of seat mobility. More research is needed to identify variables, such as tasks or anthropometrics, that might help predict these preferences.

Dwell

Dwell is a furniture industry term that can be defined as the ease or difficulty of maintaining a particular seat recline position, due to a slight resistance to reclining (static friction) that is felt only when stationary. Once the initial resistance of dwell is overcome, the chair reclines normally. Dwell differs from tilt stiffness, for both stiff and soft chairs can have either high or low dwell. Chairs currently on the market achieve dwell by methods ranging from friction to cam mechanisms.

Chairs with dwell feel stable when partially reclined, but too much dwell annoys because of the difficulty of starting the chair moving. Chairs with low dwell can feel unstable and require effort and balance to keep them reclined (if the tilt tension is stiff), or to keep them from tipping back uncontrollably (if the tilt tension is soft).

A chair with dwell requires more weight shift to start movement than to continue movement. The smoothed graphs in Figure 29.3 were obtained using the SAE H-point sitting machine, with both chairs at their softest tilt tension settings. Each chair's reclining motion was stopped at 8°. The notch in chair E shows the amount of weight shift required before the chair resumed reclining. It can be seen that chair F, which has about as much tilt stiffness as chair E, has much less dwell.

As with tilt stiffness and back/seat co-recline, one cannot evaluate such characteristics without subjective opinions. Determining the right amount of dwell, or whether the right amount of dwell changes with tension setting, will require more investigation with human subjects.

Summary and conclusion

This chapter has presented a possible approach to describing some of the key variable characteristics of modern office seating. It has shown examples suggesting that comparing data related to the tilt moion of office chairs can illuminate performance differences among them.

However, more research is needed before the graphs can be interpreted in terms of comfort or preference. When the patterns of chair dynamics are eventually linked to user variables, designers will be able to create chairs that are better tailored to the specific needs of different kinds of user, and users will be

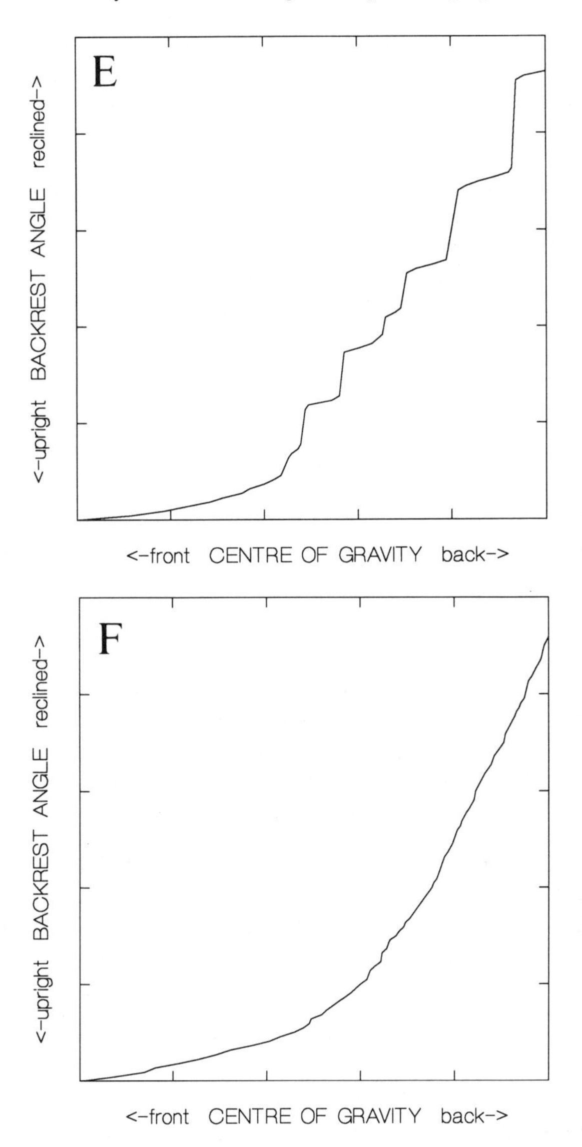

Figure 29.3 Relative amounts of dwell for two different chairs with similar tilt stiffness and fixed backrests. Each chair's recline was stopped at 8°. The notches on each curve show that different amounts of weight shift were required to get each chair moving again. The initial resistance, or dwell, of chair E is greater than that of chair F

able to choose more intelligently among chairs. Defining the variables, as we have begun to do here, is the first stage of a research agenda that promises to bring seating research, seating design, and specification practice together.

30

Office seating and movement

Ida Festervoll

Research on the relationship between body and chair provides little insight into the importance of the larger context relevant to sitting postures. Technological developments in the office affect our physical health, because they influence our everyday sitting habits. The number of people in sedentary jobs is increasing. We sit more – and many people remain in the same position throughout the working day.

The chair of the future – how will it look and function? Ergonomists, designers, chair manufacturers, researchers and technological developers must each contribute their perspectives in the development of good seating for today and tomorrow.

Technology – the problem or solution?

Technological innovation is today a subject of ongoing research and development. It introduces speech-based user interfaces, improved contrasts and contrast inter-relationships that enable us to optimize our surroundings in many ways. Much of this progress has been at the cost of computer users. Eyestrain, the incidence of neck and shoulder pains have risen with the advent of automation.

We have used our knowledge of the causes of discomfort to improve many problems. Witness the ability of new flat panel displays to reduce eyestrain by improving resolution and eliminating flicker; speech-based user interfaces to relieve wrist and shoulder discomfort; and others.

Yet many of these measures are purely cosmetic. We attend to problems created by technology – yet the cause of the problems is inherent in technology itself.

Chair manufacturers attempt to accommodate different body measurements and functional needs by making furniture more flexible. Yet is this sufficient to eliminate discomfort? It is now commonly recognized that computer work reduces natural movement and variation. If work content and design require users to assume the same position all day – and eliminate opportunities to get up and perform other activities – the work process is inadequate.

The field of preventive health care is based on the philosophy that one must address the underlying causes to produce a lasting positive effect. The structures of the body have not changed significantly throughout millenia since humans were nomads and hunter/gatherers; the body was designed for movement, and for variation between movement and rest.

Constant muscular activity, even at a low level, is not physiologically beneficial. Nor is fixation of the eyes at a given area and angle. Viewing the same object for extended durations with raised eyelids strains both the musculature and the fluid balance of the eye.

There is interest in Denmark regarding whether certain instances of spontaneous abortions can be attributed to excessively constrained postures – rather than such factors as electro-magnetic fields.

Although the march of technology cannot be stopped, it should be understood.

Technology and evolution

The evolution of technology has been guided by technocratic thinking rather than the attention to our physiological requirements for variation and movement. In addition, the decision-makers who influence this evolution have often not personally experienced the physical problems they engender. Managerial work characteristically affords variation and a high degree of postural freedom; these computer users suffer much less cost than for those forced to sit in front of a monitor all day.

One of today's paradoxes is that ergonomics guidelines for computer work artificially introduce what was once a natural feature of work – movement. Such programmes attempt to restore what technology has eliminated.

Specialists of a number of countries recommend that a session in front of a computer should not exceed four hours, beyond which the incidence of physical cost rises. How can such guidelines be accommodated when, in many workplaces, few jobs exist that do not involve computers? Will four-hour work days exist in the future, or are there alternatives?

Many regard the computer networks that link several terminals as progressive. Information flows from office to office, from work-place to work-place. This is practical, (probably) labour-saving, and often promotes productivity in the short term. But what about long-term cost to health and well-being?

In Norway the incidence of stress-induced sick leave is increasing. Although industrial work practices have improved due to our Working Environment Act, our low level of unemployment and lost time due to illness is increasing.

Stress-related disorders are the natural reaction of the body to unnatural stimuli, or rather the lack of stimuli.

Although the reasons for people's illnesses have changed, traditional methods of organizing work-places on the basis of ergonomic dimensions and distances have not. As a result, movement and variation is reduced even further. Work reorganization will not solve such problems, as alternative work-places and jobs for everyone may not exist in the future.

Humans often react negatively to change. Most people prefer to sit still, and avoid moving more than necessary. Consequently, when the available equipment enables people to sit still, they will resist changes that make work

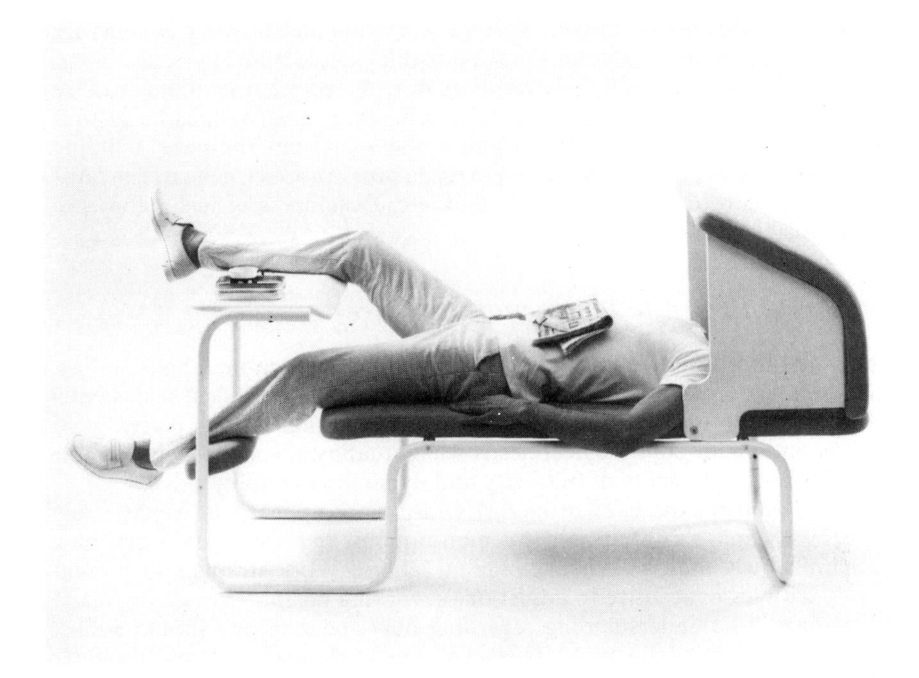

Figure 30.1

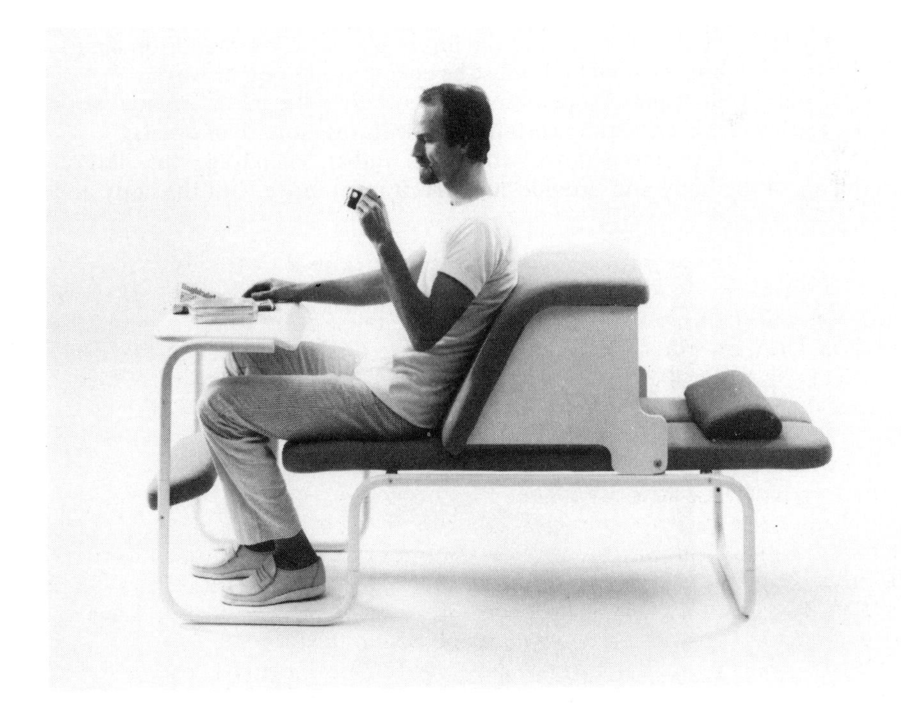

Figure 30.2

more difficult. Means of changing work organization and/or work content are not straightforward; they are not always possible, nor accepted by users.

In Norway, it is generally believed that only 10 per cent of illness can be prevented by medical staff. For example, companies must reduce hazardous effluents, local authorities must safeguard the social environment, building regulations must allow for indoor climate factors, product developers must create goods with human welfare in mind, etc. Developers of technology are also responsible.

The dilemma of the chair manufacturers

Technology and furniture must be considered in conjunction. Chairs, tables, monitors, keyboards etc. – should accommodate people's physical and mental needs. The office furniture industry – at least in Scandinavia – places great emphasis on human physiological needs. Height-adjustable tables with tiltable tops and chairs with plenty of flexibility and dynamics exemplify this.

Are such adjustments used properly? Office work appears to be physically easy. Stresses and strains caused by monotonous and sedentary work may arise more slowly than with heavy manual labour. Work strains also present the body with little incentive to adjust one's working posture. Moreover, many workers have little understanding regarding how and why one should adjust. The adverse impact of constant low-level strains to the eye, neck and shoulder muscles is frequently not recognized.

Chair manufacturers must address such factors by recognizing their contribution to the work process. The chair is more important now than ever.

The immobile society of today and the future places new demands on chairs. If technology inhibits movement, it must be encouraged in other ways.

Users that sit in front of our computers or do other monotonous work involving arms and eyes, require certain chair features more than others.

The chair must promote movement – or, rather, encourage the natural movements of the body and provide support. It must move with the body and

Figure 30.3

yet provide support and a sense of stability in all positions. Movement should not require conscious effort, or levers or wheels.

A traditional view of working chairs and sitting is that movement connotes discomfort. It is presumed that if one feels uncomfortable or sits badly, they change posture more frequently.

If movement is to be experienced as comfortable, it must not involve instability or feelings of insecurity after a certain period of adaptation.

Some research has attempted to optimize the distribution of pressure between the seat pan and buttocks/thighs. Other experiments sought to optimize the traditional 90° angle posture (cubism) by means of the shape, angle and size of the backrest in order to reduce muscular activity and maintain the natural physiological curvature of the spinal column.

Yet little agreement exists on how seating can accomplish such ends. How long should one remain in one posture? Is any position superior to others?

Many seats enable the user to lean backwards while sitting. Normally, the tilting-point is placed under the front edge of the seat pan so the legs will not be lifted from the floor when one leans backwards. Leaning back requires effort or loosening of tilt tension. If the tilt tension is too slack, upright sitting (middle position) is unstable—body weight makes the seat lean backwards. Is it possible to develop a spring tension that will promote desired movement, yet provide stability when the tilting point is far forward? Other chairs allow the seat to adjust to the forward sloping position.

Today, chairs are becoming more flexible; these tilt forward when one is working while leaning forward, and tilt back while leaning back. The sitter can

Figure 30.4

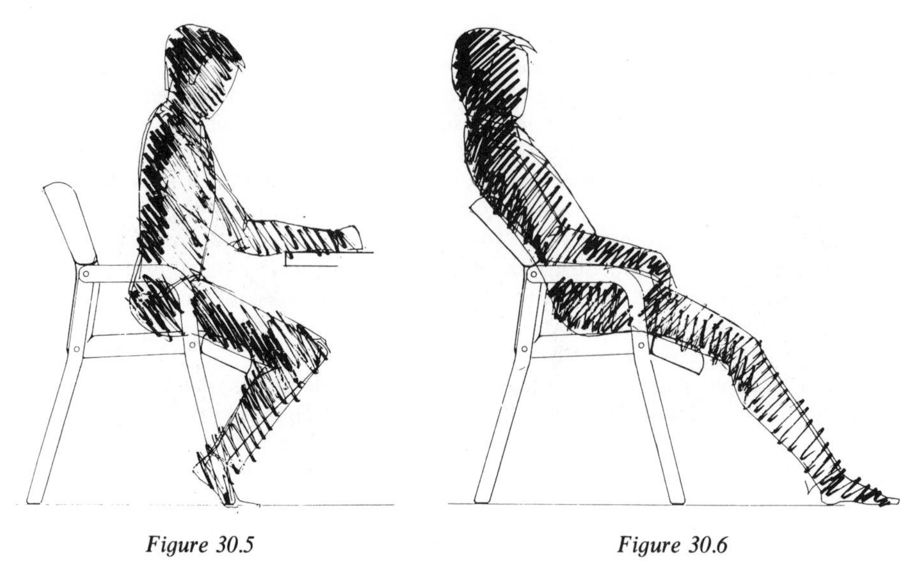

Figure 30.5 *Figure 30.6*

assume the position that is natural for their work. The effectiveness of this approach depends wholly on the overall technical design of the chair.

Most emphasis is placed on the isolated functions—whether the chair can tilt forward or backward, or both. Issues such as whether adjustment is 'automatic' or with levers, whether tilt is independent of the user's body weight and other seat adjustments, receives less attention.

In the author's experience, chairs that require manual adjustment to change position are not used as intended. People do not make the effort to adjust the chair or to operate levers to change position. They do not want to think of sitting while working.

We need the small-scale dynamics a chair can offer when it responds to shifts in weight position. Chairs should move with the user, and make it easy to move. When reaching for the telephone or to pick up a pencil, the chair should ease forward. When leaning back to talk on the phone or to take a break, the chair should make this possible—without levers. The research has ignored both the body's reactions to sitting and movement and to specific considerations associated with different chair concepts.

Most chair research has been based on specific sitting positions. Various products have been evaluated in forward-oriented positions, backward-leaning positions or while upright by measuring muscular activity, disc height, static calculations of back strain, comfort rating etc. Such research has increased our understanding of the chair and its functions in relation to the human user. However, seating research has been limited by:

1. *Ignoring the larger context of work*

   For example, forward oriented positions may require undesirable leg positions when the table is too low; the chair cannot be assessed in isolation. By limiting variables in the research, one obscures the problems—usually by setting limits to reality! Work type, table height, the desired sitting position etc. interact with desk height, terminal position and design, floor covering and the types of task.

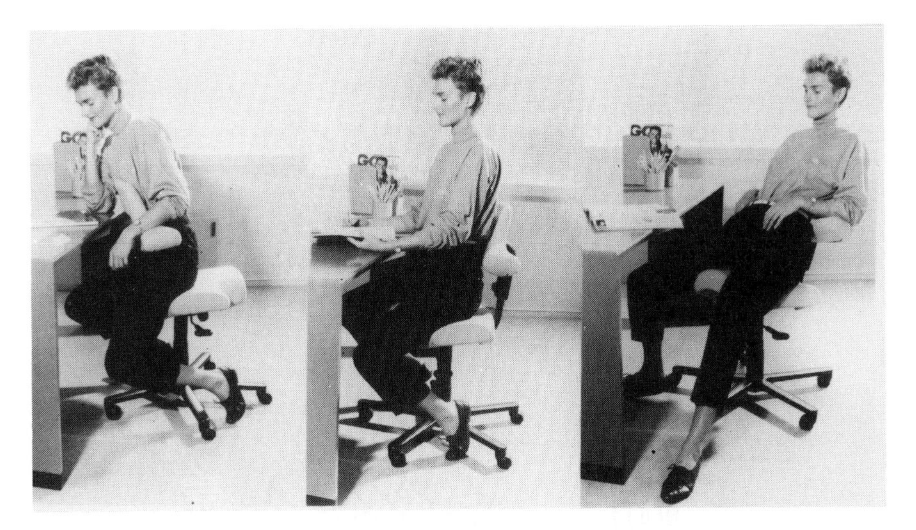

Figure 30.7

2. *Emphasis on a narrow spectrum of seating types*
   If a feature such as a tilt mechanism—is not found beneficial for one chair, it may still be effective in other chairs. Seat elements interact. As a result products differ, and will be experienced differently in terms of comfort and function. Angular displacment, seat pan shape, upholstery, tilt locations point etc. can influence the overall sitting experience.

How we work

Can products be manufactured that support the body's natural changes of posture and even promote movement automatically? How can circulation and nutrition to the discs be optimized? How much movement, and what kind of movement, is most effective? What are the relations among movement, stability and comfort? How much and what kind of muscular activity is desirable while sitting? Should time limits be imposed on sitting itself? These are some of the questions HÅG is dealing with.

HÅG a.s. has a tradition of emphasizing movement. Independent designers present their ideas in the form of sketches, prototypes or verbal concepts. These are discussed on the basis of our fundamental philosophy in a forum consisting of designers, representatives of the technical department, the marketing department and the physical therapist/ergonomist. As a physical therapist, I follow the process through the drawing-board stage, prototype development, all the way to the finished product.

This emphasis by HÅG a.s. directs the selection of such factors as the shape of the backrest, positioning of armrests, general design etc. The chair, viewed as a whole, consists of parts, all of which contribute a specific functional objective—i.e. movement. Decisions are based on critical interpretation of scientific studies of functional and anatomical factors related to sitting, and studies of various chair concepts and individualized solutions. We also

"I've been asked to test some stupid things in my time..."

Figure 30.8

develop tests and try out prototypes at all stages. We place a greater emphasis on the qualitative rather than the quantitative approaches which consider the larger context the product is used in. These seat features are evaluated in conjunction and in relation to the function of the product. This compounds the complexity of the problem, because feelings of comfort and discomfort are described relative to physiological needs, as well as to different constructing parts.

Concluding remarks

The chair, the workplace, electronic equipment, and their interaction with humans must be developed as a system on the basis of the human needs of today. This includes movement, variation and a variety of sitting postures. This requires collaboration between researchers in the fields of technology, physiology, occupational medicine, visual ergonomics and product developers and designers to provide insight into areas where more information is needed to make tomorrow's work-places good for the body and soul.

Bibliography

Andersson, B. J. G., Örtengren, R., Nachemson, A. and Elfström, G., (1971), Lumbar disc. Pressure and myo-electric back muscle activity during sitting. I. Studies on an experimental chair. II Studies on an office chair, *Scan J. Rehab, Med.*, 104–21.

Bendix, A., Jensen, C. V. and Bendix, T., (1988), Posture, acceptability and energy consumption on a tiltable and a knee-support chair, *Clinical Biomechanics*, **3**, 66–73.

Bendix, T., (1987), Adjustment of the seated workplace with special reference to heights and inclination of seat and table, *Danish Medical Bulletin*, **34** (3), 125–39.

Bendix, T. and Biering-Sørensen, F., (1983), Posture of the trunk when sitting on forward inclining seats, *Scand. J. Rehab. Med.*, **15**, 197–203.

Bennet, D. L., Gillis, D. K., Portney, L. G., Romanow, M. and Sanchez, A. S., (1989), Comparison of integrated electromyographic activity and lumbar curvature during standing and during sitting in three chairs, *Physical Therapy*, **69**(11), 902–13.

Bridger, R. S., (1988), Postural adaptions to a sloping chair and work surface, *Human Factors*, **30**(2), 237–47.

Brunswic, M., (1984), Ergonomics of seat design, *Physiotherapy*, **70**(2), 40–3.

Burton, A. K., (1984), Electro-myography and office-chair design, a pilot study, *Behaviour and Information Technology*, **3**(4), 353–7.

Corlett, E. N. and Eklund, J. A., (1986), Change of stature as an indicator of loads on the spine, in *The Ergonomics of Working Postures*, Ch. 20, Taylor & Francis, London.

Corlett, E. N., (1989), Aspect of the evaluation of industrial seating, *Ergonomics*, **32**(3), 257–69.

Diebschlag, W., (1987), Nicht nur auf, sondern in Stühlen sitzen, *Zeitscher AIT/ Architektur, Innenarchitektur, Techn. Ausbn*, **10**, 129–33.

Diebschlag, W., Müller-Limmroth, W., Baldauf, H. and Stumbaum, F., (1978), Physiological investigation for construction of seating accommodation, *Zeitschrift für Arbeitswissenschaft*, **32**, (4NF).

Drury, C. G. and Coury, B. G., (1982), A methodology for chair evaluation, *Applied Ergonomics*, **13**(3), 195–202.

Drury, C. G. and Francher, M., (1985), Evaluation of forward-sloping chair, *Applied Ergonomics*, **16**, 41–7.

Ericson, M. O. and Goldie, I., (1989), Spinal shrinkage with three different types of chair whilst performing video display unit work, *International Journal of Industrial Ergonomics*, **3**, 177–93.

Grandjean, E., Hünting, W. and Nishiyama, K., (1984), Preferred VDT work station settings, body posture and physical impairments, *Applied Ergonomics*, **15**(2), 99–104.

Grieco, A., (1986), Sitting posture: An old problem and a new one, *Ergonomics*, **29**(3), 345–62.

Hort, E., (1984), A new concept in chair design, *Behaviour and Information Technology*, **3**(4), 359–62.

Lander, C., Korbon, G. A., DeGood, O. E. and Rowlinson, J. C., (1987), The Balans chair and its semi-kneeling position. An ergonomic comparison with the conventional sitting position, *Spine*, **12**(3), 269–72.

Lueder, R., (1983), Seat comfort. A review of the construct in the office environment, *Human Factors*, **25**(6), 701–11.

Mandal, A. C., (1981), The seated man 'Homo Sedens'. The seated work position. Theory and practice, *Applied Ergonomics*, **12**(1), 19–26.

Salvendy, G., Sauter, S. L. and Hurrell, J. J. Jr., (1987), *Social Ergonomic and Stress Aspects of Work with Computers*, Elsevier Science Publishers B.V., Amsterdam.

Stumbaum, V. F. and Diebschlag, W., (1981), Multi factor study of sitting behaviour while writing at a desk, *Zeitschrift für Arbeitswissenschaft*, **35**(7NF), 179–88.

Taptagaporn, S. and Saito, S., (1990), How display polarity and lighting condition affect the pupil size of VDT operators, *Ergonomics*, **33**(2), 201–8.

Winkel, J. and Jørgensen, K., (1986), Evaluation of foot swelling and lower-limb temperatures in relation to leg activity during long-term seated office work, *Ergonomics*, **29**(2), 313–28.

31

The study of lumbar motion in seating

Hector Serber

Introduction

Traditional ergonomic chair design has been constrained by assumptions regarding posture alignment and support. Although extensive research has been conducted to define seat and back dimensioning, shapes, positioning, orientation and adjustability, issues associated with motion of lumbar segments have been ignored. This omission may well contribute to incidence of injured backs, RSI and other cumulative trauma in the work-place. Kelsey *et al.* (1979) noted that, 'It is clear that musculo skeletal conditions have a huge impact on the health of the general population and that they rank first among disease groups in the frequency with which they affect the quality of life and second in terms of their total economic cost. They also rank first in cost to Workmen's Compensation insurance carriers.'

Continuous Balanced Motion (CBM) design and a System of Postural Support (SPS), which provides continuous body alignment and safe task motion, is needed to effectively manage the support of the body masses at work. Therefore, the purpose of this paper is to present a new design tool, the Lumbar Motion Model Mannequin, as a method to analyze seated lumbar motion.

The lumbar spine is a very complex series of joints. Yet to simplify design, seated mannequins have typically assumed the lumbar is rigid, and motion is isolated at the hip joint. The Lumbar Motion Model Mannequin was developed to account for lumbar mobility. It is postulated that one joint, located at the vertebral body of L3 can represent all lumbar motion in the mid-sagittal plane with an accuracy of 99%, relative to motion by all lumbar segments. This locus, the Instantaneous Center of Lumbar Flexion (ICLF), is located at the intersection of the mid-lumbar height and trunk longitudinal centre line.

Goals of the model

1. To establish the relationship between hip angles and lumbar angles;

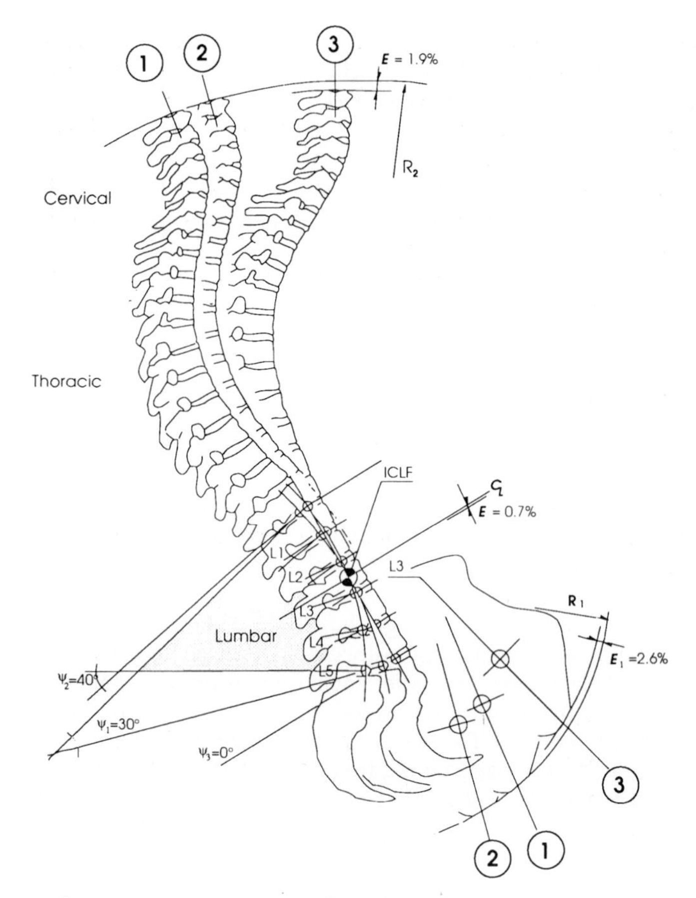

Figure 31.1 The Instantaneous Center of Lumbar Flexion (ICLF) represents all of the motion produced by the lumbar spine at a single joint

2. To allow the incorporation of this information into a template which can represent a variety of seated task postures and their positions relative to mid-range of motion and safety limits;
3. To facilitate the inclusion of lumbar motion in dynamic analysis.

Location of the ICLF

The Instantaneous Center of Lumbar Flexion (ICLF) is used to represent all of the motion produced by the lumbar spine at a single joint. This locus is located at the centre of the vertebral body of L3 (mid-lumbar). It serves as a virtual centre of motion.

The location and its associated error magnitude can be identified by opening the lumbar curve angle from 40° to 0° and by keeping the spaces between the centre of each lumbar segment constant with the change in lumbar angles.

In Figure 31.1, the motion of the joint is represented as a ball bearing. The thoracic and cervical area of the spine rotate rigidly as one, from the joint above L1. Then, tracing a circle centred at the ICLF from positions 1–3, on a geometric drawing in 60 scale, we find that the error at the ischial tuberosities is E1 = 2.6%, and at C1 the error is E = 1.9%.

Because the centres would overlap when the position of the ICLF is corrected, it is assumed that these errors can be subtracted. Using this approach, the graphic representation of the ICLF has an error of 0.7 per cent. It is assumed that this location of the ICLF will remain less than 1 per cent from its true location.

Model link system

Figure 31.2 modifies Figure 3 of Webb Associates (1978), by eliminating the illiopelvic link and joint and lumbar joints and replacing them with the Instantaneous Center of Lumbar Flexion (ICLF) joint. Similarly, the trans-sternum becomes a single joint. This yields the following: (1) Pelvic link, (2) Thoracic link, and (3) Neck link. The other links remain the same.

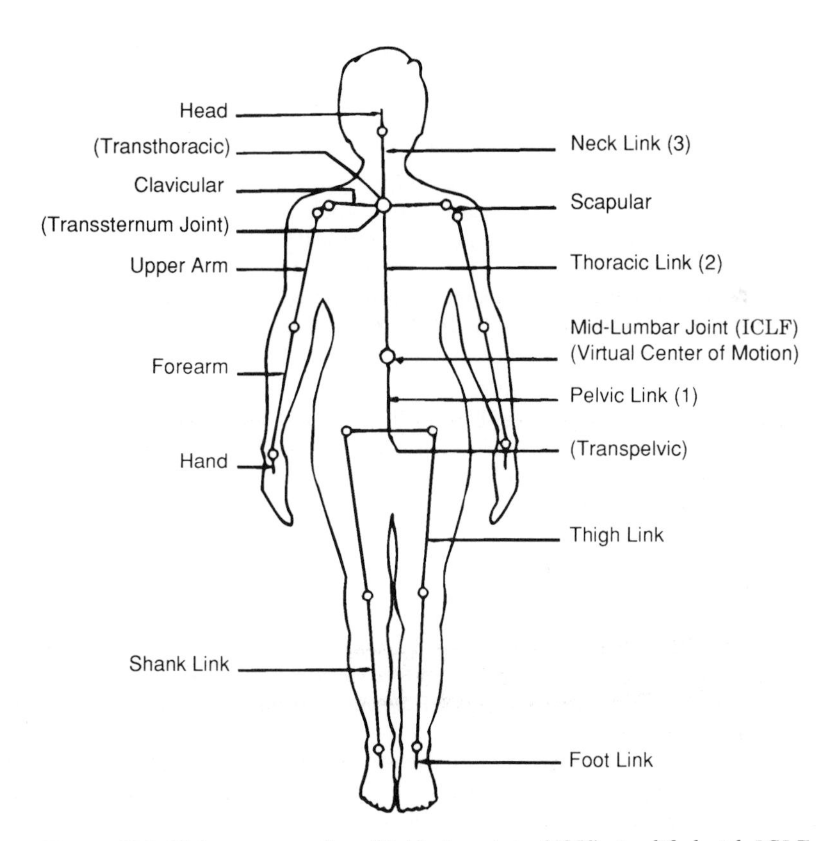

Figure 31.2 Linkage system from Webb Associates (1978) simplified with ICLF

Construction of the model

The model is based on the 50th percentile male dimensions (Kroemer *et al.*, 1993). Centres of mass and moments of inertia (Santschi, 1963), and lumbar segments (Snyder, 1972) are obtained from Webb Associates (1978) and proportionally adjusted for mass shifts due to link simplification, but not for fluid shifts. Lumbar angles associated with torso-thigh angles are extracted from Link *et al.* (1990).

To develop the hip to lumbar angular relationship, mean values of the standing lumbar curve are shown as 34.4°. Similarly, sitting at a standard chair, the mean lumbar angle is −6.5°. This implies a 40.9° of lumbar angle change for a 90° thigh-angle rotation or 0.45° of thigh rotation per 1° of lumbar angle change. This relationship is used to plot six postures of interest (1) standing, (2) NASA Neutral Body Posture, (2a, 3, 4) seating, and (5) full flexion in Table 31.1.

An experimental value can be used to support these computations. First, on a seat with 20° forward tilt (110° from vertical) mean value of the lumbar curve is 2.4°, and the computed value in position 2a at 110° yields 2.6°. According to Mandal (1981), when the thigh is 90° from the torso, the hip rotates 50°, and the lumbar angle 40°. Table 31.1 shows 49.1° hip rotation, a discrepancy of 0.9°. (*see also* Figure 31.3).

Uses of the model

The purpose of the model is to facilitate both chair design and ergonomic training. The model can be used as a template to represent an accurate silhouette for specific task posture requirements, such as how to maintain a mid-range of motion, and appropriate range of motion. The moments of inertia are used to analyze biomechanical equilibrium and constraints during seated motion as major body parts move in relationship to each other.

The lumbar motion associated with thigh-torso angles from 128° to 90° is of particular interest in the design of ergonomic chairs. The lumbar motion associated with thigh-torso angles from 110° to full flexion 35° is of particular interest to the public transportation industry in the study of seat motion and occupant dynamics during vehicle crashes. This model, translated into a free body diagram, lends itself easily to the application of dynamic equations of motion.

Discussion

The silhouette and link positions drawn from the Lumbar Motion Mannequin shown in positions 2, 2a and 3 of Figure 31.4 are superimposed at the corresponding CG's with the neck link vertical and rotated by ±9° from the mid-range posture 2a.

Note, that as the torso-thigh angle opens, from 90° to 128°, the lumbar and hip angles also open proportionately. In Table 31.1, these postures represent three equilibrium states. These are based on mean values obtained by using a physical therapist's flexible ruler. This is a relatively simple and non invasive

Table 31.1 Table of thigh, lumbar, pelvic and hip angle proportional relationships

SPECIFICATIONS: STATURE = 175.58cm WEIGHT = 78.49Kg 50%ILE MALE MANNEQUIN					
POSTURE	THIGH ANGLE	LUMBAR ANGLE θ	THIGH ROTATION	PELVIC ROTATION	HIP ROTATION
1. STANDING	180º	34.4º	0º	0º	0º
2. NASA NBP(a)	128º	10.7º	52º	23.63º	28.37º
2a. SEATING (b)	110º	2.6º	70º	31.8º	38.2º
3. SEATING (c)	90º	-6.5º	90º	40.9º	49.1º
4. CONTAIN (d)	60º	-20.1º	120º	54.53º	65.47º
5. FULL FLEXION (e)	35º	-31.5º	145º	65.5º	79.5º

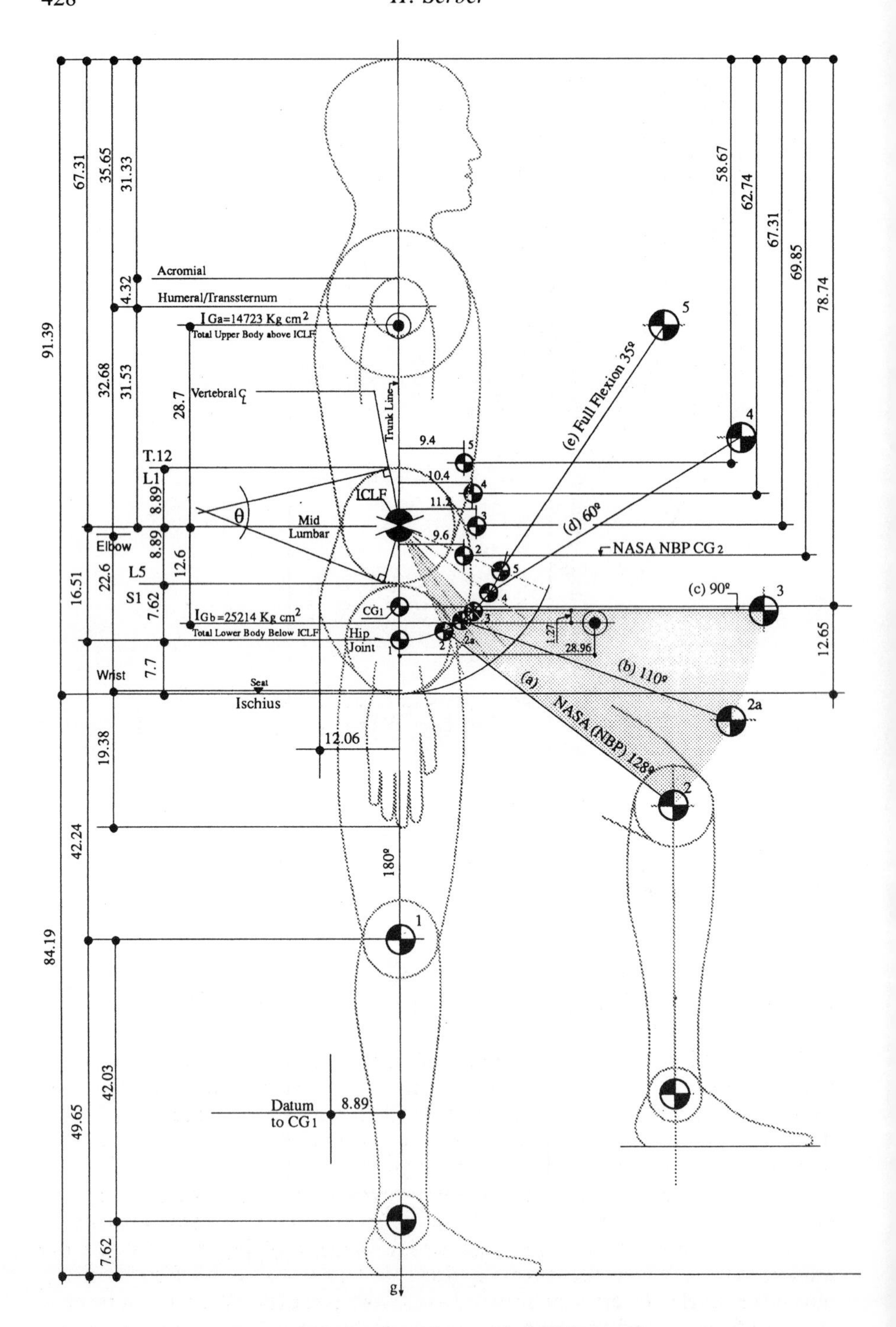

Figure 31.3 Lumbar Motion Model Mannequin

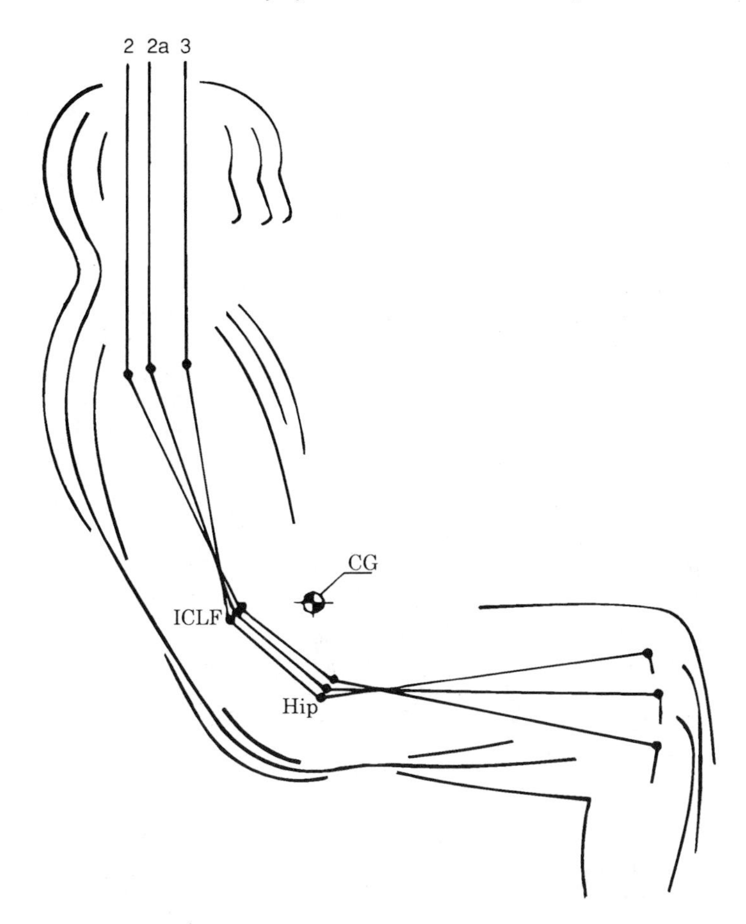

Figure 31.4 'Opening the torso-thigh angles', opens the lumbar and hip angles in proportional relationship

physical therapy tool to measure the lumbar curve by placing a flexible ruler against the spinous processes L1-S2. Hart and Rose (1986) confirmed that this method is both reliable and valid.

By using this model, the effect of loads on the lumbar spine can be analyzed, then applied to improve postural support. The author believes that the Lumbar Motion Mannequin can offer a practical method to simplify and standardize design to correlate anthropometrics, medical data and health practitioners' findings to ergonomic seating design.

It is a well known fact that the spine needs micro and macro motion to satisfy metabolic requirements and muscle movement is needed to promote blood circulation. Static seating prevents lumbar mobility and flattens the lumbar curve; when the task requires the torso to move from reclined (posterior) angles to erect or forward (anterior) postures. When the torso has to move forward and the buttocks and thigh have to remain static, stress on lumbar segments increases dramatically. When standing, intra-discal pressure is 30 per cent less than when sitting erect and decreases by 50 per cent when reclined (Nachemson and Morris, 1964). Andersson and Ortengren (1974) noted that, 'If the backrest-seat angle was increased there was always a decrease in (EMG) activity at all levels of the back'.

In an effort to incorporate balanced motion with support and alignment, the author has designed a chair mechanism which provides a Continuous Balanced Motion (CBM) seat and a System of Postural Support (SPS) (*see* Figure 31.5). The seat rotates with pelvic movement, centred proximate to the ICLF and CG of the body in mid-range posture. With a firm and adjustable back support, the lumbar depth and height adjustments are designed to support the ICLF (Mid-Lumbar), thus promoting lumbar lordosis. In conjunction with seat motion, the body is supported with both static and dynamic means. This seat motion may be stopped by the user by adjusting the seat tension. Stopping the mobility of the seat will defeat the ability of the CBM to provide lumbar mobility. The ergonomics professional should encourage use of this dynamic seat function when provided in a chair.

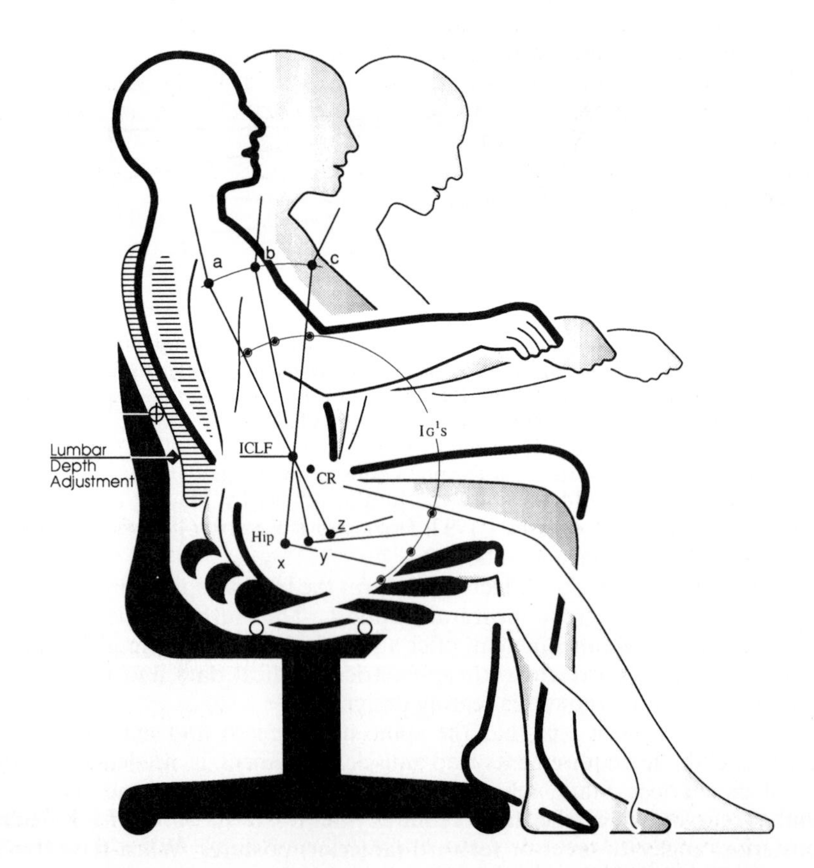

Figure 31.5 Mechanism which provides a Continuous Balance Motion (CBM) seat. The seat rotates with pelvic movement supporting with both static and dynamic means

The seat tilts $\pm 12°$ passively, and is activated by the user's motion in three ways:

1. when the torso is firmly rested against the back support (position a) and pelvic link motion is performed between hip positions x, y, and z;
2. when the torso performs familiar tasks motion between positions a, b, and c. Since the seat aligns the pelvic link with the thoracic link, the links angle can remain at mid-range throughout this motion; This eliminates excessive lumbar bending;
3. when leg motion and leg angular position affect seat tilt.

In all three cases, seat motion responds to body motion within the range of 24°. Back support adjustment ranges from 30° recline to 3° forward incline. Since the centre of rotation (CR) is fixed, the mid-lumbar spine mass will remain in place, therefore maintaining contact with the back support cushion during pelvic and hip motion.

Conclusion

Although the SPS embraces traditional ergonomic seating design principles, the Lumbar Motion Model Mannequin and the CBM seat are a new approach in ergonomic seating system design.

This material is a simplification of a more complex study. The objectives in this essay are to present a new perspective on the problem of health maintenance in task seating. The application of this perspective will affect research in 1) physical therapy, 2) occupational health and safety, 3) vehicular safety, 4) ergonomics and 5) seating design.

References

Andersson, G. B. J. and Ortengren, R., (1974), Myoelectric back muscle activity during sitting, *Scandinavian Journal of Rehabilitation Medicine*, Suppl. 3, 73–90.

Hart, D. L. and Rose, S. J., (1986), Reliability of a noninvasive method for measuring the lumbar curve, *J. Orthop. Sports Phys. Ther.*, **8**, 180.

Kelsey, J. L., White, A. A., III, Pastides, H., and Bisbee, G. E. Jr., (1979), The impact of musculoskeletal disorders on the population of the United States, *Journal of Bone and Joint Surgery*, **61A**, 959–64.

Kroemer, Kroemer and Kroemer, (1993), Ergonomics, Englewood Cliffs, NJ. Table 1-3. Body dimensions of U.S. civilian adults, female/male, in cm. Adapted from U.S. Army data reported by Gordon, Churchill, Clauser, Bradtmiller, McConville, Tebbetts, and Walker, 1989.

Link, C. S. *et al.*, (1990), Lumbar curvature in standing and sitting in two types of chairs: Relationship of hamstring and hip flexor muscle length, *Physical Therapy*, **70**(10).

Mandal, A. C., (1981), The seated man (homo sedens), the seated work position theory and practice, *Applied Ergonomics*, **12**, 19–26.

Nachemson, A. and Morris, J., (1964), In vivo measurements of intradiscal pressure, *Journal of Bone and Joint Surgery*, **46A**, 1077.

Webb Associates (eds), (1978), NASA Anthropometric Source Book, VV.I. Publication 1024, National Aeronautics and Space Administration, Technical Information Service: Washington, D.C.

PART XII

Ergonomics standards and legislative trends

32

Ergonomics standards and legislative trends for VDTs

Mark C. Volesky and Paul F. Allie

The 1990s is fast becoming the decade of ergonomics legislation. By the end of the decade all sectors of the US work force, including VDT work-places, will be covered by regulations which specify ergonomics requirements. Though roughly half of the States saw proposals of one form or another, passages into law have been rare. Some of the exceptions are the San Fransisco Ordinance and the New York City Directive. The trend to legislate standards is contrary to the intentions of the American National Standards Institute, Standards for Video Display Work Stations (ANSI/HFS 100-1988), which were drafted to provide a single, credible information source on the ergonomics of VDT work stations nationally. The breadth of current legislative proposals have expanded to include computer monitoring, adjustable furniture and electric and magnetic field (EMF) emissions protection.

Introduction

For almost 40 years, beginning primarily with the introduction of the radar screen, the ergonomics of VDT work stations has generated increasing interest. Since the VDT/CRT was introduced into the office during the early 1960s, the interest in ergonomics has evolved to the extent that a number of domestic and international interest groups have drafted ergonomics standards for the design of VDT work stations (Helander and Rupp, 1984). Such groups have included governmental bodies, general standards organizations, private enterprises and trade unions. Unfortunately, there has been, and still is, a wide degree of variance regarding the scope and content of the standards between these organizations.

The standards and guidelines can take a number of forms: guidelines issued by government agencies including the National Institute of Occupational Safety and Health (NIOSH), national voluntary standards (ANSI, DIN), trade union bargaining clauses (9–5, The National Organization of Working Women), corporate standards (IBM, Bell Labs, AT&T, etc.), and state and local legislation.

The standards and guidelines, classified as either Perceptual Ergonomics or Physical Ergonomics, are defined as follows.

1. *Perceptual:* VDT technical design specifications, which deal with radiation emissions, display colour, brightness and contrast, character size and resolution, legibility and stability.
2. *Physical:* Work-place design, which addresses seating and work-surface support of the person and the equipment, e.g., postures, operation of equipment, etc.

Fairly recent issues, such as computer/performance monitoring, mandatory rest breaks from VDT work, company-paid eye examinations, and alternative (non-VDT) work for pregnant women have also been introduced into labour-management contracts and numerous legislative proposals.

Legislative activities

Although roughly half of the states have had some form of VDT-related proposals submitted to their legislatures, few have been enacted into law. Most legislation is designed to regulate physical, rather than perceptual, ergonomics. When reviewing legislation and mandates, it is prudent to keep in mind the benefits and shortcomings of such activities. Some of the positive attributes of regulation include the following:

- assists organizations who are unsure of what their workers need;
- focuses the efforts of all interested parties;
- provides an easily accessible information source; and
- with planning, can promote uniformity of regulations between states.

Regulations can also have some negative attributes:

- may hinder incorporation of advances in ergonomics and technology;
- compliance could be costly in the short term;
- unclear language introduces confusion over compliance and auditing; and responsibilities; and
- without planning, requirements may differ both within and between states.

Some of the legislative and standard setting activities just prior to and during 1991 are listed below. Bear in mind that this summary is based on legislative sessions that ended in December, 1991. The dynamic nature of law-making makes it very likely that any bill may change during the next session.

Bills and mandates enacted

In 1988, New Mexico's governor signed an Executive Order (88-40) which covered VDT and CAD operators who are employed by the state. The Order addresses work station design, lighting, and work/rest schedules. This document is novel in that thorough explanations are provided for each work station specification and it advises that work areas should be designed with worker input. This approach should help both managers and employees gain insight into the probable causes and solutions to work-place inadequacies which could lead to repetitive injuries and stress.

In 1989, Maine passed a law that requires education and training for all employees who use VDTs for more than four consecutive hours per day. During 1991, Bill H655 was amended to cover more employees and to provide stricter language regarding education and training.

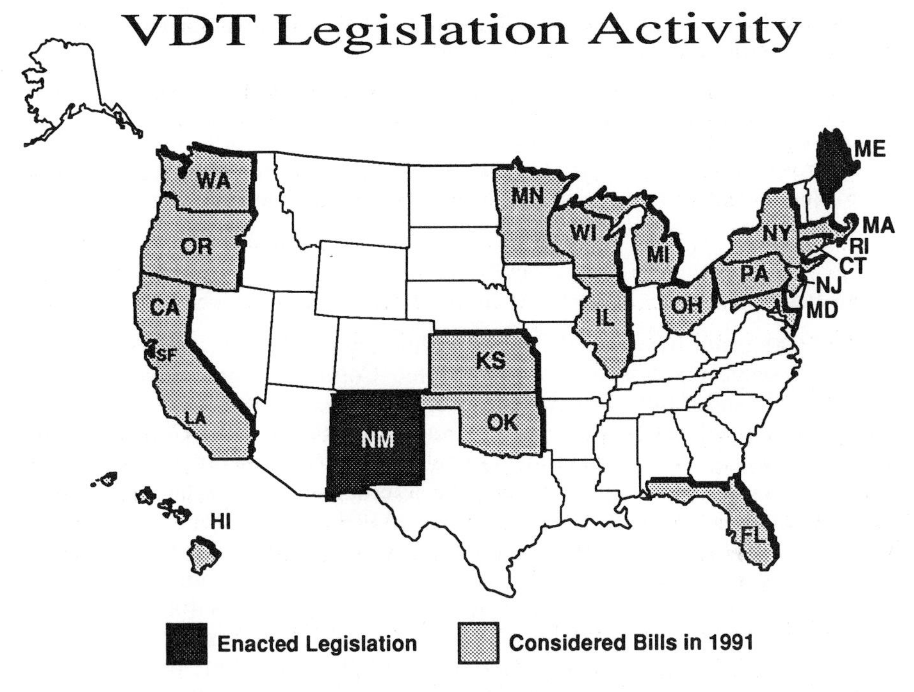

Figure 32.1 Map of U.S.

In 1991, one bill out of four was passed in the Illinois legislature. SB956 appropriated funding to study the health effects of electric and magnetic fields for VDT users. The remaining three pieces of legislation not enacted yet include two bills, HB504 and HB0504, which would have delivered strict regulations for VDT operator working conditions, and HB557, which would call for a survey regarding the incidence of repetitive strain injuries within the state, and recommend a course of action.

New York City posted a directive in January of 1991 affecting city employed VDT operators entitled *Procurement and Ergonomic Standards for Video Display Terminals and Ancillary Furniture and Equipment*. This document specifies the ANSI/HFS standards as its guideline for purchasing or leasing new VDT furniture and related equipment when designing work-spaces which will include VDTs.

The Rhode Island legislature passed HB5388, which created a special committee and directed funds for studying the causes of repetitive back strain and carpal tunnel syndrome. Other activities not enacted include: SBN88 which would require employer sponsored eye examinations, and H5363 and H5766, which would regulate the electronic monitoring of employee activities already protected by state law.

The most comprehensive measure to date, which has been enacted into law, is the 1990 San Francisco Video Display Terminal Worker Safety Ordinance (Part II, Chapter 5 of the Health Code). The Ordinance covers work station standards, eye and vision examinations, alternative work breaks, and employee education and training. The committee that developed this document apparently drew some ideas and specifications from the Suffolk County, New York VDT law. The New York State Supreme Court struck down the law soon after it was enacted, ruling that health and safety issues are the responsibility of the State Government. In a replay of history, the San Francisco Ordinance was likewise overturned under the same legal premise by California's Superior Court. In 1986, a law requiring California State colleges and universities to consider ergonomic furniture was passed.

Active and dead bills

To date California has been the most prolific regarding VDT-related proposals. VDT operators working for San Mateo County are now covered by a VDT Health and Safety Policy. This Policy was co-written with the SEIU Local 715 and addresses work station requirements. In 1991, Contra Costa County passed the Voluntary VDT Guidelines for VDT operators which also address work station requirements and alternative work schedules. 1991 brought much activity at the state level: AB644 and AB2110 were bills designed to bring work station standards into law. AB2104 mandated that CalOSHA report to the legislature on alleged VDT-related emission health hazards. CalOSHA proposed a plan to adopt the ANSI/HFS 100-1988 standards. Due to concerns raised by business groups the proposal lost political backing and a panel of advisors has been collected to redraft performance-oriented regulations that will be accepted. For now, in California, only local governments have successfully given VDT operators protective regulations.

Connecticut legislators created two bills neither of which passed into law. SB348 would have regulated break schedules. HB6229 would have created state standards for EMF emission levels equal to any standard issued by ANSI.

Florida was also unsuccessful with two bills, HB2067 and S1734. Both would have required employers to notify all employees of health risks with respect to computer display terminals.

The Georgia legislature considered bill HBN 798 which would restrict the taping of employee telephone conversations and would authorize the Public Service Commission to regulate companies that monitor employees electronically.

In Hawaii, the Department of Labour (DOL) drew up VDT work station standards with no input from business groups. After receiving many complaints the DOL has stated that it will develop a set of voluntary training programmes for VDT users. HB69, which was developed by the House to address EMF emissions, died in committee.

Indiana's SBN 308 would have required that employers disclose to employees any form of electronic monitoring that they might consider using. It would also limit the amount of monitoring.

The Kansas legislature tried to pass a bill affecting state employees. HB2418 was drafted to closely resemble the San Francisco Ordinance.

Maryland's H604 was designed to provide regulations on training and work place conditions for VDT operators working for the state.

Massachussetts legislators drew up five bills covering various topics. HBN2977 is worded to provide occupational safeguards for employees resulting from the utilization of VDTs. HBN1693 and HBN4457 both address the regulation of employer monitoring of employees. HBN3528 was written to provide protection for pregnant women using VDTs, and HBN4331 is designed to facilitate rulemaking on the biological effects of exposure to electrical and magnetic fields. None of these measures have been passed.

Legislators in Michigan have not yet drawn up any bills, but have proposed the creation of The Occupational Health Standards Commission to help develop a performance-oriented standard.

In Minnesota, two bills were introduced. HF1525, with its senate companion S1373, is an amendment to the 1990 Minnesota Statutes to include the regulation of health and safety of using VDTs. Much like the San Francisco Ordinance, HF755 is written to regulate work station design, training, alternative work schedules and eye examinations.

The New Jersey legislature developed a bill (AB3839) which goes beyond the San Francisco Ordinance in calling for more adjustability in chairs. Also, their definition of a VDT operator would allow more workers to be covered by the regulations. This bill has stalled because of consideration given to the financial conditions of both the State and private businesses in 1991. AB5099 was drafted to direct the Radiation Protection Committee to adopt EMF emission standards. A manual on healthy VDT work station design has been prepared by the state.

New York State also has five bills which cover several topics. A2136 and S1871 specify that chairs and work-surfaces for VDT operators must be adjustable. A2085 would prohibit the monitoring of phone calls by employers. A3788 and S2478 would establish VDT operator regulations.

In Ohio HB708 was created to regulate VDT furniture adjustability and work/rest breaks.

The Oklahoma legislature introduced HB1351, which quickly died. The measure would have created an advisory committee to help with VDT investigations conducted by NIOSH, and would have set up provisions for employer education and training with regard to work station design and adjustment. HBN1520 would regulate electronic monitoring of employees.

In Oregon, the Labor Committee drew up bill S840, which would have established VDT work station standards and called for a commission to study VDT emissions. Oregon's OROSHA has developed voluntary guidelines for VDTs in the work-place, which advocates compliance to ANSI/HFS 100-1988.

Pennsylvania's activities were focussed on creating a special commission to study the industrial hazards of VDTs; contained in bill SRN23.

Washington produced two companion bills, H1680 and S5493, which quickly died. These bills would have provided for the promulgation of standards requiring the use of safeguards and practices to be followed by employers of VDT operators. HB2128 was then offered in place of the companion bills and emphasized the adoption of general health and safety standards in the office environment along with worker education. HJM4007 would provide additional funding for national research into the effects of EMFs.

Wisconsin's AB608 established a grant programme in the Department of Industry, Labor and Human Relations, for which they must solicit proposals and award grants to applicants to provide VDT operator training and standard setting.

Ergonomics standards

During 1972 the West German Institute for Standardization, DIN, began a field evaluation, performed by the Technical University of Berlin, regarding standardization of VDT work-places. During the same period, other European countries such as Austria, England, Finland, France, Sweden, Switzerland, and the United States, instigated similar research. The first actual VDT standards were adopted by the Swedish Board of Occupational Health in 1979. In 1981 the DIN standards were published, and were later adopted by Austria (Helander and Rupp, 1984). Along with the creation of the European Economic Community, the EEC participants have developed the EC Directive on Screen Display Equipment, and are developing the ISO 9241 VDU guidelines.

While the EC Directive discusses organizational responsibilities for employers of VDU operators, the ISO 9241 document will be quite technical in nature, covering design specifications for input devices, VDU work stations and software parameters. Although the standard will be made up of seventeen parts, the intention is that the office work-place must be viewed in a systems sense, therefore, each of the 17 elements holds equal weight for proper ergonomic VDU work station design.

The DIN standards, although a German national standard, have had a great deal of influence on the majority of the European standards and provided a source of research for the ANSI standard, published in February of 1988. Both sets of standards address design criteria for VDT work-surfaces, including keyboard and display surfaces (height, width, depth and adjustability). Seating also encompasses a major portion of the guidelines. The Human Factors Society (the organization which drafted the standard) began a review of the ANSI/HFS 100-1988 document in 1992 to try to meet a revision deadline of mid-1994. The revision committees are focussing on the requirements that new office technologies introduce. They are also closely examining ways of accommodating individuals as well as populations within the context of the office work environment.

Standards activity in North America was most evident during the 1980s. The most recent were the establishment of the American National Standards Institute (ANSI) just mentioned above, ANSI/HFS 100-1988: *American National Standard for Human Factors Engineering of Visual Display Terminal Work Stations* and the Canadian Standards Organization (CSA). The Canadian standards were drafted by a committee comprising representatives from government, labour and private industry (CSA Standard Z412M). In 1991 an ANSI Accredited Standards Committee began working on document Z365, Control of Cumulative Trauma Disorders. This group plans to issue a broad spectrum of ergonomics standards in late 1993 or early 1994.

Since the late 1980s OSHA has been pressured to develop and adopt a set of ergonomics regulations to protect both factory and office workers from ergonomic hazards. In July of 1991, a coalition of 31 labour groups petitioned the Secretary of Labor to request OSHA to issue emergency temporary standards in order to reduce the rising number of cumulative trauma disorders. OSHA denied this request. Current OSHA activity includes the release of an advance notice for proposed rulemaking (ANPR), which is the first official step toward developing regulations. Unless Congress acts to speed up the rulemaking process, it could be five to seven years before any OSHA regulations go into effect (VDT News, Sep./Oct. 1991, p. 3). One possibility for seeing a quicker delivery date for OSHA regulations may occur if the agency adopts an already existing set of standards. Gerald Scannell, former OSHA head, has predicted that in 1994 OSHA will in fact prepare and release a set of general industry ergonomics regulations, which will include the VDT work station.

Regarding private industry, a number of major corporations have published their own standard sets which, in the end, affect the office furniture industry to the same degree as the national standards since corporate standards are the basis for purchasing office furniture. IBM, Bell Laboratories AT&T and Liberty Mutual each has its own standards. The US Government Services Administration (GSA) also has a set of standards to its procurements (Helander and Rupp, 1984).

Electric and magnetic field emissions from VDTs

Ergonomics became the buzz word of the 1980s. Research in this area is extensive but behind the demand, especially regarding the EMF emissions issue. Maximum safe level limits have been established for certain types of non-ionizing radiation, but such limits are constantly under fire because of the lack of research into long-term effects. The greatest research emphasis is currently focused on low-level electric and magnetic fields (EMFs) generated by CRT-type video display units (VDT News, March/April 1989, p. 1, May/June 1989, p. 1, Sept./Oct. 1989, p. 2). A small number of studies in the US and abroad have found statistically significant evidence of detrimental effects of EMFs on chicken embryos. However, the research and medical communities agree, for the most part, that the human foetus is quite unlike a chicken embryo, and that more research is necessary for solid conclusions to be drawn. More epidemiological studies on the effects of VDT emissions on human workers need to be performed. In 1988 the results of the Kaiser Permanente study were published. This study concluded that women who use VDTs more than half of the work week, during the first three months of pregnancy, were more likely to suffer miscarriages than those doing other types of work. Although the results were not statistically significant, or able to attribute cause (i.e., stress, smoking, etc. were unmeasured variables), the study created a high degree of concern within the labour force.

NIOSH has been researching the effects of EMFs on the human body. A team of NIOSH scientists lead by Teresa Schnorr, released its findings in March 1991 in The New England Journal of Medicine entitled, *Video Display Terminals and the Risk of Spontaneous Abortions.* After historically evaluating the pregnancies of 882 women, the researchers concluded that the use of VDTs and exposure to the accompanying electro-magnetic fields were not associated with an increased risk of spontaneous abortion. However, some critics argue for caution when interpreting NIOSH's results. They point out that both VDT and non-VDT operators were exposed to similar levels of extremely low frequency (ELF) magnetic fields in the office space. The authors defend their study by stating that the significance of the results is not diminished just because all workers were exposed to similar levels of ELF magnetic fields at work and at home. The study focuses on operators who were exposed to VDT-unique fields not emitted by other electrical sources. NIOSH will likely pursue studies to determine whether or not VDT use increases the risk of low birth weight and birth defects in humans.

IBM teamed with Ontario Hydro in a study designed to test the effects of EMFs, in the Very Low Frequency (VLF) range, on mice pregnancies. The two companies assembled an international scientific team along with third party auditors, to oversee experimental methods, for the study which was conducted at the University of Toronto (VDT News, 1989). The study concluded that when pregnant mice were exposed to magnetic fields, like those produced by VDTs, they showed no statistically significant increase in either spontaneous abortion rates or in malformations among their offspring.

The Swedes decided not to wait for irrefutable evidence of possible EMF hazards. In 1990, they adopted the emissions standards set by the Swedish National Meteorology and Testing Council. In the US, the Institute of Electrical and Electronic Engineers (IEEE) has developed a standardized measurement technique for quantifying EMF levels, which should be released in 1992. ANSI will very likely accept this document as a voluntary standard soon after it is officially released. The IEEE believes that a measurement standard is necessary to ensure that emission levels from all displays can be compared accurately. The standard also lists EMF levels for displays that are reasonably achievable with existing manufacturing technology. Interestingly, this document notes that the emission levels are not intended to imply that biological hazards exist when VDTs are used.

Despite the latest research, the EMF emissions controversy will not be resolved soon; not only because of conflicting research conclusions between the various studies, but also because the issue is emotionally charged. The majority of the studies indicate little or no relationship between VDT emissions and ill health (ranging from skin rashes to foetal deaths and birth defects (VDT News, 1989). Given the nature of the topic, only a few negative conclusions have created a measure of doubt and hence a demand, primarily from organized labour, for final, irrefutable conclusions. Most importantly, although a number of studies have found no evidence to suggest adverse effects from VDT emissions, there are no studies claiming that there are no adverse effects.

Trends, conclusions and forecasts

With so many varied groups and organizations expressing interest in ergonomics, and influencing the design of the VDT work-place, it has become extremely confusing for the average working person to sort through the information and make intelligent decisions.

The driving force behind the establishment of the ANSI standards was to provide a central, credible source of guidelines from which organized labour, legislators and private enterprises could draft their own formal mandates or guidelines (Human Factors Society, Memorandum to Don Korell, Steelcase Inc., June 1, 1985). One of the primary intentions was that a high level of uniformity would be created between groups writing guidelines and standards. We are seeing more of this type of activity today, especially in California, but it is not occurring as quickly as intended by ANSI. Even where standards or laws have been enacted, San Francisco for example, work station standards

(SFVDTWSO, 1990) have included work-surface adjustability, but have not specified an adjustment range. In other words, under the language of the law, a furniture component that has no ergonomic or comfort attributes could be acceptable simply because it has some form of adjustability, regardless of start and end of the adjustment range. In order to best assist the individuals who interpret ergonomics mandates and guidelines, a clear expression of minimum requirements must be provided for them. Unfortunately, in other states there is evidence that rushed legislation was presented without thorough exploration of the issues involved. On the national level one of the major weaknesses of the ANSI/HFS standard is that it does not fully disclose to practitioners when adjustable work-surfaces should be chosen over fixed work-surfaces. This is the type of issue that will be considered when the Human Factors Society reviews the ANSI/HFS document.

Standards activities are moving slowly, and legislative activities are on the rise. VDT/ergonomics proposals had a very limited success rate for enactment into law during the 1980s, a trend which for the most part still holds true in the early 1990s. The intense lobbying by organized labour, whose efforts have switched from the industrial to the service sector was very influential in passing the San Francisco Video Terminal Worker Safety Ordinance. The economic downturn in the early 1990s acted to slow or stall much legislative activity at the state level. One should fully expect to see continued legislative efforts to pass laws that regulate furniture and work-place standards for the office environment.

Another prominent topic for the early 1990s is VDT EMF emissions. In addition to the research, a number of products, such as filters and actual low-emission displays are being introduced into the market. Tandberg Data A/S, a Norwegian company, and the Safe Computing Company of Boston, have introduced controlled-emission CRTs and liquid crystal displays, respectively. In early 1992 the Institute for Electrical and Electronics Engineers (P-1140 panel) is scheduled to release a measurement protocol for VDT electric and magnetic field emissions. IBM, NEC and Compac, among others, have begun marketing low EMF VDTs throughout the world.

The Steelcase World Office Environment Index, which polled more than 1000 office workers around the world, found that 36 per cent of those polled think EMF radiation is a serious problem; 49 per cent of heavy VDT users say that it is a serious problem. Even though two recent studies (NIOSH and IBM/Ontario Hydro) have shown no correlation between VDT use and increased risk of miscarriages, VDT operators are still unsure about working within electric and magnetic fields. As long as the CRT is utilized in the work-place, this issue will cause concern among VDT users.

Changes in technology are also affected by health and safety issues. According to Frost and Sullivan Inc. (VDT News, 1989), sales for flat screens, primarily liquid crystal displays (LCDs), are expected to reach $2.5 billion by 1992, and $5.7 billion by 1997. Active matrix liquid crystal displays have a number of distinct advantages over CRTs, including low power requirements and much weaker EMF emissions. High resolution colour LCD displays are now being introduced in Japan (Portable with Colour LCD First of More to Come, Byte Magazine, Oct. 1989, p. 22). As the costs of these units decrease, the LCD will gradually replace the CRT. Research into High Definition monitors has opened the possibility that other, low emission display technologies will emerge. The coming of such new displays will diminish the current concern over EMFs and their effect on the human body.

We always expect that new technology will bring new products to market. What is less certain though, is how industry, ergonomics practitioners, legislators and even the end user will accept and prepare for these advancements. VDT operators depend on these people for expert application of ergonomics and corporate acceptance of advancements, for their health, productivity and well being. In the 1990s, the office environment will become the venue for worker health and safety.

References

Berns, T., (1984), Proceedings of the World Conference on Ergonomics in Computer Systems, Helsinki, 235–55.

Burch, J. L., (1984), *Ergonomics: The Science of Productivity and Health*, The Report Store, Lawrence, Kansas, ix–x.

Canadian Standards Association, (1989), *A Guideline on Office Ergonomics*, Available through CSA in Toronto, Canada.

Cohen, B. G. F., (1984), Human aspects in office automation, National Institute of Occupational Safety and Health, Division of Biomedical and Behavioral Science, 71–4.

Helander, M. G. and Rupp, B. A., (1984), An overview of standards guidelines for visual display terminals, *Applied Ergonomics*, **15**(3), 185–95.

Human Factors Society, American National Standards Institute, (1988), *American National Standard for Human Factors Engineering of Visual Display Terminal Workstations*, Available from the Human Factors Society in Santa Monica, CA.

Human Factors Society, 1985, Memorandum to Don Korell, Steelcase Inc., June 1.

Korell, D., (1984), Memorandum: B.I.F.M.A. A proposed section on VDT work stations, October 16, 1984.

Portable with Color LCD First of More to Come, (1989), *Byte Magazine*, October, 22.

Roel, R. E. and Spindle, W., (1989), VDT Law Overturned, *Newsday Magazine*, December 28, pp. 3 and 27.

San Francisco Video Display Terminal Worker Safety Ordinance, (1990), Article 23 of Part II, Chapter 5 in the San Francisco Municipal Code, Ordinance Number 405–90.

Schnorr, T. M., *et al.*, (1991), Video display terminals and the risk of spontaneous abortion, NIOSH study, *New England Journal of Medicine*, **324**, 727–33.

The Bureau of National Affairs, Inc., (1984), *VDTs in the Work-place: A Study of the Effects on Employment*, 1–13, 65–89.

VDT News, (1989), March/April, p. 1, May/June, p 1, Sept/Oct, p. 2.

VDT News, (1991), Sept/Oct, p. 3.

Westin, A. F., (1984), *The Changing Work-place*, Knowledge Industry Publications, 11.1, 12.6.

PART XIII

Overview

33

Office 2000

Leonard B. Kruk

Emerging computer and communication technologies will continue to drive change in offices in the 1990s. The World Future Society predicts 'all the technological knowledge we work with today, will represent only 1 per cent of the knowledge that will be available in 2050'.

Such advances in technology will affect both our working postures, changes in position, freedom of movement, and psychological and organizational stresses that we are subjected to. New technologies will also contribute to innovations to tomorrow's seating.

As such, a thorough understanding of seating requires a corresponding analysis of how it will be affected by changes in technology. The aim of this chapter is to review these technologies to provide a framework for reviewing the corresponding implications for tomorrow's office seating.

New technologies will allow organizations to transcend geography, operate trans-globally as though they were local, and disperse co-operative activities. Teams of managers and professionals working together on short- and long-term projects will no longer be restricted by geography, travel time or budgets. The numbers and kinds of communications and computing tools that will emerge into the market will explode. Our major dilemma will be how to adapt individual and group work processes to take advantage of technology's potential. We are now experiencing the first stage of technological overload. Therefore, the first half of this decade will be a period of introduction and the second half a period of assimilation.

Supportive technologies which may affect office workers' postures and the extent of postural "fix" in future work stations or conference rooms include:

1. Facsimile. A low-tech solution that is widely accepted will continue to proliferate in offices through the late 1990s (Roemi, 1990). The market is expected to reach 12 billion by 1995 (Market Intelligence Research Co., 1991). By the end of this century, this paper-based technology will give way to paperless, electronic technologies. In the interim, fax boards installed inside personal computers will proliferate. Such fax boards can potentially alleviate confidentiality problems associated with free standing fax machines and allow laptop users to fax from

wherever they are. Emerging fax machines are able to accommodate blue print drawings and colour.

2. CD-ROM and optical storage technologies storing massive amounts of information will allow researchers to subscribe to over 300 specialized data bases (Orr, 1989). Computer Library exemplifies a popular CD-ROM data base which provides both complete text and abstracted articles relating to the computer industry. One monthly disc contains over 60000 articles and abstracts. Electronic libraries will also make information available in a variety of media: text, images, sound and full-motion video. An example is LiberNet, a prototype electronic library network being developed by Pacific Bell.
3. Electronic records retrieval systems using optical disc storage may both eliminate the need for large, centralized file rooms and promote a paperless office environment (Harrington, 1991).
4. High-end, colour work stations will transmit graphic images. Demand is growing for larger screens. Increases in your monitor size from 14 to 17 inches actually increases the display area by about 50 per cent; this is particularly beneficial for window environments, desktop publishing and CAD. Twenty and 27 inch screens were recently demonstrated at Comdex. The ergonomics of interfacing with larger screens needs further research.
5. Thin flat-panel, colour and monochrome, low power demand display screens, such as LCD (Light Crystal Display), will begin to replace the CRT screens in the next few years. These screens, currently available on laptop computers, radiate less heat and electro-magnetic radiation emission than with CRTs. LCDs may also facilitate multiple screen (4 to 6) viewing.
6. Pen-based input may humanize the way we interact with computers by transforming computers into electronic paper (Dickinson, 1991). The capacity of reading handprinting into electronic forms now exists; GO Inc., Microsoft and GRID Systems are amongst the major developers of this new technology. Pen-input uses a stylus or pen to draw and write on a flat, LCD screen. As one handprints, the information is converted into typeface. In addition, handwriting, such as signatures, can be stored. Supplementary keyboards may be used to input large quantities of information since it is faster than handwriting. Stylus input will most likely be used by executives for editing, drawing and notetaking when travelling, as well as, in conference rooms.
7. Several alternative, built-in pointing devices will replace mice in laptop and notebook computers (Reinhardt, 1991). The shift to graphical user interfaces and the need for more ergonomic and efficient input tools will drive their development. These alternative devices include trackballs, touchpads, touchscreens, and a home row key mouse (with sensors under the letter J). The development of these devices is driven by the need for user comfort.
8. High-tech is also invading the meeting room. Video conferencing equipment, projectional VCRs, LCD projection screens, visualizers, electronic information display boards, fax machines and scanners, personal computers, and electronic white boards will become commonplace. Such rooms are being redefined as work areas for supporting group/technology interactions.
9. Video-conferencing on individual desktops will allow users to conduct a video conference in one window on the PC while running interactive multimedia applications in another. Both individual work areas and meeting rooms will evidence a breakthough in the next few years (Marshak, 1990).
10. Wireless communications are now being introduced and experimented with (Bradshor, 1990). This will take place on two fronts: wireless LANs using radio frequency or infra-red technologies and cable-free sharing of peripherals. With the later, a cable-free peripheral-sharing device, roughly the size of a portable CD player will allow your PC to communicate with any peripheral or terminal through the building's AC circuitry.

11. Laptops will represent an increasing share of the personal computer market and impact both executive offices and conferences rooms and account for 70 per cent of all personal computer sales by the end of the decade (Quick, 1991). In both settings, ease of accessibility to both power and communication ports will be required. Many notebooks, according to analyst Tim Bajarin of Creative Strategies Research International of San Jose, CA, will use infrared light transmitters to communicate with wireless networks.
12. Although voice recognition systems will be utilized for input to the personal computer, its applications will be limited through the mid-1990s (Rifkin, 1991). Today's commercial voice recognition applications provide command and control capabilities under which users can issue spoken commands to select menu items, launch macros and click buttons. Speaker-independence and continuous-speech recognition represent two major hurdles developers must cross to create the next generation of voice applications.
13. Technologies are increasingly becoming available to assist disabled people in the office. These include the Dragon Dictate-30K voice recognition system, from Dragon Systems Inc., the IBM PhoneCommunicator, for the hearing impaired; and DEC's DEC talk voice synthesizer. The new American Disabilities Act mandates that facilities accommodate handicapped workers; this requires an understanding of such technologies.

As we approach the turn of the century, organizations will begin to learn how to efficiently use and integrate the technologies. They will be linked to the information and communication technologies of their corporate offices and boardrooms; their field personnel, consultants, and trainers; their homes, schools and communities.

Research will be required on the implications of these technologies to future seating design. How will people work and sit in offices and homes while utilizing these new technologies? Further, will some of these technologies be incorporated into office and conference room chairs in the future? What kinds of special purpose high-tech will become available?

By the end of the century, technology will become much more than a tool. Rather it will become an integral element of a human technology system associated with use of new forms of artificial intelligence (AI), voice recognition, neural networks and other emerging technologies. Interactive three-dimensional video, very high-bandwidth communication channels, virtual reality communications, shape recognition cameras and team work stations may tie multimedia tools to the human visual system.

Many specialists maintain that office automation enhances productivity in tomorrow's offices. However, such benefits will not be achieved without an understanding of how they will affect the organization's work processes, and different kinds of user. Technology will affect posture and postural requirements.

References

Bradshor, K., (1990), *Wireless Signals Link Computers to Future*, Grand Rapids Press, November 25, p. E12.

Dickinson, J., (1991), The handwriting is on the wall for pen-based computers, *Computer World*, **4**(3), 48.

Harrington, M., (1991), Write-once technology still too costly for most, *Computer World*, **25**(8), 6.

Market Intelligence Research Co., US and World Facsimile Markets, Mountain View, CA., 1991.
Marshak, D., (1990), Video-conferencing meets multimedia, Patricia Seybold's *Network Monitor*, **5**(10), 18.
Orr, J., (1989), Brainstorming, Column, Andrew Seybold's *Outlook on Professional Computing*, **7**(10), 17.
Quick, G., (1991), Notebook changes to come, study shows improved portables will spur market growth, *Computer Reseller News*, **406**, 61.
Reinhardt, A., (1991), Touch and feel interfaces: Built-in pointing devices replace mice in notebooks and laptops, *Byte*, **16**(2), 223–4.
Rifkin, G., (1991), Toward the voice-literate computer, leading Japan into a keyboardless future where PCs multiply, Column, *The New York Times*, Feb. 24, 9.
Roemi, L., (1990), Growth opportunities abound in all segments, *Modern Office Technology*, January, 94.

Index

HETERICK MEMORIAL LIBRARY
3 5111 00397 4049